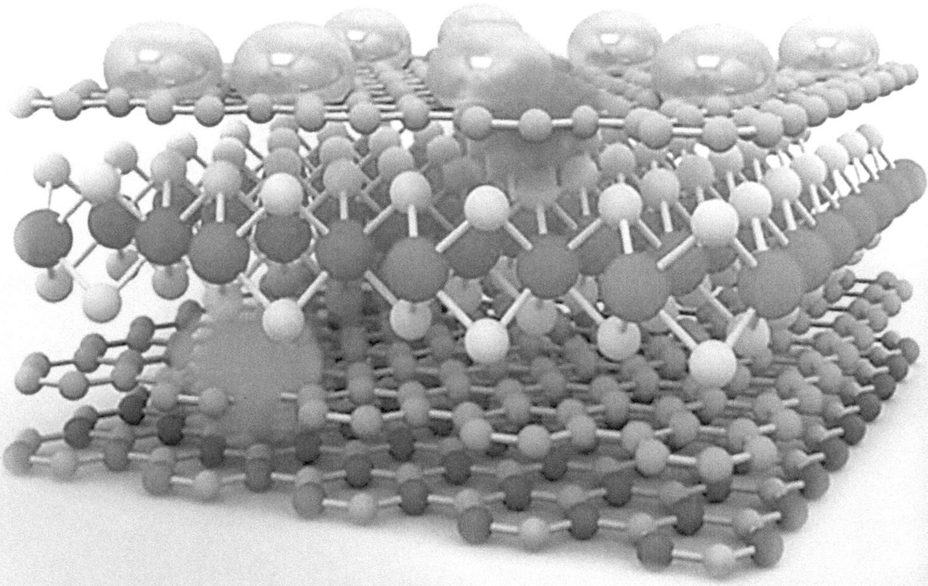

Hybrid Nanocomposites

Hybrid Nanocomposites

Fundamentals, Synthesis, and Applications

edited by

Kaushik Pal

PAN STANFORD PUBLISHING

Published by

Pan Stanford Publishing Pte. Ltd.
Penthouse Level, Suntec Tower 3
8 Temasek Boulevard
Singapore 038988

Email: editorial@panstanford.com
Web: www.panstanford.com

British Library Cataloguing-in-Publication Data
A catalogue record for this book is available from the British Library.

Hybrid Nanocomposites: Fundamentals, Synthesis, and Applications

ISBN 978-981-4800-34-1 (Hardcover)
ISBN 978-0-429-00096-6 (eBook)

Contents

Preface

I undertook the task of editing *Hybrid Nanocomposites: Fundamentals, Synthesis, and Applications*, and all the authors investigated materials science and nanotechnology concepts to contribute to this book. I have maintained both the novelty of the findings and the number of exercises and applications. Current research trends and potential applications in future advances such as nanomaterials, nanometrology, nanoelectronics, optoelectronics, transparent conducing and flexible thin films, nanobiotechnology, and surfaces and interfaces are a key challenge for those working on hybrid nanomaterials, graphene, and liquid crystals, where new imaging and analysis spectroscopy/electron microscopy responses are vital.

The book contains the following significant topics:

- Fabrication and applications of nanomaterials
- Synthesis of hybrid nanocomposites, current research trends, and potential applications
- Key challenges for those working on hybrid nanomaterials, graphene, liquid crystals, and spectroscopic responses
- Variability and site recognition of biopolymers and biosensors
- 1D nanostructures consisting of biopolymers and inorganic compounds

The book also identifies what environmental, health and safety, ethical, or societal implications or uncertainties may arise from the use of the technology, both current and future. It explains how technology might be used in the future, estimating the likely timescales in which the most far-reaching applications of technology might become a reality.

This book highlights two things: the novelty of the detailed research methodology and the scholars' hands-on expertise. Beginner research scholars will be able to use the information provided in this book and perform experimental analysis. Scientists will be able to augment their experimental proficiency using several strategies and various techniques. Not only is there an abundance of skill-building expertise, there also is a wide variety of realistic applications using real data, so "dreaded approaches" will be seen as a useful and practical extension of what researchers have learned. Every chapter ends in simple and innovative ways that PhD beginners will be comfortable with to fit in their study material. I am confident that researchers will appreciate having a textbook written for them and their research interest in the fields in materials science and nanotechnology will be stimulated.

I would like to extend my appreciation to the people at Pan Stanford Publishing for their whole-hearted support in producing this book. Many thanks to all of them in advance for their efforts to make this book a bestseller when there are already many good books for faculty, researchers, and scientists to choose from. I would like to thank the many students, scholars, scientists, and faculty members who are using this book. I also sincerely appreciate the efforts of the reviewers who gave many helpful suggestions to improve the content of this book. I specifically want to thank the board of advisors who contributed feedback throughout the process.

Prof. (Dr.) Kaushik Pal

Research Professor (Independent Scientist & Group Leader)

2019

Chapter 1

Graphene-Based Polymer Nanocomposites for Sensor Applications

Srinivasan Krishnan,[a,b] Ravisankar Tadiboyina,[c] Murthy Chavali,[b,d] Maria P. Nikolova,[e] Ren-Jang Wu,[f] Da Bian,[g] Yeau-Ren Jeng,[g] P.T.S.R.K. Prasada Rao,[h] Periasamy Palanisamy,[i] and Sudhakar Reddy Pamanji[j]

[a]*College of Chemistry and Chemical Engineering, Henan University, Kaifeng 475004, Henan Province, P.R. China*

[b]*MCETRC, Tenali, Guntur 522201, Andhra Pradesh, India*

[c]*Aakash Educational Services Ltd., No. 2, AB-Block, 2nd Avenue, Anna Nagar, Chennai 600040, Tamil Nadu, India*

[d]*Shree Velagapudi Rama Krishna Memorial College (PG Studies), Affiliated to Acharya Nagarjuna University, Nagaram 522268, Guntur District, Andhra Pradesh, India*

[e]*Department of Material Science and Technology, University of Ruse "Angel Kanchev," 8 Studentska Str., POB 7017, Ruse, Bulgaria*

[f]*Department of Applied Chemistry, College of Science, Providence University, 200, Sec. 7, Taiwan Boulevard, Shalu District, Taichung City 43301, Taiwan*

[g]*Department of Mechanical Engineering, National Chung Cheng University, Ming-Hsiung, Chia-Yi 621, Taiwan*

[h]*Department of Chemistry, P B Siddhartha College of Arts & Science, A S Rama Rao Road, Moghalrajpuram, Siddhartha Nagar, Vijayawada 520010, Andhra Pradesh, India*

[i]*Department of Physics, Gnanamani College of Engineering, Pachai, Namakkal 637018, Tamil Nadu, India*

[j]*Department of Zoology, VSU PG Centre Kavali, Peddapavani Road, Kavali 524201, Andhra Pradesh, India*

ChavaliM@gmail.com; wurenjang@gmail.com

Hybrid Nanocomposites: Fundamentals, Synthesis, and Applications
Edited by Kaushik Pal

ISBN 978-981-4800-34-1 (Hardcover), 978-0-429-00096-6 (eBook)
www.panstanford.com

Graphene is a unique 2D nanostructured material with particles composed of few-nanometers thickness. Owing to its strength (130 GPa), Young's modulus (1 TPa), enhanced thermal conductivity, and superior electronic properties, graphene acts as an outstanding reinforcing material today. Graphene is a multifunctional reinforcement that can improve the properties of polymers, even at extremely low loading. In addition, mechanical strength, a large surface area, easy functionalization, and other attractive physiochemical characteristics of graphene endorse its broad exploits in sensing/biosensing applications. A graphene-based polymer nanocomposite possesses exceptional properties such as mechanical, electrical, gas barrier, thermal, and flame-retardant properties than a neat polymer. Graphene-reinforced polymer composite–based biosensors have the advantage of higher sensitivity, with selectivity, fast response time, stability, and a low limit of detection (LOD). The exceptional properties of graphene-reinforced polymers can demonstrate superior performance in numerous applications such as electronic devices, memory devices, and semiconductive sheets in transistors, hydrogen storage, flexible packaging, and structural components for transportation or energy, aerospace, and printable electronics. This chapter highlights several graphene-based polymer nanocomposites for sensing applications, which include sensing of dopamine (DA), glucose, ammonia, hydrazine, nitric oxide, guanine, adenine, hemoglobin, methane, and formaldehyde gas.

1.1 Introduction

Nowadays nanotechnology drives the technology, especially surrounded by nanoparticles (NPs) and 2D graphene sheets. Graphene sheets became a priority and subject of scientific interest for several research groups due to their exceptional electron transport, mechanical properties, and high surface area. Graphene is a one-atom-thick single layer of graphite and was first produced by mechanical exfoliation and is the lightest, thinnest, and strongest material ever discovered.

Graphene is a 2D atomically thick and sheet-like material composed of carbon (sp^2) atoms in an arrangement of a honeycomb-like structure (Fig. 1.1). Graphene is renowned as a basic building block for other graphitic allotropes of carbon: (i) graphite (3D carbon allotrope) is made of stacked graphene sheets that are separated by 3.37 Å, (ii) carbon nanotubes (CNTs; 1D carbon allotropes) can be made by rolled-up graphene sheets and slicing, and (iii) fullerene (buckyballs, 0D carbon allotrope) is made by a wrapped-up part of a graphene sheet [1–3]. It has attracted considerable interest due to its unique electrical conductivity, high flexibility, high surface area, chemical stability, low thermal noise, strong mechanical properties, high optical transparency, and thermal properties [4–13]. Graphene has a high surface-to-volume ratio, is free from catalytic impurities, and has high charge mobility than that of CNTs [14–16]. In addition, graphene can be easily synthesized on a large scale via reduction of graphene oxide (GO). As a result of these remarkable properties, it has been extensively used in numerous fields such as polymer nanocomposites, sensors, and energy storage (e.g., batteries, solar cells, supercapacitors and fuel cells, and optical devices) [17–20]. Graphene-based polymeric nanocomposites also show a very low percolation threshold of electrical conductivity and improved mechanical, thermal, and barrier properties. In this chapter, from the field of materials science, graphene-based polymer nanocomposites are highlighted and reviewed for their applications and for most promising developments. Also, potential applications of various graphene-based polymer nanocomposite materials, based on GO and reduced graphene oxide (rGO), have been discussed. Due to the high diversity, properties, and advantages of graphene, a multitude of nanocomposite-based applications have been envisioned to be practical.

Graphene is a single sheet made of polycyclic aromatic carbon dislocated from a 3D graphite structure. Graphene can be synthesized by two primary approaches, top-down and bottom-up methods (Fig. 1.2). The top-down method generally involves the exfoliation of individual sheets of graphene made from graphite. The first isolation of graphene was demonstrated by Novoselov and Gaim in 2004 by using exfoliation of simple adhesive tape [3]. The exfoliation techniques are mainly processed by weakening the van der Waals forces between the layers of graphene sheets through (i) mechanical (scotch tape method) and liquid-phase exfoliation of graphite, (ii) chemical (oxidation of graphite, solution-based exfoliation, exfoliation/reduction of graphite oxide), (iii) electrochemical (exfoliation and oxidation/thermal reduction) methods, or (iv) unzipping the CNTs (plasma treatment, acid reactions, liquid NH_3/Li intercalation–exfoliation, and catalytic approaches). The top-down method is extensively used for the synthesis of graphene. However, graphene synthesized through the top-down approach results in a large number of defects and also the usage of highly toxic reagents limits its applications [3, 13, 21–27].

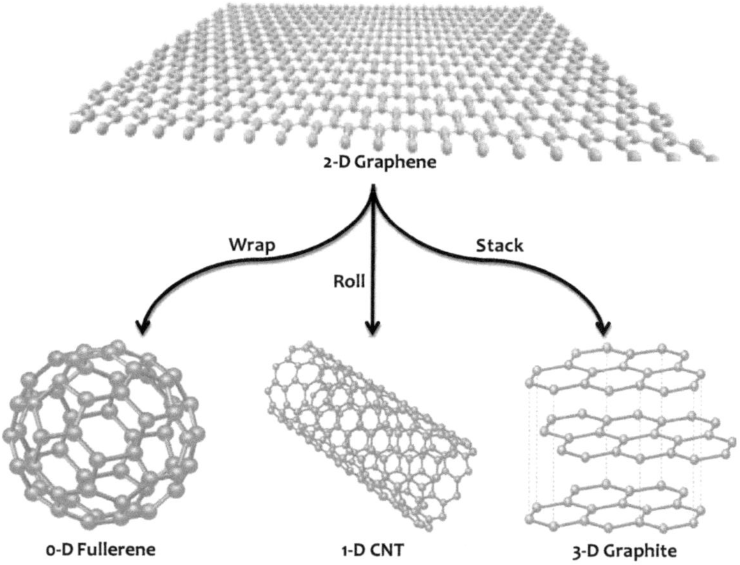

Figure 1.1 Graphene, the mother building block of all other carbon allotropes, can be wrapped to form the 0D fullerene/buckyballs, rolled to form 1D nanotubes, and stacked to form the 3D graphite.

Besides, a completely different approach for the synthesis of graphene is a bottom-up method in which graphene is made to generate directly on a surface. In this approach, graphene is generated by building small carbon molecular blocks into a large or single-layer graphene structure via (i) organic synthesis (chemical vapor/solid deposition), (ii) chemical vapor deposition (CVD; catalytic process), (iii) epitaxial growth or decomposition of silicon carbide SiC at temperatures above 1100°C (thermal). As compared to the top-down approach, this method results in fewer defects on the surface of graphene. The application of graphene is significantly dependent on the nature of the materials, type of defects, and substrates [13, 23, 28–30]. The advantages and disadvantages of various methodologies for the synthesis of graphene are also summarized in Table 1.1.

Table 1.1 Methodologies for the production of graphene

S. no.	Method	Advantage	Disadvantage	Reference(s)
1.	Confined self-assembly	Thickness control	Existence of defects	[31]
2.	CVD	Coverage of large areas, potential cost-effectiveness, promising plasma-coupled CVD techniques	Transfer steps needed, high energy consumption, toxic chemical sources	[32–38]
3.	Arc discharge	Can produce ~10 g/h of graphene	Low yield of graphene, carbonaceous impurities	[39, 40]
4.	Epitaxial growth on SiC	Direct growth in isolating substrates, single few layers obtained, reproducibility, order, clean	High cost of the SiC wafers, high temperatures	[31––96]
5.	Unzipping of CNTs	Size controlled by selection of the starting nanotubes	Expensive starting material, oxidized graphene	[97–99]
6.	Reduction of CO	Unoxidized sheets	Contamination with α-Al_2O_3 and α-Al_2S	[100]
7.	Liquid exfoliation of graphite	Processability, good amounts	Very small flakes, high amounts of edge defects	[14]
8.	Exfoliation via GO	Massive production	Very low quality	[14]
9.	Electrochemicalintercalation methods	Simplicity, low cost, good amounts with good quality, control of oxidation	Expansion/exfoliation steps needed, specific equipment	[14]

(*Continued*)

Table 1.1 (*Continued*)

S. no.	Method	Advantage	Disadvantage	Reference(s)
10.	Micromechanical exfoliation	Large size and unmodified graphene sheets, low complexity, free of defects	Very small-scale production, high cost	[2]
11.	Direct sonication of graphite	Unmodified graphene, inexpensive	Low yield, separation	[101, 102]
12.	Electrochemical exfoliation/ functionalization of graphite	Single-step functionalization and exfoliation, high electrical conductivity of the functionalized graphene	Cost of ionic liquids	[103]
13.	Superacid dissolution of graphite	Unmodified graphene, scalable	Use of hazardous chlorosulfonic acid, cost of acid removal	[104]
14.	From graphite derivatives (graphite oxide or graphite fluoride) Li alkylation of graphite fluoride	Large size, functionalizedd sheets, no oxygen functionality	Cost of the starting material, restacking after annealing	[105]
15.	Chemical reduction of colloidal GO in water	Large sheet size, some routes use only water	Use of hazardous chemicals, only dispersed in hydrophilic polymers	[106–112]
16.	Chemical reduction of organically treated GO	Colloidal stability in organic solvents, better exfoliation	Low thermal stability, in situ chemical reduction degrades some polymers	[106, 113, 114]
17.	Thermal exfoliation/ reduction of GO	One-step exfoliation/reduction, short heating time, dry basis	High heating temperature, smaller sheet size compared to chemically reduced sheets	[115, 116]

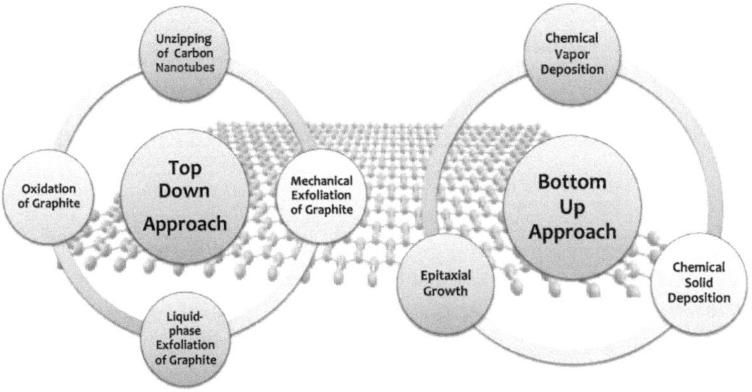

Figure 1.2 Production methods for the synthesis of graphene.

1.2 Graphene-Based Polymer Nanocomposites

The high aspect ratio, surface area, superior thermal and electrical conductivity, tensile strength, optical transparency, electromagnetic interference (EMI) shielding, and flexibility possessed by graphene facilitate it to be the most promising candidate as a nanofiller for a polymer matrix. Further, the precursor for the synthesis of graphene is also abundantly available, thus making it the most prominent nanofiller than customary nanofillers like CNTs, carbon black (CB), sodium montmorillonite (Na-MMT), exfoliated graphite (EG), layered silicates, and carbon nanofibers (CNFs). Moreover, graphene is a multifunctional reinforcement that can improve the properties (such as mechanical, electrical, gas barrier, and thermal properties) of polymers even at extremely low loading [117–123].

Compared to polymers, graphene possesses extraordinary properties that are reflected in graphene-based polymer composites. As a consequence, the tailor-made graphene-based polymer nanocomposite shows superlative properties such as mechanical, electrical, gas barrier, thermal, and flame-retardant properties than a neat polymer. In addition, graphene-based polymer composites possess much better improvement than other carbon filler–based polymer composites [124–128]. The exceptional properties of graphene-reinforced polymers can demonstrate superior performance in numerous applications such as electronic devices, memory devices, semiconductive sheets in transistors, hydrogen storage, flexible packaging, structural components for transportation or energy, aerospace and printable electronics, etc. [14]. Similarly, the successive bulk synthesis of graphene also promotes the fabrication of graphene-reinforced polymer composites and hybrid materials.

1.3 Synthesis of Graphene-Assembled Polymer Nanocomposites

In general, the properties of polymer nanocomposites mainly depend on the dispersion, and the methodology depends on the molecular weight, polarity, hydrophobicity, and reactive groups present in graphene and the polymer [129a]. It is vital to disperse the graphene in the given polymer matrix to synthesize composites with improved properties. Depending on the polarity, molecular weight, hydrophobicity, and reactive groups of the polymer, graphene, and solvent [129b], there are three general methodologies for fabricating/preparing graphene-filled/graphene-reinforced polymer composites (See Table 1.2):

a. *Solution blending*: Both the polymer and graphene or modified graphene layers are allowed to swell in a suitable solvent system (water, acetone, chloroform, tetrahydrofuran [THF], dimethylformamide [DMF], toluene, etc.) and thereafter mixed. The polymer adsorbs on to the delaminated graphene sheets, and finally, the solvent is evaporated [130] to obtain the nanocomposites.
b. *Melt blending*: This allows mixing the filler and polymer in a molten state. Melt intercalation allows mixing the filler and polymer in a molten state. Usually, thermoplastic polymers are mixed at elevated temperatures using conventional methods like extrusion and injection molding.
c. *In situ polymerization*: Graphene or modified graphene is first swollen within the liquid monomer by adding an initiator that initiates the polymerization either by heat or by radiation. Numerous polymer nanocomposites such as polystyrene (PS)/graphene have been prepared by this method.

1.3.1 Solution Blending

Solution blending or solution mixing is the most efficient method for the production of polymer-based nanocomposites by creating a strong interface between the polymer and the filler. In this method, the polymer is dissolved in aqueous or organic solvents such as water, acetone, chloroform, toluene, DMF, and dichloromethane (DCM), subsequently mixing the resulting solution with a dispersed solution of graphene using magnetic agitation, mechanical mixing, or ultrasonication treatment. The main advantage of this methodology is low-, high-, or even nonpolar polymers can be used for the preparation of graphene-based polymer composites. In addition, this method works well even for semicrystalline as well as amorphous polymers and also independent of the polymer structure. However, some polymers like polyolefin and polyamides (PAs) are insoluble in common solvents. As a result, environmentally compromised solvents (*o*-dichlorobenzene or *m*-cresol) are used to dissolve these polymers at high temperature [13, 14, 131, 132]. The polymers, including PS, polyimides (PIs), polycarbonate (PC), polyacrylamide, and poly(methyl methacrylate) (PMMA), have been successfully mixed with graphene by this technique [124, 125, 133–135].

Table 1.2 Methodology for the production of graphene/polymer nanocomposites

S. no.	Graphene polymer composites	Fabrication method	Reference(s)
1.	Graphene/PVP	Electrochemical reduction and electropolymerization	[48]
2.	rGO/PEDOT	Electrodeposition and electrochemical reduction	[136]
3.	Graphene/PS	Υ-irradiation-induced graft polymerization, solvothermal	[49]
4.	Graphene/PANI	In situ polymerization, electrochemical apta sensor	[50]
5.	rGO/PPy	In situ chemical oxidative polymerization	[50]
6.	Graphene/porphyrin	Condensation and electrochemical reduction	[51]

(*Continued*)

Table 1.2 (*Continued*)

S. no.	Graphene polymer composites	Fabrication method	Reference(s)
7.	Graphene/chitosan	Layer-by-layer assembly	[52]
8.	EDTA–silane/rGO/Nafion	Silanization and ultrasonication	[53]
9.	GO/PANI	In situ chemical polymerization	[54]
10.	Graphene/chitosan	Chemical reduction and ultrasonication	[55]
11.	rGO/PpPD	Heat treatment	[56]
12.	Graphene/PoPD	Electrochemical reduction and electropolymerization	[57]
13.	Graphene sheets/GSCR–MIPs	Free-radical polymerization	[58]
14.	Graphene/chitosan	Blending and chemical reduction	[59]
15.	GO/PS	Solvent blending + hydrazine	[60, 61, 124]
16.	GO/vinyl chloride/vinyl acetate copolymer	Solvent blending + hydrazine	[62]
17.	Graphite/PS	Solvent blending + ionic liquid	[103]
18.	GO/epoxy	In situ polymerization at 250°C	[63, 64]
19.	GO/PMMA	In situ polymerization, solution mixing	[65–67]
20.	Graphene/rGO/PU	Solvent blending, in situ polymerization, melt compounding	[68–71]
21.	rGO/PEN	Melt compounding	[72]
22.	rGO/PC	Melt compounding	[73]
23.	rGO/natural rubber	Melt or solvent blending + vulcanization	[137]

S. no.	Graphene polymer composites	Fabrication method	Reference(s)
24.	Graphene/PS-PI-PS	Melt or solvent blending	[1]
25.	Graphene/PDMS	Oligomer blending + polymerization	[1]
26.	Graphene/PVDF	Solvent blending	[74]
27.	Graphene /SAN	Preblending using solvents, followed by melt compounding	[75]
28.	GO/PVA	Solution mixing	[76–78]
29.	GO/CNT/PVA	Solution mixing	[79]
30.	Graphene/PVC	Solution mixing	[80]
31.	Graphene/hydrogenated carboxylated nitrile butadiene rubber	Solution mixing	[81]
32.	GO/PMMA foam	Blending and foaming	[82]
33.	Graphene/PA12	Melt blending	[83]
34.	GO/PBS	Solution mixing and then melt blending	[84]
35.	GO/PA/polyphenylene (PA/PPO, 90/10)	Solution mixing and then melt blending	[85]
36.	GO/PCL film	Solution casting	[86]
37.	GO/PCL nanofibrous membranes	Electrospin	[86]
38.	GO/ PI	In situ polymerization	[87]
39.	GO/polyester	Solution mixing	[88]
40.	GO/PLA	Melt blending	[89]

PVP, poly(vinyl pyrrolidone); PEDOT, poly(3,4-ethylene dioxythiophene); PpPD, poly(*p*-phenylenediamine); PoPD, poly(*o*-phenylenediamine); GSCR–MIP, graphene sheets/Congo red–molecularly imprinted polymer; PU, polyurethane; PEN, poly(ethylene naphthalate); PS-PI-PS, poly(styrene-co-isoprene-co-styrene) triblock copolymer; PDMS, poly(dimethylsiloxane); PVDF, poly(vinylidene fluoride); SAN, poly(styrene-ran-acrylonitrile); PVA, poly(vinyl alcohol); PVC, poly(vinyl chloride); PPO, polyphenylene oxide; PCL, poly (ε-caprolactone); PLA, poly(lactic acid).

1.3.2 Melt Blending

The melt blending or compounding technique is extensively used for the large-scale production of thermoplastic composites and nanocomposites. This methodology is preferred by most of the processing industry due to its cost benefits. In this approach, graphene/modified graphene is directly mixed with the thermoplastic polymer in the molten state and then molded by extrusion molding or injection molding.The polymer composites are then exfoliated or intercalated to form nanocomposites [138–140]. In this method, solvents are not in use during processing, and therefore huge amounts of specimens with different shapes can be produced in a short span of time. However, the high-temperature mechanically assisted processing could be harmful to both the polymer and the graphene sheets, which results in a decreased aspect ratio of graphene sheets and the molecular weight of the polymers. A broad range of graphene-based polymer nanocomposites such as PMMA/graphene, polypropylene (PP)/graphite, poly(ethylene-2,6-naphthalate)/graphene, high-density polyethylene (HDPE)/graphite, PC/graphene, polyphenylene sulfide (PPS)/graphite, and polyamide (PA6)/graphite have been prepared by this methodology [139, 141–144].

1.3.3 In situ Polymerization

In this fabrication strategy, the dispersed solution of graphene or modified graphene is mixed with neat monomers and is polymerized (initiated either by heat or by radiation) to get graphene-based polymer nanocomposites [13, 23, 130, 145]. The in situ polymerization methodology has produced polymer composites with a covalent crosslink between the filler and the polymer matrix. Further, in situ polymerization can also result in the production of noncovalent polymer composites such as polyethylene, PMMA, and polypyrrole (PPy). In situ polymerization methods have successively produced a huge number of graphene-based polymer nanocomposites, which include PS/graphene [130, 145–150], polystyrene sulfonate (PSS)/layered double hydroxide (LDH) [151], PMMA/graphite [149, 152, 153], polyurethane (PU)/graphene [154], PI/LDH [155], epoxy/graphene [156], poly(acrylic acid-co-acrylamide)/graphene [157], polyethylene terephthalate (PET)/LDH [158], poly(sodium methacrylic acid) (PMANa)/graphene [159], poly(dimethylsiloxane) (PDMS)/graphene [160, 161], and polyaniline (PANI)/GO nanocomposites [162, 163].

1.4 Varieties of Graphene-Based Polymer Nanocomposites

Graphene and its derivatives (nanofiller)-reinforced polymer nanocomposites have shown promising potential for a variety of important industrial and point-of-care applications such as sensors (chemical and biosensors), aerospace, electronics, electrostatic discharge (ESD) and EMI shielding, green energy, and automotive industries. The wide range of polymer nanocomposites based on graphene nanofillers are presented in Fig. 1.3. As mentioned earlier, 2D graphene possesses an enhanced high aspect ratio, a larger specific surface area, and better mechanical, electrical, and thermal properties than other reinforcements (CNTs, carbon, and Kevlar fibers).

1.4.1 Graphene/Polyaniline Nanocomposites

Graphene/PANI nanocomposites have attracted remarkable interest owing to their immense applications such as enhanced conductivity, ease of processing, cost-effectiveness, biocompatibility, superior electrocatalytic activity, and a cost-effective source material for the fabrication of sensors [164, 165]. Graphene/PANI composite paper was synthesized by in situ anodic electropolymerization of aniline over graphene paper [166]. Anodic electropolymerization of aniline was carried out by a three-electrode cell (counter-Pt plate, reference-standard calomel electrode [SCE], and working electrode-graphene paper). PANI was electropolymerized over graphene paper at a constant potential rate of 0.75 V. In another method, Wang et al. prepared a high-performing graphene/PANI composite electrode by the spin-coating method [167]. In this preparation, an aqueous solution of GO was coated on a quartz glass substrate and then reduced thermally to obtain graphene film. Subsequently, n-methylpyrrolidone (NMP)-dispersed solution of PANI was then deposited over graphene films. The as-prepared graphene/PANI electrode is more appropriate for the designing of electrochromic devices. Similarly, chemically modified graphene and PANI composites were also fabricated by in situ polymerization of aniline in the presence of an acidic solution of GO [168].

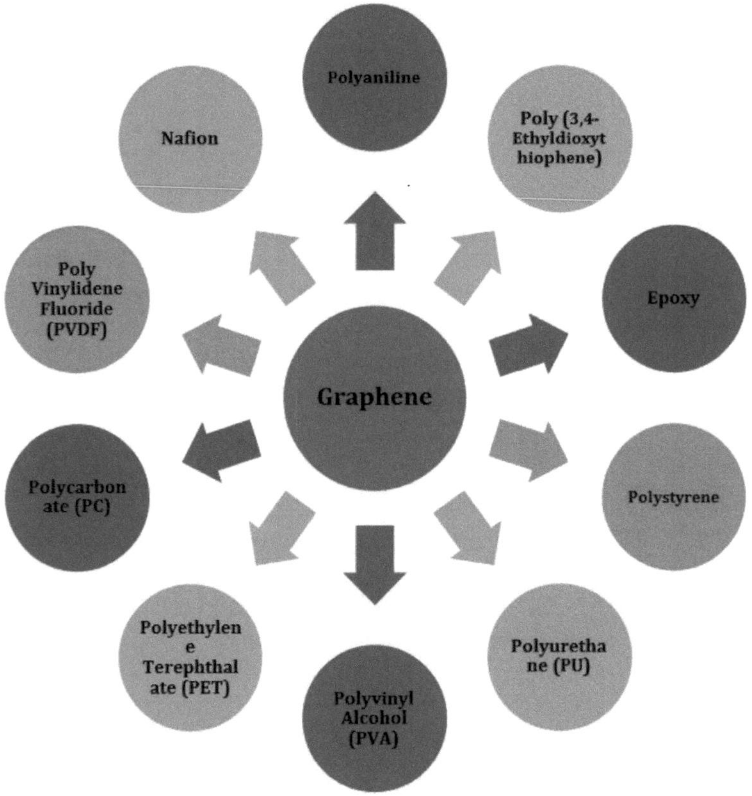

Figure 1.3 Different types of graphene-assembled polymer composites.

1.4.2 Graphene/Poly(3,4-Ethylene Dioxythiophene)

The essential properties such as high conductivity, improved stability, low density, enhanced catalytic activity, and convenient processing of poly(3,4-ethylene dioxythiophene) (PEDOT) paves the way to exploiting it as a conducting polymer in various electrochemical applications such as supercapacitors, sensors, and solar cells [169–172]. Very recently, PEDOT and its composites have been reported to disclose its outstanding thermoelectric performance [173]. In 2013, Xu et al. reported a thermoelectric composite made of PEDOT and graphene synthesized using in situ polymerization [174]. The graphene/PEDOT composites possess higher thermal stability and it show very little weight loss (below 297°C). The graphene/PEDOT composite is thermally more stable than the PSS/PEDOT composite.

1.4.3 Graphene/Epoxy Nanocomposites

Epoxy and its composites are multipurpose materials for numerous industrial applications such as automobiles and aerospace applications. Though epoxy composites have some limitations, graphene paves the ways to overcome those limitations. Graphene/epoxy nanocomposites can be prepared via in situ polymerization [175–177], and they can be utilized as an effective lightweight material for EMI shielding. Hence, the thermal conductivity of epoxy resin was found to be very poor; however, the incorporation of graphene offered a significant enhancement. Graphene can extensively improve the physical and chemical properties of an epoxy matrix even at very low loadings [178, 179]. As compared to neat epoxy resin, 5 wt% of GO-reinforced epoxy resin showed higher thermal conductivity (four times) [180, 181]. Consequently, graphene/epoxy composites explore a potential thermal interface for heat dissipation application.

1.4.4 Graphene/Polystyrene Nanocomposites

PS is a thermoplastic material that has various applications in a variety of fields like protective packaging, toner inks, construction, and automotive and consumer products. Graphene/PS nanocomposites can be prepared by the solution-blending method, reversible addition-fragmentation (RAFT) polymerization, and in situ emulsion polymerization. [182]. The thin films of graphene/PS nanocomposites are naturally semiconducting materials and can exhibit an ambipolar field effect [129]. Similarly, the conductivity of the composite is found to be directly proportional to the volume percentage of filler loading [124]. Liu et al. reported the synthesis of a PS/ionic liquid–functionalized graphene composite using the solution-blending method, which possessed higher electrical conductivity than neat polymers [103, 183]. Similarly, the thermal stability of the nanocomposite was found to be higher than that of pure PS [145].

1.4.5 Graphene/Polyurethane Nanocomposites

PUs are the most versatile synthetic polymers and have prominent industrial applications such as microcellular foam, synthetic fibers, insulation panels, elastomeric wheels, automotive suspension, bushings, tires, sealants, seals, and gaskets [184, 185]. There are numerous techniques to prepare graphene/PU nanocomposites, such as in situ polymerization, solventless method, solution route, melt method, sol–gel, etc. [186]. Graphene/PU nanocomposites possess several applications, for example, shape memory effect, gas barrier, oil adsorbent, dye-sensitized solar cells, EMI-shielding materials, and fuel cells [187].

1.4.6 Graphene/Poly(Vinyl Alcohol) Nanocomposites

Owing to its efficient film forming, gas (oxygen) barrier properties,solubility in water, biocompatibility, adhesiveness, and nonhazardous/toxic properties, poly(vinyl alcohol) (PVA) has been extensively used in protective coating and packaging applications [188–191]. However, PVA contains a number of hydroxyl groups in its structure and it has very poor water vapor barrier properties. Consequently, it is readily plasticized by water and the effect on swelling properties extremely limits its application [192, 193]. The incorporation of graphene or GO paves the way to overcome these limitations, since graphene has been demonstrated as an efficient nanofiller to improve the mechanical properties, thermal stability, and water vapor barrier properties [194, 195]. Due to the hydrogen bonding between graphene and PVA, the dispersion of graphene (at the molecular level) in the PVA matrix was improved. As a result,mechanical properties of the GO/PVA nanocomposite were found to be superior to those of bare PVA [125, 181]. There are so many routes that have been developed to fabricate GO/PVA nanocomposites, among which the most commonly used one is the solution-mixing method [196–200]. The thermal stability of the nanocomposite was improved significantly by the addition of graphene (0.2 wt%) into the PVA matrix [201].

1.4.7 Graphene/Polyethylene Terephthalate Nanocomposites

PET is a semicrystalline thermoplastic polymer with exceptional molecular structure, high thermal stability, low melt viscosity, excellent mechanical properties, and chemical resistance [202]. However, the crystallization behavior of PET limits its applications [203]. The addition of nanofillers into the PET matrix helps to alter the crystallization behavior of PET [204]. Liu et al. demonstrated reduced GO/PET composites using in situ melt polycondensation. The enhanced tensile strength of more than 60% was observed at 0.5% of GO loadings [205]. Similarly, Zhang et al. reported the synthesis of GO/PET nanocomposites via the melt-compounding method. The electrical conductivity of GO/PET nanocomposites improved rapidly with the addition of graphene [129].

1.4.8 Graphene/Polycarbonate Nanocomposites

The fabrication of graphene/PC nanocomposites is highly desirable because of the optical transparency, enhanced mechanical strength, temperature resistance, high impact strength, dimensional stability, and good thermal stability of the matrix polymer [206, 207]. Functionalized graphene sheet (FGS)/PC and graphite/PC nanocomposites were prepared using the melt-compounding method [137]. The electrical conductivity and tensile modulus of the graphene/PC nanocomposites were found to be higher than those of neat PC.

1.4.9 Graphene/Poly(Vinylidene Fluoride) Nanocomposites

Poly(vinylidene fluoride) (PVDF) is a well-known ferroelectric semicrystalline polymer. Owing to its unique features, PFDF-based nanocomposites have been extensively used in numerous applications such as antistatic shielding, sensors, self-regulated heaters, actuators, nonvolatile memories, overcurrent protectors, lithium batteries, and energy harvesters [208, 209]. Song et al. prepared graphene/PVDF composites using ultrasonic processing and mechanical mixing [210]. FGS/PVDF nanocomposites were prepared by solution processing and compression molding using GO and expanded graphite [208]. The mechanical properties of both composites are found to be higher than those of neat PVDF. In addition, the thermal stability of the FGS/PVDF nanocomposite was found to be higher than that of the expanded graphite/PVDF nanocomposite.

1.4.10 Graphene/Nafion Nanocomposites

Nafion has comparably better mechanical strength, ion exchange capacity, and chemical stability and higher ionic conductivity. Thus, it has been widely used in various applications such as electrochemical sensors and fuel cells [211]. A modified electrode based on tris(2,21-bipyridyl) ruthenium (II) $[Ru(bpy)_3]^{2+}$, Nafion, and graphene was prepared by the solution-mixing method of graphene and Nafion [212]. The modified electrode offers superior sensitivity, specificity, and stability.

1.4.11 Graphene/Carbon Nanotube–Polymer Nanocomposites

The combination of CNTs and graphene with polymers provides a route to the creation of materials with countless applications almost in a combinatorial manner because it does not only refer to the combination of two compounds but in the assembly of two families of materials. Polymer–CNT and polymer–graphene-based sensors have demonstrated their great potential in a wide variety of challenging chemical sensing and biosensing applications. The synergistic effect of the intrinsic properties of both carbon nanomaterials such as near-infrared (NIR) fluorescence or fluorescence quenching, high electrical and thermal conductivity, chemical stability, and mechanical strength with the tuneable properties of polymers in terms of their chemical structure and functionality, combined with their low cost, easy processability, recyclability, and sustainability, makes these polymer composites ideal for the development of new types of chemical sensors [213].

1.4.12 Typical Graphene-Based Polymer Composites

There are several other graphene-based polymer composites reported. For example, a GO/poly(ε-caprolactone) (PCL) composite was synthesized by in situ polymerization and the resulting nanocomposite possesses excellent mechanical properties and robustness [214]. Similarly, GO/poly(lactic acid) (PLA) nanocomposites were fabricated using the response surface methodology [215]. The addition of graphene into the PLA polymer resulted in improved tensile strength. Preparation of a graphene/PMMA nanocomposite via in situ polymerization was reported by Mohammadi et al. [216]. Pan et al. fabricated graphene/PA composite coatings using the spraying method.The tribological results of the resulting composite coatings explored that the wear life of composite coatings was found to be superior to that of neat coatings [217]. Pang and coworkers prepared a novel conductive composite made of ultrahigh-molecular-weight polyethylene (HMWPE) with a double-percolated and segregated structure [218]. The graphene/polydiacetylene (PDA) nanocomposites were synthesized by Liang et al. using the solution-processing method [219]. The excellent actuation character with fast response rate, controllable motion, and high-frequency resonance was observed in the resulting nanocomposites. A graphene/PPS composite was fabricated by spraying methodology and the resulting nanocomposite possessed seven times greater wear life than that of the neat polymer matrix [220]. Graphene as a fine filler with high strength can improve the load-carrying capacity of the composite coating and make wear life increase significantly.

1.5 Applications of Graphene-Based Polymer Composites

Owing to the enhanced performance of graphene-functionalized polymer nanocomposites, these materials have been extensively utilized for a wide range of industrial and practical applications such as electronic devices, ESD and EMI shielding, energy storage, sensors, electronic devices, and biomedical applications (Fig. 1.4) [13, 132]. In addition, graphene-based nanocomposites also employed electrodes materials for organic light-emitting diodes, dye-sensitized solar cells, organic solar cells, liquid crystals, and field emission devices.

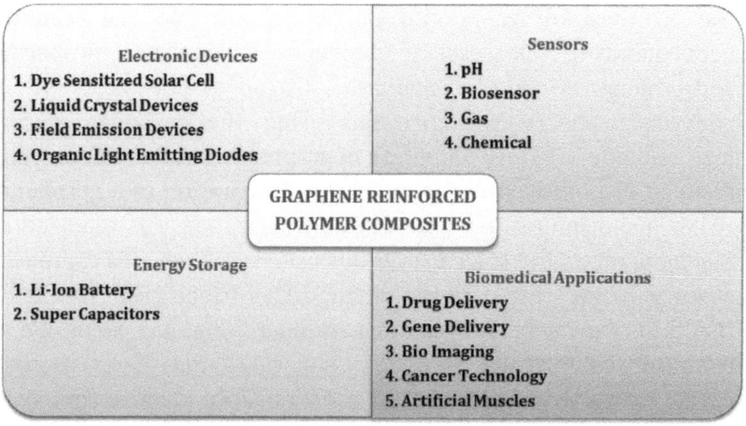

Figure 1.4 Applications of graphene-assembled polymer composites.

1.5.1 Sensors Applications

Graphene has a large specific area, low Johnson noise, and conductance-changing behavior as a function of surface adsorption. Graphene has been demonstrated as a promising candidate for the detection of a variety of target molecules. Sensors can be fabricated with the combination of nanofiller and conducting polymers. As a result, graphene-functionalized or graphene-reinforced polymer nanocomposite materials can be used for various sensor applications (see Table 1.3) (e.g., temperature, biomolecules, pressure, pH, and strain sensors) owing to their 2D (atom-thick) conjugated structures, higher conductivity, and large specific surface areas [132, 208, 221–224]. In addition, graphene-based polymer composite films have superior electrocatalytic activity, enhanced electrochemical stability, and faster charge transfer between the components. With all these advantages and performance, graphene-based polymer composite materials became a promising candidate for numerous sensing applications [225–227]. Moreover, graphene is also impermeable to gaseous molecules, thus paving the way for gas sensor applications [228–231a].

Table 1.3 Comparison of various graphene-based polymer nanocomposites for sensing applications

S. no.	Graphene–polymer composites	Target	LOD	Reference(s)
1.	PANI/GO/ZnO	NH_3	300 ppm	[246]
2.	PPy/GO	NH_3	1300 ppm	[247]
3.	PANI/rGO	NH_3	50 ppm	[248]
4.	PPy/ZnO	LPG	1400 ppm	[271]
5.	rGO/SnO_2/PANI	NH_3	20 ppm	[249]
6.	rGO/Au	NO_2	5 ppm	[272]
7.	Graphene quantum dots	NH_3	10 ppm	[273]
8.	1D ZnO NR/PP	NH_3	1000 ppm	[274]
9.	PANI/ZnO	NH_3	1000 ppm	[275]
10.	PANI/Cu	NH_3	50 ppm	[276]
11.	Graphene/PVDF	Temperature	NA	[208]
12.	HPCD-GO/TPP	Hemoglobin	5.0×10^{-9} M	[260]
13.	Graphene/PEI/GOD	Glucose	2 mM	[261]
14.	Graphene/GOD/chitosan	Glucose	0.02 mM	[262]
15.	GOD/Pt/graphene/chitosan	Glucose	0.6 μM	[263]
16.	GOD/Au/graphene/Nafion	Glucose	1 μM	[264]

(Continued)

Table 1.3 (*Continued*)

S. no.	Graphene–polymer composites	Target	LOD	Reference(s)
17.	Graphene/Nafion/GC	Guanine and adenine	0.58 M and 0.75 M	[267]
18.	Graphene/Chitosan	UA	NA	[268]
19.	Graphene/silver/silver oxide/Nafion	Thrombonodulin	31.5 pg/mL	[269]
20.	Graphene/Nafion/GCE	Methyl parathion	1.6 ng/ml	[270]
21.	Pd-decorated graphene	Hydrogen gas	0.1%	[277]
22.	Graphene/PANI	Hydrogen gas	1%	[164]
23.	Graphene/ZnO	Hydrogen gas	200 ppm	[278]
24.	Polyaniline/graphene	Hydrazine	~15.38 mM	[257]
25.	Graphene/rubber	Motion sensors	NA	[279]
26.	Au@PPy/rGO	DA	18.92 pM	[280]
27.	PEDOT-rGO composite	NO_2	1 ppm	[245]
28.	Pt NP ensemble-on-graphene hybrid nanosheet (PNEGHNS)	H_2O_2	80 nM	[281]
29.	Polymer film at chitosan–platinum NPs/graphene–gold NPs double nanocomposites	Erythromycin	2.3×10^{-8} mol/L	[282]
30.	NiCPNP/rGO nanocomposites	Glucose sensing	–	[283]
31.	GSCR-MIPs	DA	1.0×10^{-7} M	[284]

S. no.	Graphene–polymer composites	Target	LOD	Reference(s)
32.	Graphene/copper phthalocyanine/PANI nanocomposites	AA	24.46 μA mM^{-1}	[285]
33.	Graphene/PANI nanocomposites	NH_3	1 ppm	[286]
34.	MIP-coated graphene quantum dots	*p*-nitrophenol	9.00 ng mL^{-1}	[287]
35.	GO/poly(diallyl-dimethyl ammonium chloride) (PDDA) nanocomposites	Humidity sensor	NA	[288]
36.	SnO_2/rGO, CuO/rGO	HCHO and NH_3 Gas	0.8 ppm	[289]
37.	rGO/PDDA nanocomposite	Humidity	NA	[290]
38.	rGO MIPs	*p*-nitrophenol	0.005 μM	[291]
39.	GO/SiO_2 MIPs	DA	3.0×10^{-8} M	[292]
40.	rGO/P NFs	NO_2	0.15 ppm	[293]
41.	rGO bonded to Au electrode	NO_2	1.00 ppm	[294]
42.	rGO/SnO_2	NO_2	1.00 ppm	[295]
43.	rGO/Au NPs	NO_2	0.20 ppm	[296]
44.	rGO/polymer 3D	NO_2	NA	[297]
45.	Porous PEDOT on rGO	NO_2	2 ppm	[298]
46.	rGO/PANI nanocomposites and AuNPs@MIPs	Serotonin (5-hydroxytryptamine, 5-HT)	11.7 nmol/L	[299]

(Continued)

Table 1.3 (*Continued*)

S. no.	Graphene–polymer composites	Target	LOD	Reference(s)
47.	GO, and a cationic conjugated polymer, poly[(9,9-bis (6'-*N,N,N*-trimethylammonium)hexyl)-fluorenylene phenylene dibromide] (PFP)	DNA	40 pM	[300]
48.	Graphene/PANI	Methane	~200 ppm	[301]
49.	Graphene/PANI	NH_3	1 ppm	[286]
50.	PEDOT/GO	DA	39 nM	[136]
51.	MIP/graphene–Au NPs	Glycoprotein	7.5×10^{-12}g/mL	[302]
52.	PDA/grapheme	THF, $CHCl_3$, CH_3, OH, and DMF	0.01%	[303]
53.	PPy graphene nanocomposite decorated with TiO_2 NPs	NH_3	50 ppm	[304]
54.	Graphene-PEDOT:PSS Graphene poly(3,4-ethylenedioxythiophene):PSS	NH_3	500 ppm	[305]
55.	PPy/rGO	NH_3	3 ppm	[305]
56.	PEDOT/GO film	Hydroquinone (HQ) and catechol(CT)	1.6 μM	[306]
57.	Graphene/PANI	Toluene	100 ppm	[307]
58.	Graphene–PEDOT:PSS	NH_3	25 ppm	[308]
59.	GO–magnetite–MIPs	Epinephrine	1.09×10^{-9} mol/L	[309]
60.	Pt/PANI/graphene NS	H_2O_2 and glucose	NA	[310]

S. no.	Graphene–polymer composites	Target	LOD	Reference(s)
61.	rGO/PI	Strain	NA	[311]
62.	Graphene/PANI/PS	Pb^{2+} and Cd^{2+}	3.30 μg L^{-1} and 4.43 μg L^{-1} for Pb^{2+} and Cd^{2+}	[312]
63.	Graphene/copper(II)phthalocyanine-tetrasulfonic acid tetra sodium salt (CuPc)/PANI	AA	6.3×10^{-8} M	[285]
64.	BQD-PAA-GO, OQD-P2VP-GO.poly(acrylic acid) (PAA) and poly(2-vinyl pyridine) (P2VP)	pH	NA	[313]
65.	GSCR-MIPs	DA	10^{-7}	[284]
66.	PVA/GO	Water vapor	NA	[41]
67.	Graphene-MIPs	Bovine hemoglobin (BHb)	2.0×10^{-10} mg mL^{-1}	[42]
68.	Graphene–chitosan MIPs	DA	10^{-11}M	[43]
69.	Chitosan–Pt NPs and graphene–Au NPs / graphene–Au NPs/chitosan–Pt NPs/Au electrode	Erythromycin	2.3×10^{-8} mol L^{-1}	[282]
70.	PANI/rGO/GCE	Clonazepam	4.63 ng mL^{-1}	[44]
71.	PANI/GO/GCE	Calcium channel blocker drug levamlodipine (LAMP)	1.07 ng mL^{-1}	[45]
72.	Poly-DA-treated GO/PVA	Humidity	NA	[46]
73.	Phenylenediamine (PPD) rGO	Dimethyl methylphosphonate (DMMP)	10 ppm	[47]
74.	PANI/GO, PANI/GO/ZnO	NH_3	1000 ppm	[246]

1.5.2 Gas Sensors

The monitoring, alerting, and rapid detection of toxic gases [231b] become more predominant to prevent or minimize accidents that involve explosions or poisoning. Toxic or bad odors, such as hydrogen sulfide (H_2S), ammonia (NH_3), carbon monoxide (CO), chlorine (Cl_2), bromine (Br_2), hydrogen chloride (HCl), hydrogen fluoride (HF), nitric oxide (NO), nitrogen dioxide (NO_2), sulfur dioxide (SO_2), hydrogen cyanide (HCN), phosgene ($COCl_2$), benzene (C_6H_6), formaldehyde (HCHO), methyl bromide (CH_3Br), arsine (AsH_3), phosphine (PH_3), boranes (BH_3), silane (SiH_4), ozone (O_3), propane (C_3H_8), methane (CH_4), liquefied petroleum gas (LPG), and germane (GeH_4) gases, are repeatedly encountered in livelihood circumstances, industries (chemical, petroleum, food, electronic, coal mines), warehouses, vehicles, enclosed parking areas, waste disposal, sewerage, and even battle fields. Recently, the prerequisite of sensors that can detect air pollutants (NO_x, SO_x, and CO_2) in environments has increased significantly. Great efforts have been made to reduce and control the exhausts of pollutants from industrial stationary facilities and automobiles. These gaseous elements are mostly present at very trace levels and often mixed with various disturbing gases. Consequently, the development of highly sensitive with enhanced-specificity gas sensors for the same (depending on the conditions of operation) and different gaseous targets becomes highly desirable [232, 233]. The conventional methods (calorimetric and chromatographic methods) are time consuming and tedious. As a result, fast exact detection methods were evolved during these decades [234–243].

For instance, Al-Mashat et al. reported the preparation of a graphene/PANI nanocomposite as a sensitive layer and its application toward the development of a hydrogen (H) gas sensor. In this method, the performance of the developed sensor was compared with that of sensors based on only graphene sheets and PANI nanofibers separately. Owing to the high surface-to-volume ratio of graphene, the developed sensor has a higher sensitivity of 16.57% toward 1% of H gas than that of sensors based on only graphene sheets (0.83%) and PANI nanofibers (9.38%) [164]. Alizadeh et al. demonstrated the blending of chemically exfoliated graphene with PMMA and their utilization as a chemiresistor sensor for the sensitive detection of formaldehyde vapor with a detection limit of 0.01 ppm [244]. Dunst et al. fabricated an electrochemical sensor for NO_2 detection based on a PEDOT/rGO composite via electrodeposition [245]. The effect of GO on gas-sensing properties of the conducting polymer (PANI) and metal oxide (ZnO) was studied by Gaikwad et al. The PANI/GO/ZnO nanocomposite showed a response of 5.706 for 1000 ppm of NH_3, which has a 10.3 times enhanced response than that of the PANI sensor [246].

In another study, Patil et al. reported the in situ polymerization of a PPy/GO composite and its gas-sensing application. The gas-sensing properties of PPy/GO nanocomposites were tested with H_2S, LPG, CO_2, and NH_3 at room temperature. It was shown that PPy/GO nanocomposites with varying weight ratios of GO (5%, 10%, and 20%) had enhanced sensitivity and selectivity towards the detection of NH_3 [247]. Huang et al. demonstrated an NH_3 gas sensor based on rGO-anchored PANI hybrids. The effective detection of ammonia gas with a positive synergetic effect was achieved by the combination of graphene with PANI [248]. Ye et al. fabricated an NH_3 sensor based on GO@SiO_2-reinforced PANI composites via in situ chemical oxide polymerization to improve the sensing potential of PANI. As compared to bare PANI and GO@SiO_2, GO@SiO_2–PANI composites showed efficient response toward the detection ofNH_3 [249].

One-dimensional nanostructures of metal oxides such as ZnO, SnO_2, and Cu_2O nanowires (NWs) or nanorods (NRs) have been widely explored for sensing applications, mainly due to their large specific surface areas, high length-to-width ratios, and excellent mechanical flexibility [250–252]. However, the low conductivities of these nanostructures usually limit their performances. Blending them with 2D graphene sheets to form hybrid architectures can improve their sensing behaviors. For example, Kohl et al. developed a wet method to grow vertically aligned ZnO NRs on a chemically converted graphene (CCG) film. The resulting ZnO/graphene hybrid can be used to detect H_2S at room temperature [252]. In this case, the adsorption of oxygen on the surface of ZnO NRs was crucial for achieving excellent sensing performance, possibly due to the fact that the adsorbed oxygen was converted to ionic species by capturing electrons from ZnO. Therefore, the sensor exhibited a resistance increase in an oxygen environment. After introducing H_2S, the electron concentration on the surface of ZnO NRs increased due to the interaction between H_2S and the adsorbed oxygen ions. Consequently, the resistance of the ZnO NR/graphene composite-based sensor decreased [253].

1.5.3 Applications of Biosensors, Optical Sensors, and Calorimetric Sensors

Eswaraiah et al. demonstrated the real-time strain sensor based on graphene-functionalized poly(vinylidene fluoride) (f-GPVDF) nanocomposite films using the solvent-casting method [254]. In that, the surface functionalization of graphene facilitates the development of a 3D crosslinked network of graphene with the polymer matrix. Similarly, noncovalent functionalization of graphene by pyrene carboxylic acid was reported by Xia et al. A number of optical and molecular sensing properties (that are not present in pristine graphene films) were achieved by this method without any change in the conducting nature of graphene [255]. Han et al. successfully developed a modified electrode using chitosan-functionalized graphene for the determination of ascorbic acid (AA), dopamine (DA), and uric acid (UA), with a detection range of 50–1200 mM (AA), 1.0–24 mM (DA), and 2.0–45 mM (UA) [256]. Sadia Ameen et al. reported the fabrication of a modified electrode using PANI/graphene composites (via in situ electrochemical synthesis) for hydrazine sensors. This electrode shows very high sensitivity ($\sim 32.54 \times 10^{-5}$ A cm^{-2} mM^{-1}) with a detection limit of ~15.38 mM [257].

Similarly, Liu et al. reported a PVP/graphene-modified glassy carbon electrode (GCE) for the sensitive electrochemical detection of DA in the presence of AA [258]. Wang et al. fabricated rGO-doped conducting polymer–PEDOT nanocomposites for the detection of DA. In this method, rGO/PEDOT nanocomposites were prepared via electrochemical reduction. The limit of detection (LOD) was found to be 39 nM without any interference from AA and UA [136]. Liu et al. developed a sensor of a tetracycline antibiotic with poly(*o*-phenylenediamine) (PoPD), molecularly imprinted polymer (MIP), and rGO [259]. This sensor utilizes the response application of rGO and the special recognition of the MIP. Xu and coworkers demonstrated a hemoglobin (Hb) sensor using hydroxypropyl-b-cyclodextrin (HPCD)-modified GOs and tetra-phenylporphyrin

(TPP). Due to the interactions (photoinduced electron transfer) between HPCD and TPP, HPCD-GO/TPP/glassy carbon (GC) acts as an excellent biosensor that possesses exceptional electrocatalytic activity toward the oxidation as well as reduction of Hb. The HPCD-GO/TPP-modified electrode exhibited a detection limit of 5×10^{-9} M [260]. The first graphene-based electrochemical biosensor for the detection of glucose was constructed by Shan et al. The novel PVP-decorated graphene/polyethyleneimine (PEI)-functionalized ionic liquid/glucose oxidase (GOD) was used as a modified electrode [261]. Similarly, Kang's group reported the fabrication of a GCE with graphene/chitosan film for the detection of glucose [262], and their LOD is found in the range from 0.08 to 12 mM. In another study, a GOD/Pt/graphene/chitosan-based nanocomposite film was prepared by Wu et al. As-prepared nanocomposites offered quick and sensitive determination with a detection limit of 0.6 μM of glucose [263]. Baby et al. reported GOD/Au/graphene/Nafion-based amperometric glucose biosensors, which exhibited superior sensing applications with a detection limit of 1 μM [264a].

The many important sensing properties of graphene are attributed to its capability to detect biomolecules. It was Lu et al. [264b] who reported a graphene-based biosensor along with a dye-labeled as a DNA probe that could be found and quenched by GO for the first time, resulting from the fluorescent energy transfer between the dye and GO. In addition, the sensitive charge carrier modulation of chemically modified graphene has allowed the development of biodevices that can detect a single bacterium/sense DNA by Jang et al. [264c].

Hydrogen peroxide, which is a general enzymatic production of oxidases and peroxidases, has been applied in the food industry as a mediator. So it is necessary to detect hydrogen peroxide. To detect hydrogen peroxide, Zhou and coworkers developed a graphene-modified sensor to investigate the electrochemical feature of hydrogen peroxide on this sensor [265]. The graphene-modified sensor shows a remarkable increase in the electron transfer rate compared to a graphite-modified sensor. On the basis of the high electrocatalytic activity of graphene toward hydrogen peroxide, graphene could be an excellent electrode material for oxidase biosensors. Shan et al. [266a] designed a graphene/PEI-functionalized ionic liquid nanocomposite–modified biosensor to detect glucose. This graphene sensor possesses a wide response from 2 to 14 mM ($R = 0.994$), good reproducibility (relative standard deviation of the current response to 6 mM glucose at −0.5 V was 3.2% for 10 successive measurements), and high stability (response current + 4.9% after 1 week).

Nowadays, optical and colorimetric biosensors have attracted plentiful attention because of their low cost, simplicity, ease of use, and sensibleness [266b]. Since the changes in color can be read by the naked eye, colorimetric biosensors do not require expensive or advanced instrumentation and can be applied to field investigation and point-of-care diagnosis [266c]. The basic strategy of colorimetric biosensors is translating the detection events into color changes. There are many smart materials including graphene-based polymer nanomaterials. Itis worth mentioning here that there are only a of handful reports describing optical biosensors based on graphene and polymer nanocomposites, which is very much open for young researchers to investigate.

To detect guanine and adenine in milk powder, urine, and DNA samples, Yin et al. designed a graphene/Nafion–composite film–modified GC electrode with excellent anti-interference, stability, and reproducibility [267]. Similarly, for the detection of UA, Lian and coworkers fabricated a highly sensitive graphene-doped chitosan-based molecularly imprinted electrochemical sensor [268]. Yang et al. reported the development of a Nafion-dispersed graphene solution and silver–silver oxide nanoparticles (Ag-Ag_2O NPs)-derived label-free electrochemical immune sensor for the determination of serum thrombomodulin (TM). The detection limit of TM was found as 31.5 pg/mL [269]. Further, Xue et al. developed an electrochemical sensor based on a graphene/Nafion-modified GC electrode for the detection of nitroaromatic/organophosphorus pesticides (OPS). Methyl parathion presented in the vegetable samples was successfully determined using this sensor system [270].

1.6 Conclusions, Outlook, and Future Scope

Materials science and engineering are playing a very important role in our day-to-day life, especially in the development of new materials. There have been several exciting advancements in the field of materials science in the past quarter century, many researchers have generated enthusiasm and excitement in the material known as graphene, the material of the future, which is pure carbon in the form of sheets one atom thick. Graphene is estimated to be as flexible as rubber, while conducting heat and electricity with enormous efficiency. Further, because it is only an atom thick, it is nearly 2D, imbuing it with many interesting light-related and water-related properties. According to Segal et al. [314], there has been a significant scale-up in the production of GO, and similarly, graphene platelets have also gathered interest as primary nanofillers, which led to their mass production since it is the onset of graphene-based nanocomposites [315]. Polymer nanocomposites are the result of continuous effort since the 1990s. Graphene-based polymer nanocomposites represent one of the most technologically promising developments to emerge from the interface of graphene-based materials and polymer materials.

Applications of graphene-based polymer nanocomposites are multidirectional, in almost all fields. Several branches such as chemistry, physics, and biology to chemical, mechanical, electrical, and civil engineering can allow the rise of graphene and its polymer nanocomposites to attain their true potential. Meaningful advancements to bridge the gap between graphene and its polymer nanocomposites (GPNC) research and its applications are likely to occur only if a broader scientific and engineering perspective is taken.

To optimize the properties of graphene–polymer composites, the control of graphene-based materials' orientation and dispersion during processing is critical. Much more efforts in nanoengineering are needed to understand the behavior of graphene-based materials in processing. In coating applications, coating methods have a profound effect on the properties and morphology of the resulted coating. Two or more coating techniques may be used at the same time to produce a multilayer graphene-based coating system to meet industrial requirements, as the current techniques are difficult to satisfy industry standards when they are used separately. The development of new technologies for graphene-based materials' fabrication and processing is still essential to face the demands and challenges of industries nowadays.

Graphene-based polymeric nanocomposites also show a very low percolation threshold of electrical conductivity and improved mechanical, thermal, and barrier properties. In this chapter, from the field of materials science, graphene-based polymer nanocomposites were highlighted and reviewed for their applications and for most promising developments. Also, potential applications of graphene in various polymer nanocomposite materials, based on GO and rGO, were highlighted. Due to the high diversity, properties, and advantages of graphene, a multitude of nanocomposite-based applications have been envisioned to be practical. These multifunctional graphene composites coupled with affordable cost will soon be seen in the global market.

Graphene-based polymer nanocomposites show promising growth in technology and applications and are emerging as one of the priority materials, both as a material and as a composite, but still lots of challenges lie on the path to be addressed and resolved to realize mature graphene/graphene-based polymer nanocomposites' utilization to their fullest potential regarding synthesis methods, costs, and applications. However, many challenges must be addressed for these nanocomposites in order to exploit their full potential. It is evidenced that graphene exhibits many extraordinary properties, while graphene-based materials have a wide range of potential applications such as flexible transparent electrodes, sensors, and electronic components. Defect-free graphene is the perfect material for being used in all kinds of applications, but the fabrication techniques are still not mature. Moreover, the scale-up of fabricating graphene-based materials with acceptable cost is still extremely challenging. Essentially, the potential health risks of graphene-based materials need to be evaluated before large-scale utilization. Though graphene or graphene-based polymer nanocomposites do not show up in the top 10 breakthroughs in materials science, some of the biggest areas of discovery, application, and commercial interest include the following:

- Many governments allocate research funding for innovations that have the potential to have a tangible economic impact.
- Graphene production can be one of the top five trends in mechanical engineering, along with other key materials' R&D by 2020.

- Because of various applications, research in materials science is performed and funded by a diverse group of players.
- Militaries looking at the novelty in utilizing graphene-based metamaterials, for example, experiments surrounding invisibility cloaks.
- CNTs with their present commercial applications, in combination with graphene-based polymer nanocomposites, likely play a central role in emerging nanotechnologies.
- Carbon fiber–reinforced plastics are used in racecars and bikes, among other applications.
- Materials for long-life Li-ion batteries are used in laptops and cellular phones.
- Academics often lead the way in research with breakthroughs.
- Outside of governments, industries, and institutions of higher education, direct investment in materials science research comes from corporations.

The dominance of graphene over CNTs as reinforcement stems from easy admission to the graphitic precursor material, scalable methods, cost, and orientation flexibility (morphology). Graphene's current and potential applications are astounding and could revolutionize many products, markets, and fields. These include the following:

- Electronics, computing. sensing, biomedical science, medicine, waterproofing, energy storage, water purification, and air purification are some applications.
- Composites of graphene plus plastics or polymers can be an industry standard, from bikes to wind turbines in the years to come.
- A potential material for 3D printing.
- Graphene composites are making their way to production in many industries, such as aerospace and automobiles, for example, in the production of lightweight planes (~3700 kg on average), indirectly contributing toward a reduction in CO_2 emission.
- Further properties can be enhanced in graphene-based composites through improved morphological control.

- Graphene's incorporation into other materials, such as paint, plastic, and polymer production, for novel properties is also being explored.

However, the three major challenges in the upcoming future are fresh air, pure water, and green energy. Hopefully the use of graphene-based polymer nanocomposite materials can solve all these three problems, and one can hope for the material of the future to not only undoubtedly solve these problems but also revolutionize the future usage of composite materials.

References

1. Mukhopadhyay, P. and Gupta, R. K. (2013). Graphene/polymer nanocomposites, in *Graphite, Graphene, and Their Polymer Nanocomposites*, eds. Kim, H., Abdala, A. A. and Macosko, C. W. (CRC Press, Boca Raton, FL), Chapter 16, pp. 513–556.
2. Geim, A. K. (2007). The rise of graphene, *Nat. Mater.*, **6**, pp. 183–191.
3. Novoselov, K. S., Geim, A. K., Morozov, S. V., Jiang, D., Zhang, Y., Dubonos, S. V., Grigorieva, I. V. and Firsov, A. A. (2004). Electric field effect in atomically thin carbon films, *Science*, **306**, pp. 666–669.
4. Xiaoqing, Y., Wensi, Z., Panpan, Z. and Zhiqiang, S. (2017). Fabrication technologies and sensing applications of graphene-based composite films: advances and challenges, *Biosens Bioelectron.*, **89**, pp. 79–84.
5. Xia, J., Masaki, N., Jiang, K. and Yanagida, S.(2007). The influence of doping ions on poly (3, 4-ethylene dioxythiophene) as a counter electrode of a dye-sensitized solar cell, *J. Mater. Chem.*, **17**, pp. 2845–2850.
6. Papageorgiou, N., Maier, W. F. and Grätzel, M. (1997). An iodine/triiodide reduction electrocatalyst for aqueous and organic media, *J. Electrochem. Soc.*, **144**, pp. 876–884.
7. Kay, A. and Grätzel, M. (1996). Low-cost photovoltaic modules based on dye sensitized nanocrystalline titanium dioxide and carbon powder, *Sol. Energy Mater. Sol. Cells*, **44**, pp. 99–117.
8. He, Q., Wu, S., Gao, S., Cao, X., Yin, Z., Li, H., Chen, P. and Zhang, H. (2011). Transparent, flexible, all-reduced graphene oxide thin film transistors, *ACS Nano*, **5**, pp. 5038–5044.
9. Liu, G., Jin, W. and Xu, N. (2015). Graphene-based membranes, *Chem. Soc. Rev.*, **44**, pp. 5016–5030.

10. Lee, J. H., Lee, E. K., Joo, W. J., Jang, Y., Kim, B. S., Lim, J. Y., Choi, S. H., Ahn, S. J., Ahn, J. R., Park, M. H., Yang, C. W., Choi, B. L., Hwang, S. W. and Whang, D. (2014). Wafer-scale growth of single-crystal monolayer graphene on reusable hydrogen-terminated germanium, *Science*, **344**, pp. 286–289.

11. Pumera, M. (2010). Graphene-based nanomaterials and their electrochemistry, *Chem. Soc. Rev.*, **39**, pp. 4146–4157.

12. Yang, Z., Ren, J., Zhang, Z., Chen, X., Guan, G., Qiu, L., Zhang, Y. and Peng, H. (2015). Recent advancement of nanostructured carbon for energy applications, *Chem. Rev.*, **115**, pp. 5159–5223.

13. Abraham, J., Arify, M. and Thomas, S. (2016). A comprehensive study of surface modified graphene-based polymer nanocomposites for multifunctional electronic applications, in *Young Researchers in Vacuum Micro/Nano Electronics*, pp. 1–13.

14. Salavagione, H. J., Martínez, G. and Ellis, G. (2011). Graphene-based polymer nanocomposites, in *Physics and Applications of Graphene: Experiments*, ed. Mikhailov, S. (InTech), doi:10.5772/14665.

15. Ding, J., Sun, W., Wei, G. and Su, Z. (2015). Cuprous oxide microspheres on graphene nanosheets: an enhanced material for non-enzymatic electrochemical detection of H_2O_2 and glucose, *RSC Adv.*, **5**, pp. 35338–35345.

16. Zhang, P., Lu, X., Huang, Y., Deng, J., Zhang, L., Ding, F., Su, Z., Wei, G. and Schmidt, O. G. (2015). MoS_2 nanosheets decorated with gold nanoparticles for rechargeable Li–O_2 batteries, *J. Mater. Chem. A*, **3**, pp. 14562–14566.

17. Tjong, S. C. (2011). Polymer nanocomposite bipolar plates reinforced with carbon nanotubes and graphite nanosheets, *Energy Env. Sci.*, **4**, pp. 605–626.

18. Chiacchiarelli, L. M., Rallini, M., Monti, M. and Puglia, D. (2013). The role of irreversible and reversible phenomena in the piezoresistive behaviour of graphene epoxy nanocomposites applied to structural health monitoring, *Compos. Sci. Technol.*, **80**, pp. 73–79.

19. Ren, L., Qiu, J. and Wang, S. (2012). The thermo-adaptive functionality of graphene/polydimethylsiloxane nanocomposites, *Smart Mater. Struct.*, **21**, p. 105032.

20. Chen, Y., Qi, Y., Tai, Z., Yan, X., Zhu, F. and Xue, Q. (2012). Preparation, mechanical properties and biocompatibility of graphene oxide/ ultrahigh molecular weight polyethylene composites, *Eur. Polym. J.*, **48**, pp. 1026–1033.

21. Dreyer, D. R., Park, S., Bielawski, C. W. and Ruoff, R. S. (2010). The chemistry of graphene oxide, *Chem. Soc. Rev.*, **39**, p. 228.

22. Novoselov, K. S., Fal'ko, V. I., Colombo, L., Gellert, P. R., Schwab, M. G. and Kim, K. (2012). A roadmap for graphene, *Nature*, **490**, pp. 192–200.

23. Krane, N. (2011). Preparation of graphene, *Selected Topics in Physics: Physics of Nanoscale Carbon* (Freie Univ., Berlin).

24. Kosynkin, D. V., Higginbotham, A. L., Sinitskii, A., Lomeda, J. R., Dimiev, A., Price, B. K. and Tour, J. M. (2009). Longitudinal unzipping of carbon nanotubes to form graphene nanoribbons, *Nature*, **458**, pp. 872–876.

25. Jiao, L., Zhang, L., Wang, X., Diankov, G. and Dai, H. (2009). Narrow graphene nanoribbons from carbon nanotubes, *Nature*, **458**, pp. 877–880.

26. Cano-Márquez, A. G., Rodíguez-Macías, F. J., Campos-Delgado, J., Espinosa-González, C. G., Tristán-López, F., Ramírez-González, D., Cullen, D. A., Smith, D. J., Terrones, M. and Vega-Cantú, Y. I. (2009). Ex-MWNTs: graphene sheets and ribbons produced by lithium intercalation and exfoliation of carbon nanotubes, *Nano Lett.*, **9**, pp. 1527–1533.

27. Elıas, A. L., Botello-Mendez, A. S. R., Meneses-Rodríguez, D., Jehova González, V., Ramírez-González, D., Ci, L., Munoz-Sandoval, E., Ajayan, P. M., Terrones, H. and Terrones, M. (2010). Longitudinal cutting of pure and doped carbon nanotubes to form graphitic nanoribbons using metal clusters as nanoscalpels, *Nano Lett.*, **10**, pp. 366–372.

28. Berger, C., Song, Z., Li, X., Wu, X., Brown, N., Naud, C., Mayou, D., Li, T., Hass, J., Marchenkov, A. N. and Conrad, E. H. (2006). Electronic confinement and coherence in patterned epitaxial graphene, *Science*, **312**, pp. 1191–1196.

29. Kim, K. S., Zhao, Y., Jang, H., Lee, S. Y., Kim, J. M., Kim, K. S., Ahn, J. H., Kim, P., Choi, J. Y. and Hong, B. H. (2009). Large-scale pattern growth of graphene films for stretchable transparent electrodes, *Nature*, **457**, pp. 706–710.

30. Li, X. S., Cai, W. W., An, J., Kim, S., Nah, J., Yang, D., Piner, R., Velamakanni, A., Jung, I., Tutuc, E. and Banerjee, S. K. (2009). Large-area synthesis of high-quality and uniform graphene films on copper foils, *Science*, **324**, pp. 1312–1314.

31. Zhang, W., Cui, J., Tao, C. A., Wu, Y., Li, Z., Ma, L., Wen, Y. and Li, G. (2009). A strategy for producing pure single-layer graphene sheets based on a confined self-assembly approach, *Angew. Chem. Int. Ed.*, **48**, pp. 5864–5868.

32. Wang, X., You, H., Liu, F., Li, M., Wan, L., Li, S., Li, Q., Xu, Y., Tian, R., Yu, Z., Xiang, D. and Cheng, J. (2009). Large-scale synthesis of few-layered graphene using CVD, *Chem. Vap. Deposition*, **15**, pp. 53–56.

33. Wang, Y., Chen, X., Zhong, Y., Zhu, F. and Loh, K. P. (2009). Large area, continuous, few-layered graphene as anodes in organic photovoltaic devices, *Appl. Phys. Lett.*, **95**, pp. 063302/1–063302/3.

34. Xianbao, W., Haijun, Y., Fangming, L., Mingjian, L., Li, W., Shaoqing, L., Qin, L., Yang, X., Rong, T., Ziyong, Y., Dong, X. and Jing, C. (2009). Large-scale synthesis of few-layered graphene using CVD, *Chem. Vap. Deposition*, **15**, pp. 53–56.

35. Dervishi, E., Li, Z., Watanabe, F., Biswas, A., Xu, Y., Biris Alexandru, R., Saini, V. and Biris Alexandru, S. (2009). Large-scale graphene production by the RF-CCVD method, *Chem. Commun.*, **27**, pp. 4061–4063.

36. Li, X., Cai, W., An, J., Kim, S., Nah, J., Yang, D., Piner, R., Velamakanni, A., Jung, I., Tutuc, E., Banerjee, S. K., Colombo, L. and Ruoff, R. S. (2009). Large-area synthesis of high-quality and uniform graphene films on copper foils, *Science*, **324**, pp. 1312–1314.

37. Chong-an, D., Dacheng, W., Gui, Y., Yunqi, L., Yunlong, G. and Daoben, Z. (2008). Patterned graphene as source/drain electrodes for bottom-contact organic field effect transistors, *Adv. Mater.*, **20**, pp. 3289–3293.

38. Chae, S. J., Gunes, F., Kim, K. K., Kim, E. S., Han, G. H., Kim, S. M., Shin, H. J., Yoon, S. M., Choi, J. Y., Park, M. H., Yang, C. W., Pribat, D. and Lee, Y. H. (2009). Synthesis of large-area graphene layers on a poly-nickel substrate by chemical vapour deposition: wrinkle formation, *Adv. Mater.*, **21**, pp. 2328–2333.

39. Li, N., Wang, Z., Zhao, K., Shi, Z., Gu, Z. and Xu, S. (2009). Large-scale synthesis of n-doped multi-layered graphene sheets by the simple arc-discharge method, *Carbon*, **48**, pp. 255–259.

40. Karmakar, S., Kulkarni, N. V., Nawale, A. B., Lalla, N. P., Mishra, R., Sathe, V. G., Bhoraskar, S. V. and Das, A. K. (2009). A novel approach towards the selective bulk synthesis of few-layer graphenes in an electric arc, *J. Phys. D: Appl. Phys.*, **42**, pp. 115201–115214.

41. Ma, J., Li, Y., Yin, X., Xu, Y., Yue, J., Bao, J. and Zhou, T. (2016). Poly (vinyl alcohol)/graphene oxide nanocomposites prepared by in situ polymerization with enhanced mechanical properties and water vapour barrier properties, *RSC Adv.*, **6**, pp. 49448–49458.

42. Luo, J., Jiang, S. and Liu, X. (2014). An electrochemical sensor for bovine haemoglobin based on a novel graphene-molecular imprinted

polymers composite as a recognition element, *Sens. Actuators, B*, **203**, pp. 782–789.

43. Liu, B., Lian, H. T., Yin, J. F. and Sun, X. Y. (2012). Dopamine molecularly imprinted electrochemical sensor based on graphene–chitosan composite, *Electrochim. Acta*, **75**, pp. 108–114.
44. Jain, R., Sinha, A. and Khan, A. L. (2016). Polyaniline–graphene oxide nanocomposite sensor for quantification of calcium channel blocker levamlodipine, *Mater. Sci. Eng., C*, **65**, pp. 205–214.
45. Jain, R., Sinha, A., Kumari, N. and Khan, A. L. (2016). A polyaniline/graphene oxide nanocomposite as a voltammetric sensor for electroanalytical detection of clonazepam, *Anal. Methods*, **8**, pp. 3034–3045.
46. Hwang, S. H., Kang, D., Ruoff, R. S., Shin, H. S. and Park, Y. B. (2014). Poly (vinyl alcohol) reinforced and toughened with poly (dopamine)-treated graphene oxide, and its use for humidity sensing, *ACS Nano*, **8**, pp. 6739–6747.
47. Hu, N., Wang, Y., Chai, J., Gao, R., Yang, Z., Kong, E. S. W. and Zhang, Y. (2012). Gas sensor based on p-phenylenediamine reduced graphene oxide, *Sens. Actuators, B*, **163**, pp. 107–114.
48. Liu, Q., Zhu, X., Huo, Z., He, X., Liang, Y. and Xu, M. (2012). Electrochemical detection of dopamine in the presence of ascorbic acid using PVP/graphene modified electrodes, *Talanta*, **97**, pp. 557–562.
49. Liu, W., Xiao, J., Wang, C., Yin, H., Xie, H. and Cheng, R. (2013). Synthesis of polystyrene-grafted-graphene hybrid and its application intheelectrochemical sensor of dopamine, *Mater. Lett.*, **100**, pp. 70–73.
50. Liu, S., Xing, X., Yu, J., Lian, W., Li, J., Cui, M. and Huang, J. (2012). A novel label-free electrochemical aptasensor based on graphene-polyaniline composite film for dopamine determination, *Biosens. Bioelectron.*, **36**, pp. 186–191.
51. Han, H. S., Lee, H. K., You, J. M., Jeong, H. and Jeon, S. (2014). Electrochemical biosensor for simultaneous determination of dopamine and serotonin based on electrochemically reduced GO-porphyrin, *Sens. Actuators, B*, **190**, pp. 886–895.
52. Weng, X., Cao, Q., Liang, L., Chen, J., You, C., Ruan, Y., Lin, H. and Wu, L. (2013). Simultaneous determination of dopamine and uric acid using layer-by-layer graphene and chitosan assembled multilayer films, *Talanta*, **117**, pp. 359–365.
53. Hou, S., Kasner, M. L., Su, S., Patel, K. and Cuellari, R. (2010). Highly sensitive and selective dopamine biosensor fabricated with silanized graphene, *J. Phys. Chem. C*, **114**, pp. 14915–14921.

54. Liu, C., Zhang, J., Yifeng, E., Yue, J., Chen, L. and Li, D. (2014). One-pot synthesis of graphene–chitosan nanocomposite modified carbon paste electrode for selective determination of dopamine, *Electron. J. Biotechnol.*, **17**, pp. 183–188.

55. Han, D., Han, T., Shan, C., Ivaska, A. and Niu, L. (2010). Simultaneous determination of ascorbic acid, dopamine and uric acid with chitosan-graphene modified electrode, *Electroanalysis*, **22**, pp. 2001–2008.

56. Liu, S., Yu, B. and Zhang, T. (2013). Preparation of crumpled reduced graphene oxide–poly (p-phenylenediamine) hybrids for the detection of dopamine, *J. Mater. Chem. A*, **1**, pp. 13314–13320.

57. Liu, X., Zhu, H. and Yang, X. (2014). An electrochemical sensor for dopamine based on poly (o-phenylenediamine) functionalized with electrochemically reduced graphene oxide, *RSC Adv.*, **4**, pp. 3706–3712.

58. Qin, X., Xiao-Ya, H. and Shi-Rong, H. (2011). Electrochemical sensors based on electropolymerized films, in *Electropolymerization* (InTech), pp. 187–198.

59. Liao, C., Zhang, M., Niu, L., Zheng, Z. and Yan, F. (2014). Organic electrochemical transistors with graphene-modified gate electrodes for highly sensitive and selective dopamine sensors, *J. Mater. Chem. B*, **2**, pp. 191–200.

60. Ren, P. G., Yan, D. X., Chen, T., Zeng, B. Q. and Li, Z. M. (2011). Improved properties of highly oriented graphene/polymer nanocomposites, *J. Appl. Polym. Sci.*, **121**, pp. 3167–3174.

61. Fang, M., Wang, K., Lu, H., Yang, Y. and Nutt, S. (2009). Covalent polymer functionalization of graphene nanosheets and mechanical properties of composites, *J. Mater. Chem.*, **19**, pp. 7098–7105.

62. Fornes, T. D. and Paul, D. R. (2003). Modeling properties of nylon 6/clay nanocomposites using composite theories, *Polymer*, **44**, pp. 4993–5013.

63. Vermant, J., Ceccia, S., Dolgovskij, M. K., Maffettone, P. L. and Macosko, C. W. (2007). Quantifying dispersion of layered nanocomposites via melt rheology, *J. Rheol.*, **51**, pp. 429–450.

64. Qiu, J. and Wang, S. (2011). Enhancing polymer performance through graphene sheets, *J. Appl. Polym. Sci.*, **119**, pp. 3670–3674.

65. Lee, C., Wei, X., Kysar, J. W. and Hone, J. (2008). Measurement of the elastic properties and intrinsic strength of monolayer graphene, *Science*, **321**, pp. 385–388.

66. Zosel, A. (1982). Rheological properties of disperse systems at low shear stresses, *Rheol. Acta*, **21**, pp. 72–80.

67. Wang, J., Hu, H., Wang, X., Xu, C., Zhang, M. and Shang, X. (2011). Preparation and mechanical and electrical properties of graphene nanosheets–poly (methyl methacrylate) nanocomposites via in situ suspension polymerization, *J. Appl. Polym. Sci.*, **122**, pp. 1866–1871.

68. Liang, J., Xu, Y., Huang, Y., Zhang, L., Wang, Y., Ma, Y., Li, F., Guo, T. and Chen, Y. (2009). Infrared-triggered actuators from graphene-based nanocomposites, *J. Phys. Chem. C*, **113**, pp. 9921–9927.

69. Yang, Y., Wang, J., Zhang, J., Liu, J., Yang, X. and Zhao, H. (2009). Exfoliated graphite oxide decorated with PDMAEMA chains and polymer particles, *Langmuir*, **25**, pp. 11808–11814.

70. Garboczi, E. J., Snyder, K. A., Douglas, J. F. and Thorpe, M. F. (1995). Geometrical percolation threshold of overlapping ellipsoids, *Phys. Rev. E: Stat. Phys. Plasmas Fluids Relat. Interdiscip. Top.*, **52**, pp. 819–828.

71. Celzard, A., McRae, E., Deleuze, C., Dufort, M., Furdin, G. and Mareche, J. F. (1996). Critical concentration in percolating systems containing a high-aspect-ratio filler, *Phys. Rev. B: Condens. Matter*, **53**, pp. 6209–6214.

72. Wakabayashi, K., Pierre, C., Dikin, D. A., Ruoff, R. S., Ramanathan, T., Brinson, L. C. and Torkelson, J. M. (2008). Polymer-graphite nanocomposites: effective dispersion and major property enhancement via solid-state shear pulverization, *Macromolecules*, **41**, pp. 1905–1908.

73. Kim, H. and Macosko, C. W. (2008). Morphology and properties of polyester/exfoliated graphite nanocomposites, *Macromolecules*, **41**, pp. 3317–3327.

74. Ren, J., Silva, A. S. and Krishnamoorti, R. (2000). Linear viscoelasticity of disordered polystyrene-polyisoprene block copolymer based layered-silicate nanocomposites, *Macromolecules*, **33**, pp. 3739–3746.

75. Steurer, P., Wissert, R., Thomann, R. and Muelhaupt, R. (2009). Functionalized graphenes and thermoplastic nanocomposites based upon expanded graphite oxide, *Macromol. Rapid Commun.*, **30**, pp. 316–327.

76. Bao, C., Guo, Y., Song, L. and Hu, Y. (2011). Poly (vinyl alcohol) nanocomposites based on graphene and graphite oxide: a comparative investigation of property and mechanism, *J. Mater. Chem.*, **21**, pp. 13942–13950.

77. Liang, J., Huang, Y., Zhang, L., Wang, Y., Ma, Y., Guo, T. and Chen, Y. (2009). Molecular-level dispersion of graphene into poly (vinyl alcohol) and effective reinforcement of their nanocomposites, *Adv. Funct. Mater.*, **19**, pp. 2297–2302.

78. Wang, J., Wang, X., Xu, C., Zhang, M. and Shang, X. (2011). Preparation of graphene/poly (vinyl alcohol) nanocomposites with enhanced mechanical properties and water resistance, *Polym. Int.*, **60**, pp. 816–822.

79. Li, Y., Yang, T., Yu, T., Zheng, L. and Liao, K. (2011). Synergistic effect of hybrid carbon nanotube-graphene oxide as nanofiller in enhancing the mechanical properties of PVA composites, *J. Mater. Chem.*, **21**, pp. 10844–10851.

80. Vadukumpully, S., Paul, J., Mahanta, N. and Valiyaveettil, S. (2011). Flexible conductive graphene/poly (vinyl chloride) composite thin films with high mechanical strength and thermal stability, *Carbon*, **49**, pp. 198–205.

81. Bai, X., Wan, C., Zhang, Y. and Zhai, Y. (2011). Reinforcement of hydrogenated carboxylated nitrile–butadiene rubber with exfoliated graphene oxide, *Carbon*, **49**, pp. 1608–1613.

82. Zhang, H. B., Yan, Q., Zheng, W. G., He, Z. and Yu, Z. Z. (2011). Tough graphene– polymer microcellular foams for electromagnetic interference shielding, *ACS Appl. Mater. Interfaces*, **3**, pp. 918–924.

83. Chatterjee, S., Nüesch, F. A. and Chu, B. T. (2011). Comparing carbon nanotubes and graphene nanoplatelets as reinforcements in polyamide 12 composites, *Nanotechnology*, **22**, p. 275714.

84. Wang, X., Song, L., Yang, H., Lu, H. and Hu, Y. (2011). Synergistic effect of graphene on anti-dripping and fire resistance of intumescent flame retardant poly (butylene succinate) composites, *Ind. Eng. Chem. Res.*, **50**, pp. 5376–5383.

85. Cao, Y., Zhang, J., Feng, J. and Wu, P. (2011). Compatibilization of immiscible polymer blends using graphene oxide sheets, *ACS Nano*, **5**, pp. 5920–5927.

86. Wan, C. and Chen, B. (2011). Poly (ε-caprolactone)/graphene oxide biocomposites: mechanical properties and bioactivity, *Biomed. Mater.*, **6**, p. 055010.

87. Wang, J. Y., Yang, S. Y., Huang, Y. L., Tien, H. W., Chin, W. K. and Ma, C. C. M. (2011). Preparation and properties of graphene oxide/polyimide composite films with low dielectric constant and ultrahigh strength via in-situ polymerization, *J. Mater. Chem.*, **21**, pp. 13569–13575.

88. Liu, K., Chen, L., Chen, Y., Wu, J., Zhang, W., Chen, F. and Fu, Q. (2011). Preparation of polyester/reduced graphene oxide composites via in situ melt polycondensation and simultaneous thermo-reduction of graphene oxide, *J. Mater. Chem.*, **21**, pp. 8612–8617.

89. Kim, I. H. and Jeong, Y. G. (2010). Polylactide/exfoliated graphite nanocomposites with enhanced thermal stability, mechanical modulus, and electrical conductivity, *J. Polym. Sci., Part B: Polym. Phys.*, **48**, pp. 850–858.

90. Rollings, E., Gweon, G. H., Zhou, S. Y., Mun, B. S., McChesney, J. L., Hussain, B. S., Fedorov, A. V., First, P. N., de Heer, W. A. and Lanzar, A. (2006). Synthesis and characterization of atomically thin graphite films on a silicon carbide substrate, *J. Phys. Chem. Solids*, **67**, pp. 2172–2177.

91. de Heer, W. A., Berger, C., Wu, X., First, P. N., Conrad, E. H., Li, X., Li, T., Sprinkle, M., Hass, J., Sadowski, M. L., Potemski, M. and Martinez, G. (2007). Epitaxial graphene, *Solid State Commun.*, **143**, pp. 92–100.

92. Mattausch, A. and Pankratov, O. (2008). Density functional study of graphene overlayers on SiC, *Phys. Status Solidi B*, **245**, pp. 1425–1435.

93. Ni, Z. H., Chen, W., Fan, X. F., Kuo, J. L., Yu, T., Wee, A. T. S. and Shen, Z. X. (2008). Raman spectroscopy of epitaxial graphene on a SiC substrate, *Phys. Rev. B: Condens. Matter*, **77**, p. 115416.

94. Sutter, P. W., Flege, J. I. and Sutter, E. A. (2008). Epitaxial graphene on ruthenium, *Nat. Mater.*, **7**, pp. 406–411.

95. Seyller, T., Bostwick, A., Emtsev, K. V., Horn, K., Ley, L., McChesney, J. L., Ohta, T., Riley, J. D., Rotenberg, E. and Speck, F. (2008). Epitaxial graphene: a new material, *Phys. Status Solidi B*, **245**, pp. 1436–1446.

96. Sprinkle, M., Soukiassian, P., de Heer, W. A., Berger, C. and Conrad, E. H. (2009). Epitaxial graphene: the material for graphene electronics, *Phys. Status Solidi RRL*, **3**, pp. A91–A94.

97. Kosynkin, D. V., Higginbotham, A. L., Sinitskii, A., Lomeda, J. R., Dimiev, A., Price, B. K. and Tour, J. M. (2009). Longitudinal unzipping of carbon nanotubes to form graphene nanoribbons, *Nature*, **458**, pp. 872–876.

98. Hirsch, A. (2009). Unzipping carbon nanotubes: a peeling method for the formation of graphene nanoribbons, *Angew. Chem. Int. Ed.*, **48**, pp. 6594–6596.

99. Jiao, L., Zhang, L., Wang, X., Diankov, G. and Dai, H. (2009). Narrow graphene nanoribbons from carbon nanotubes, *Nature*, **458**, pp. 877–880.

100. Kim, C. D., Min, B. K. and Jung, W. S. (2009). Preparation of graphene sheets by the reduction of carbon monoxide, *Carbon*, **47**, pp. 1610–1612.

101. Bourlinos, A. B., Georgakilas, V., Zboril, R., Steriotis, T. A. and Stubos, A. (2009). Liquid-phase exfoliation of graphite towards solubilized graphenes, *Small*, **5**, pp. 1841–1845.

102. Hernandez, Y., Nicolosi, V., Lotya, M., Blighe, F. M., Sun, Z., De, S., McGovern, I. T., Holland, B., Byrne, M., Gun'Ko, Y. K., Boland, J. J., Niraj, P., Duesberg, G., Krishnamurthy, S., Goodhue, R., Hutchison, J., Scardaci, V., Ferrari, A. C. and Coleman, J. N. (2008). High-yield production of graphene by liquid-phase exfoliation of graphite, *Nat. Nanotechnol.*, **3**, pp. 563–568.

103. Liu, N., Luo, F., Wu, H., Liu, Y., Zhang, C. and Chen, J. (2008). One step ionic liquid assisted electrochemical synthesis of ionic-liquid-functionalized graphene sheets directly from graphene, *Adv. Funct. Mater.*, **18**, pp. 1518–1525.

104. Behabtu, N., Lomeda, J. R., Green, M. J., Higginbotham, A. L., Sinitskii, A., Kosynkin, D. V., Tsentalovich, D., Parra-Vasquez, A. N. G., Schmidt, J., Kesselman, E., Cohen, Y., Talmon, Y., Tour, J. M. and Pasquali, M. (2010). Spontaneous high-concentration dispersions and liquid crystals of graphene, *Nat. Nanotechnol.*, **5**, pp. 406–411.

105. Worsley, K. A., Ramesh, P., Mandal, S. K., Niyogi, S., Itkis, M. E. and Haddon, R. C. (2007). Soluble graphene derived from graphite fluoride, *Chem. Phys. Lett.*, **445**, pp. 51–56.

106. Stankovich, S., Dikin, D. A., Piner, R. D., Kohlhaas, K. A., Kleinhammes, A., Jia, Y., Wu, Y., Nguyen, S. T. and Ruoff, R. S. (2007). Synthesis of graphene-based nanosheets via chemical reduction of exfoliated graphite oxide, *Carbon*, **45**, pp. 1558–1565.

107. Wang, H., Robinson, J. T., Li, X. and Dai, H. (2009). Solvothermal reduction of chemically exfoliated graphene sheets, *J. Am. Chem. Soc.*, **131**, pp. 9910–9911.

108. Wang, G., Yang, J., Park, J., Gou, X., Wang, B., Liu, H. and Yao, J. (2008). Facile synthesis and characterization of graphene nanosheets, *J. Phys. Chem. C*, **112**, pp. 8192–8195.

109. Williams, G., Seger, B. and Kamat, P. V. (2008). TiO_2-graphene nanocomposites. UV-assisted photocatalytic reduction of graphene oxide, *ACS Nano*, **2**, pp. 1487–1491.

110. Zhou, Y., Bao, Q., Tang, L. A. L., Zhong, Y. and Loh, K. P. (2009). Hydrothermal dehydration for the "green" reduction of exfoliated graphene oxide to graphene and demonstration of tunable optical limiting properties, *Chem. Mater.*, **21**, pp. 2950–2956.

111. Nethravathi, C. and Rajamathi, M. (2008). Chemically modified graphene sheets produced by the solvothermal reduction of colloidal dispersions of graphite oxide, *Carbon*, **46**, pp. 1994–1998.

112. Fan, Z., Wang, K., Wei, T., Yan, J., Song, L. and Shao, B. (2010). An environmentally friendly and efficient route for the reduction of graphene oxide by aluminium powder, *Carbon*, **48**, pp. 1670–1692.

113. Wang, G., Shen, X., Wang, B., Yao, J. and Park, J. (2009). Synthesis and characterization of hydrophilic and organophilic graphene nanosheets, *Carbon*, **47**, pp. 1359–1364.

114. Lomeda, J. R., Doyle, C. D., Kosynkin, D. V., Hwang, W. F. and Tour, J. M. (2008). Diazonium functionalization of surfactant-wrapped chemically converted graphene sheets, *J. Am. Chem. Soc.*, **130**, pp. 16201–16206.

115. Schniepp, H. C., Li, J. L., McAllister, M. J., Sai, H., Herrera-Alonso, M., Adamson, D. H., Prud'homme, R. K., Car, R., Saville, D. A. and Aksay, I. A. (2006). Functionalized single graphene sheets derived from splitting graphite oxide, *J. Phys. Chem. B*, **110**, pp. 8535–8539.

116. Aksay, I. A., Milius, D. L., Korkut, S. and Prud'homme, R. K. (2009). Functionalized graphene sheets having a high carbon to oxygen ratios for polymer composite, W.O. Patent 2009/134492 A2.

117. Ambrosi, A., Chua, C. K., Bonanni, A. and Pumera, M. (2014). Electrochemistry of graphene and related materials, *Chem. Rev.*, **114**, pp. 7150–7188.

118. Peigney, A., Laurent, C., Flahaut, E., Bacsa, R. R. and Rousset, A. (2001). Specific surface area of carbon nanotubes and bundles of carbon nanotubes, *Carbon*, **39**, pp. 507–514.

119. Huang, J. C. (2002). A carbon black filled conducting polymers and polymer blends, *Adv. Polym. Technol.*, **21**, pp. 299–313.

120. Moniruzzaman, M. and Winey, K. I. (2006). Polymer nanocomposites containing carbon nanotubes, *Macromolecules*, **39**, pp. 5194–5205.

121. Okamoto, M. and Ray, S. S. (2004). Polymer/clay nanocomposites, in *Encyclopedia of Nanoscience and Nanotechnology*, ed. Nalwa, H. S. (American Scientific, Stevenson Ranch, CA), **8**, pp. 1–52.

122. Wang, M., Yan, C. and Ma, L. (2012). *Graphene Nanocomposites, Composites, and Their Properties*, ed. Ning, H. (InTech), doi:10.5772/50840. Available from: https://www.intechopen.com/books/composites-and-their-properties/graphene-nanocomposites.

123. Wang, X., Zhi, L. and Mullen, K. (2008). Transparent, conductive graphene electrodes for dye-sensitized solar cells, *Nano Lett.*, **8**, pp. 323–327.

124. Stankovich, S., Dikin, D. A., Dommett, G. H. B., Kohlhaas, K. M., Zimney, E. J., Stach, E. A., Piner, R. D., Nguyen, S. T. and Ruoff, R. S. (2006). Graphene-based composite materials, *Nature*, **442**, pp. 282–286.

125. Ramanathan, T., Abdala, A. A., Stankovich, S., Dikin, D. A., Alonso, M. H., Piner, R. D., Adamson, D. H., Schniepp, H. C., Chen, X. R. R. S., Ruoff, R. S. and Nguyen, S. T. (2008). Functionalized graphene sheets for polymer nanocomposites, *Nat. Nanotechnol.*, **3**, pp. 327–331.

126. Lee, Y. R., Raghu, A. V., Jeong, H. M. and Kim, B. K. (2009). Properties of waterborne polyurethane/functionalized graphene sheet nanocomposites prepared by an in situ method, *Macromol. Chem. Phys.*, **210**, pp. 1247–1254.

127. Xu, Y., Wang, Y., Jiajie, L., Huang, Y., Ma, Y., Wan, X. and Chen, Y. (2009). A hybrid material of graphene and poly (3,4-ethyl dioxythiophene) with high conductivity, flexibility, and transparency, *Nano Res.*, **2**, pp. 343–348.

128. Quan, H., Zhang, B., Zhao, Q., Yuen, R. K. K. and Li, R. K. Y. (2009). Facile preparation and thermal degradation studies of graphite nanoplatelets (GNPs) filled thermoplastic polyurethane (TPU) nanocomposites, *Composites Part A*, **40**, pp. 1506–1513.

129. (a) Zhang, H. B., Zheng, W. G., Yan, Q., Yang, Y., Wang, J., Lu, Z. H., Ji, G. Y. and Yu, Z. Z. (2010). Electrically conductive polyethylene terephthalate/graphene nanocomposites prepared by melt compounding, *Polymer*, **51**, pp. 1191–1196; (b) Muralidharan, M. N. and Ansari, S. (2013). Thermally reduced graphene oxide/thermoplastic polyurethane nanocomposites as photomechanical actuators, *Adv. Mater. Lett.*, **4**, pp. 927–932.

130. Zheng, W., Lu, X. and Wong, S. C. (2004). Electrical and mechanical properties of expanded graphite-reinforced high-density polyethylene, *J. Appl. Polym. Sci.*, **91**, pp. 2781–2788.

131. Wakabayashi, K., Pierre, C., Dikin, D. A., Ruoff, R. S., Ramanathan, T., Brinson, L. C. and Torkelson, J. M. (2008). Polymer–graphite nanocomposites: effective dispersion and major property enhancement via solid-state shear pulverization, *Macromolecules*, **41**, pp. 1905–1908.

132. Das, T. K. and Prusty, S. (2013). Graphene-based polymer composites and their applications, *Polym. Plast. Technol. Eng.*, **52**, pp. 319–331.

133. Higginbotham, A. L., Lomeda, J. R., Morgan, A. B. and Tour, J. M. (2009). Graphite oxide flame-retardant polymer nanocomposites, *ACS Appl. Mater. Interfaces*, **1**, pp. 2256–2261.

134. Chen, D., Zhu, H. and Liu, T. (2010). In situ thermal preparation of polyimide nanocomposite films containing functionalized graphene sheets, *ACS Appl. Mater. Interfaces*, **2**, pp. 3702–3708.

135. Lee, S., Kim, Y. J., Kim, D. H., Ku, B. C. and Joh, H. I. (2012). Synthesis and properties of thermally reduced graphene oxide/polyacrylonitrile composites, *J. Phys. Chem. Solids*, **73**, pp. 741–743.

136. Wang, W., Xu, G., Cui, X. T., Sheng, G. and Luo, X. (2014). Enhanced catalytic and dopamine sensing properties of electrochemically reduced conducting polymer nanocomposite doped with pure graphene oxide, *Biosens. Bioelectron.*, **58**, pp. 153–156.

137. Kim, H. and Macosko, C. W. (2009). Processing-property relationships of polycarbonate/ graphene composites, *Polymer*, **50**, pp. 3797–3809.

138. Wanga, W. P. and Pana, C. Y. (2004). Preparation and characterization of the polystyrene/graphite composite prepared by cationic grafting polymerization, *Polymer*, **45**, pp. 3987–3995.

139. Kalaitzidou, K., Fukushima, H. and Drzal, L. T. (2007). A new compounding method for exfoliated graphite-polypropylene nanocomposites with enhanced flexural properties and lower percolation threshold, *Compos. Sci. Technol.*, **67**, pp. 2045–2051.

140. Kim, S. K., Kim, N. H. and Lee, J. H. (2006). Effects of the addition of multiwalled carbon nanotubes on the positive temperature coefficient characteristics of carbon-black-filled high-density polyethylene nanocomposites, *Scr. Mater.*, **55**, pp. 1119–1122.

141. Kim, S., Do, I. and Drzal, L. T. (2009). Thermal stability and dynamic mechanical behaviour of exfoliated graphite nanoplatelets-LLDPE nanocomposites, *Polym. Compos.*, **31**, pp. 755–761.

142. Chen, G., Wu, C., Weng, W., Wu, D. and Yan, W. (2003). Preparation of polystyrene= graphite nanosheet composites, *Polymer*, **44**, pp. 1781–1784.

143. Weng, W., Chen, G. and Wu, D. (2005). Transport properties of electrically conducting nylon 6/foliated graphite nanocomposites, *Polymer*, **46**, pp. 6250–6257.

144. Kulia, T., Bose, S., Khanra, P., Kim, N. H., Rhee, K. Y. and Lee, J. H. (2011). Characterization and properties of in situ emulsion polymerized poly (methyl methacrylate)/graphene nanocomposites, *Composites Part A*, **42**, pp. 1856–1861.

145. Hu, H. T., Wang, J. C., Wan, L., Liu, F. M., Zheng, H., Chen, R. and Xu, C. (2010). Preparation and properties of graphene nanosheets – polystyrene nanocomposites via in situ emulsion polymerization, *Chem. Phys. Lett.*, **484**, pp. 247–253.

146. Salavagione, H. J., Martinez, G. and Ellis, G. (2011). Recent advances in the covalent modification of graphene with polymers, *Macromol. Rapid Commun.* **32**, pp. 1771–1789.

147. Fang, M., Wang, K. G., Lu, H. B., Yang, Y. L. and Nutt, S. (2010). Single-layer graphene nanosheets with controlled grafting of polymer chains, *J. Mater. Chem.*, **20**, pp. 1982–1992.

148. Fang, M., Wang, K. G., Lu, H. B., Yang, Y. L. and Nutt, S. (2009). Covalent polymer functionalization of graphene nanosheets and mechanical properties of composites, *J. Mater. Chem.*, **19**, pp. 7098–7105.

149. Lee, S. H., Dreyer, D. R., An, J. H., Velamakanni, A., Piner, R. D., Park, S., Zhu, Y. W., Kim, S. O., Bielawski, C. W. and Ruoff, R. S. (2010). Polymer brushes via controlled, surface-initiated atom transfer radical polymerization (ATRP) from graphene oxide, *Macromol. Rapid Commun.*, **31**, pp. 281–288.

150. Ye, L., Meng, X. Y., Ji, X., Li, Z. M. and Tang, J. H. (2009). Synthesis, and characterization of expandable graphite-poly (methyl methacrylate) composite particles and their application to flame retardation of rigid polyurethane foams, *Polym. Degrad. Stab.*, **94**, pp. 971–979.

151. Kornmann, X. (2001). Synthesis and characterization of thermoset-layered silicate nanocomposites, PhD thesis, Lulea Tekniska Universitet, Sweden.

152. Potts, J. R., Lee, S. H., Alam, T. M., An, J., Stoller, M. D., Piner, R. D. and Ruoff, R. S. (2011). Thermomechanical properties of chemically modified graphene/poly(methyl methacrylate) composite made by in situ polymerizations, *Carbon*, **49**, pp. 2615–2623.

153. Chen, G., Wu, D., Weng, W. and Wu, C. (2003). Exfoliation of graphite flakes and its nanocomposites, *Carbon*, **41**, pp. 619–621.

154. Kim, H., Miura, Y. and Macosko, C. W. (2010). Graphene/polyurethane nanocomposites for improved gas barrier and electrical conductivity, *Chem. Mater.*, **22**, pp. 3441–3450.

155. Moujahid, E. M., Besse, J. P. and Leroux, F. (2003). Poly (styrene sulfonate) layered double hydroxide nanocomposites Stability and subsequent structural transformation with changes in temperature, *J. Mater. Chem.*, **13**, pp. 258–264.

156. Wang, S. R., Tambraparni, M., Qiu, J. J., Tipton, J. and Dean, D. (2009). Thermal expansion of graphene composites, *Macromolecules*, **42**, pp. 5251–5255.

157. Huang, Y., Zeng, M., Ren, J., Wang, J., Fan, L. and Xu, Q. (2012). Preparation and swelling properties of graphene oxide/poly (acrylic acid-co-acrylamide) super-absorbent hydrogel nanocomposites, *Colloids Surf., A*, **401**, pp. 97–106.

158. Hsueh, H. B. and Chen, C. Y. (2003). Preparation and properties of LDHs/polyimide nanocomposites, *Polymer*, **44**, pp. 1151–1161.

159. Zhang, B., Yu, B., Zhou, F. and Liu, W. (2013). Polymer brush stabilized amorphous MnO_2 on graphene oxide sheets as novel electrode materials for high-performance supercapacitors, *J. Mater. Chem. A*, **1**, pp. 8587–8592.

160. Kuilla, T., Bhadra, S., Yao, D., Kim, N. H., Bosed, S. and Lee, J. H. (2010). Recent advances in graphene based polymer composites, *Prog. Polym. Sci.*, **35**, pp. 1350–1375.

161. Verdejo, R., Bernal, M. M., Romasanta, L. J. and Lopez-Manchado, M. A. (2011). Graphene filled polymer nanocomposites, *J. Mater. Chem.*, **21**, pp. 3301–3310.

162. Wang, H., Hao, Q., Yang, X. and Wang, X. (2009). Graphene oxide doped polyaniline for the supercapacitor, *Electrochem. Commun.*, **11**, pp. 1158–1161.

163. Mashat, L. A., Shin, K., Zader, K. K., Plessis, J. D., Han, S. H., Kojima, R. W., Kaner, R. B., Li, D., Gou, X., Ippolito, S. J. and Ki, W. W. (2010). Graphene/polyaniline nanocomposite for hydrogen sensing, *J. Phys. Chem.*, **114**, pp. 16168–16173.

164. Liu, T., Su, H., Qu, X., Ju, P., Cui, L. and Ai, S. (2011). Acetylcholinesterase biosensor based on 3-carboxyphenylboronic acid/reduced graphene oxide–gold nanocomposites modified electrode for amperometric detection of organophosphorus and carbamate pesticides, *Sens. Actuators, B*, **160**, pp. 1255–1261.

165. Li, B., Liu, T., Wang, Y. and Wang, Z. (2012). ZnO/graphene-oxide nanocomposite with remarkably enhanced visible-light-driven photocatalytic performance, *J. Colloid Interface Sci.*, **377**, pp. 114–121.

166. Peponi, L., Tercjak, A., Verdejo, R., Lopez-Manchado, M. A., Mondragon, I. and Kenny, J. M. (2009). Confinement of functionalized graphene sheets by triblock copolymers, *J. Phys. Chem.*, **113**, pp. 17973–17978.

167. Wang, D. W., Li, F., Zhao, J., Ren, W., Chen, Z. G., Tan, J., Wu, Z. S., Gentle, I., Lu, G. Q. and Cheng, H. M. (2009). Fabrication of graphene/polyaniline composite paper via in situ anodic electropolymerizations for the high-performance flexible electrode, *ACS Nano*, **7**, pp. 1745–1752.

168. Zhang, K., Zhang, L. L., Zhao, X. S. and Wu, J. (2010). Graphene/polyaniline nanofiber composites as supercapacitor electrodes, *Chem. Mater.*, **22**, pp. 1392–1401.

169. (a) Rice, A. H., Giridharagopal, R., Zheng, S. X., Ohuchi, F. S., Ginger, D. S. and Luscombe, C. K. (2011). Controlling vertical morphology within the active layer of organic photovoltaics using poly (3-hexylthiophene) nanowires and phenyl-C61-butyric acid methyl ester, *ACS Nano*, **5**,

pp. 3132–3140; (b) Kirchmeyer, S. and Reuter, K. (2005). Scientific importance, properties and growing applications of poly (3, 4-ethylenedioxythiophene), *J. Mater. Chem.*, **15**, pp. 2077–2088; (c) Xu, H., Pang, X., He, Y., He, M., Jung, J., Xia, H. and Lin, Z. (2015). An unconventional route to monodisperse and intimately contacted semiconducting organic-inorganic nanocomposites, *Angew. Chem. Int. Ed.*, **54**, pp. 4636–4640.

170. Han, Y. and Lu, Y. (2008). Easy fabrication and excellent electrical conductivity of graphite oxide/poly (3, 4-ethylene dioxythiophene) nanocomposites, *Synth. Met.*, **158**, 744–748.

171. Ryu, K. S., Lee, Y. G., Hong, Y. S., Park, Y. J., Wu, X., Kim, K. M., Kang, M. G., Park, N. G. and Chang, S. H. (2004). Poly (ethylene-dioxythiophene) (PEDOT) as polymer electrode in redox supercapacitor, *Electrochim. Acta*, **50**, pp. 843–847.

172. Jang, J., Chang, M. and Yoon, H. (2005). Chemical sensors based on highly conductive poly (3,4-ethylene dioxythiophene) nanorods, *Adv. Mater.*, **17**, pp. 1616–1620.

173. Li, X., Liang, L., Yang, M., Chen, G. and Guo, C. Y. (2016). Poly (3, 4-ethylenedioxythiophene)/graphene/carbon nanotube ternary composites with improved thermoelectric performance, *Org. Electron.*, **38**, pp. 200–204.

174. Xu, K., Chen, G. and Qiu, D. (2013). Convenient construction of poly (3, 4-ethylenedioxythiophene)–graphene pie-like structure with enhanced thermoelectric performance, *J. Mater. Chem. A*, **1**, pp. 12395–12399.

175. Wei, J., Vo, T. and Inam, F. (2015). Epoxy/graphene nanocomposites–processing and properties: a review, *RSC Adv.*, **5**, pp. 73510–73524.

176. Balandin, A. A. (2011). Thermal properties of graphene and nanostructured carbon materials, *Nat. Mater.*, **10**, pp. 569–581.

177. Yu, J., Lu, K., Sourty, E., Grossiord, N., Koning, C. E. and Loos, J. (2007). Characterization of conductive multiwall carbon nanotube/polystyrene composites prepared by latex technology, *Carbon*, **45**, pp. 2897–2903.

178. Kumar, R., Singh, R. K., Singh, D. P., Joanni, E., Yadav, R. M. and Moshkalev, S. A. (2017). Laser-assisted synthesis, reduction and micro-patterning of graphene: recent progress and applications, *Coord. Chem. Rev.*, **342**, pp. 34–79.

179. Awasthi, S., Awasthi, K., Kumar, R. and Srivastava, O. N. (2009). Functionalization effects on the electrical properties of multi-walled

carbon nanotube-polyacrylamide composites, *J. Nanosci. Nanotechnol.*, **9**, pp. 5455–5460.

180. Yu, A., Ramesh, P., Sun, X., Bekyarova, E., Itkis, M. E. and Haddon, R. C. (2008). Enhanced thermal conductivity in a hybrid graphite nanoplatelet-carbon nanotubes filler for epoxy composites, *Adv. Mater.*, **20**, pp. 4740–4744.
181. Yu, A., Ramesh, P., Itkis, M. E., Elena, B. and Haddon, R. C. (2007). Graphite nanoplatelet-epoxy composite thermal interface materials, *J. Phys. Chem. C*, **111**, pp. 7565–7569.
182. Eda, G. and Chhowalla, M. (2009). Graphene-based composite thin films for electronics, *Nano Lett.*, **9**, pp. 814–818.
183. Zhao, X., Zhang, Q. and Chen, D. (2010). Enhanced mechanical properties of graphene-based poly (vinyl alcohol) composites, *Macromolecules*, **43**, pp. 2357–2363.
184. Ahmed, N., Kausar, A. and Muhammad, B. (2015). Advances in shape memory polyurethanes and composites: a review, *Polym. Plast. Technol. Eng.*, **54**, pp. 1410–1423.
185. Rahnama, M. R., Barikani, M., Barmar, M. and Honarkar, H. (2014). An investigation into the effects of different nanoclays on polyurethane nanocomposites properties, *Polym. Plast. Technol. Eng.*, **53**, pp. 801–810.
186. Khan, D. M., Kausar, A. and Salman, S. M. (2016). Exploitation of nanofiller in polymer/graphene oxide–carbon nanotube, polymer/graphene oxide–nanodiamond, and polymer/graphene oxide–montmorillonite composite: a review, *Polym. Plast. Technol. Eng.*, **55**, pp. 744–768.
187. Zhu, Y., Murali, S., Cai, W., Li, X., Suk, J. W., Potts, J. R. and Ruoff, R. S. (2010). Graphene and graphene oxide: synthesis, properties, and applications, *Adv. Mater.*, **22**, pp. 3906–3924.
188. Tripathi, S., Mehrotra, G. K. and Dutta, P. K. (2010). Preparation and physicochemical evaluation of chitosan/poly (vinyl alcohol)/pectin ternary film for food-packaging applications, *Carbohydr. Polym.*, **79**, pp. 711–716.
189. DeMerlis, C. C. and Schoneker, D. R. (2003). Review of the oral toxicity of polyvinyl alcohol (PVA), *Food Chem. Toxicol.*, **41**, pp. 319–326.
190. Hu, X. Q., Ye, D. Z., Tang, J. B., Zhang, L. J. and Zhang, X. (2016). From waste to functional additives: thermal stabilization and toughening of PVA with lignin, *RSC Adv.*, **6**, pp. 13797–13802.

191. Zhang, B., Xu, S., Tang, H. and Wu, P. (2013). Crosslinked acetylacetonated poly (vinyl alcohol-co-vinyl acetate) nanocomposites with graphene oxide and reduced graphene oxide: a new way to modify the property of nanocomposites, *RSC Adv.*, **3**, pp. 8372–8379.

192. Wang, J., Qiu, W., Wang, N. and Li, L. (2015). Influence of hydroxyapatite onthethermoplastic foaming performance of water-plasticized poly (vinyl alcohol), *RSC Adv.*, **5**, pp. 84578–84586.

193. Cui, Y., Kumar, S., Kona, B. R. and van Houcke, D. (2015). Gas barrier properties of polymer/clay nanocomposites, *RSC Adv.*, **5**, pp. 63669–63690.

194. Wang, J., Wang, X., Xu, C., Zhang, M. and Shang, X. (2011). Preparation of graphene/poly (vinyl alcohol) nanocomposites with enhanced mechanical properties and water resistance, *Polym. Int.*, **60**, pp. 816–822.

195. Milosavljević, N. B., Kljajević, L. M., Popović, I. G., Filipović, J. M. and Kalagasidis Krušić, M. T. (2010). Chitosan, itaconic acid and poly (vinyl alcohol) hybrid polymer networks ofahigh degree of swelling and good mechanical strength, *Polym. Int.*, **59**, pp. 686–694.

196. Zhang, J., Hu, P., Wang, X., Wang, Z., Liu, D., Yang, B. and Cao, W. (2012). CVD growth ofthelarge area and uniform graphene on tilted copper foil for high performance flexible transparent conductive film, *J. Mater. Chem.*, **22**, pp. 18283–18290.

197. Zubair, N. A., Rahman, N. A., Lim, H. N., Zawawi, R. M. and Sulaiman, Y. (2016). Electrochemical properties of PVA–GO/PEDOT nanofibers prepared using electrospinning and electropolymerization techniques, *RSC Adv.*, **6**, pp. 17720–17727.

198. Li, J., Shao, L., Zhou, X. and Wang, Y. (2014). Fabrication of high strength PVA/rGO composite fibres by gel spinning, *RSC Adv.*, **4**, pp. 43612–43618.

199. Xu, Y., Hong, W., Bai, H., Li, C. and Shi, G. (2009). Strong and ductile poly (vinyl alcohol)/graphene oxide composite films with a layered structure, *Carbon*, **47**, pp. 3538–3543.

200. Liang, J., Huang, Y., Zhang, L., Wang, Y., Ma, Y., Guo, T. and Chen, Y. (2009). Molecular-level dispersion of graphene into poly (vinyl alcohol) and effective reinforcement of their nanocomposites, *Adv. Funct. Mater.*, **19**, pp. 2297–2302.

201. Guo, J., Ren, L., Wang, R., Zhang, C., Yang, Y. and Liu, T. (2011). Water-dispersible graphene noncovalently functionalized with tryptophan and its poly (vinyl alcohol) nanocomposite, *Composites Part B*, **42**, pp. 2130–2135.

202. Panasyuk, G. P., Izotov, A. D., Azarova, L. A., Shabalin, D. G. and Voroshilov, I. L. (2015). New methods for utilization of waste polyethylene terephthalate, *Theor. Found. Chem. Eng.*, **49**, pp. 580–583.

203. Jabarin, S. A. (1987). Crystallization kinetics of polyethylene terephthalate. II. Dynamic crystallization of PET, *J. Appl. Polym. Sci.*, **34**, pp. 97–102.

204. Wang, Y., Gao, J., Ma, Y. and Agarwal, U. S. (2006). Study on mechanical properties, thermal stability and crystallization behaviour of PET/MMT nanocomposites, *Composites Part B*, **37**, pp. 399–407.

205. Liu, K., Chen, L., Chen, Y., Wu, J., Zhang, W., Chen, F. and Fu, Q. (2011). Preparation of polyester/reduced graphene oxide composites via in situ melt polycondensation and simultaneous thermo-reduction of graphene oxide, *J. Mater. Chem.*, **21**, pp. 8612–8617.

206. Brunelle, D. J. and Korn, M. R. eds. (2005). Advances in polycarbonate: an overview, in *Advances in Polycarbonates*, ACS Symposium Series, Vol. 898 (American Chemical Society, Washington, DC), Chapter 1, pp. 1–5.

207. Yoon, S. H. and Jung, H. T. (2017). Grafting polycarbonate onto graphene nanosheets: synthesis and characterization of high-performance polycarbonate–graphene nanocomposites for ESD/EMI applications, *RSC Adv.*, **7**, pp. 45902–45910.

208. Ansari, S. and Giannelis, E. P. (2009). Functionalized graphene sheet-poly (vinylidene fluoride) conductive nanocomposites, *J. Polym. Sci., Part B: Polym. Phys.*, **47**, pp. 888–897.

209. Cho, S., Lee, J. S. and Jang, J. (2015). Poly (vinylidene fluoride)/NH2-treated graphene nanodot/reduced graphene oxide nanocomposites withanenhanced dielectric performance for ultrahigh energy density capacitor, *ACS Appl. Mater. Interfaces*, **7**, pp. 9668–9681.

210. Song, H. S. and Liu, D. B. (2011). Preparation of graphene and the research of dielectric properties of graphene/PVDF composites [J], *Chem. Eng.*, **8**, pp. 1–3.

211. Aragaw, B. A., Su, W. N., Rick, J. and Hwang, B. J. (2013). Highly efficient synthesis of reduced graphene oxide–nafion nanocomposites with strong coupling for enhanced proton and electron conduction, *RSC Adv.*, **3**, pp. 23212–23221.

212. Li, H., Chen, J., Han, S., Niu, W., Liu, X. and Xu, G. (2009). Electrochemiluminescence from tris (2, 2′-bipyridyl) ruthenium (II)–graphene–nafion modified electrode, *Talanta*, **79**, pp. 165–170.

213. Salavagione, H. J., Díez-Pascual, A. M., Lázaro, E., Vera, S. and Gómez-Fatou, M. A. (2014). Chemical sensors based on polymer composites with carbon nanotubes and graphene: the role of the polymer, *J. Mater. Chem. A*, **2**, pp. 14289–14328.

214. Wang, R., Wang, X., Chen, S. and Jiang, G. (2012). In situ polymerization approach to poly (ε-caprolactone)-graphene oxide composites, *Des. Monomers Polym.*, **15**, pp. 303–310.

215. Chieng, B. W., Ibrahim, N. A. and Yunus, W. W. (2012). Optimization ofthetensile strength of poly (lactic acid)/graphene nanocomposites using response surface methodology, *Polym. Plast. Technol. Eng.*, **51**, pp. 791–799.

216. Mohammadi, S., Sanjani, N. S. and Mahdavi, H. (2011). Functionalization of graphene sheets via chemically grafting of PMMA chains through in situ polymerization, *J. Macromol. Sci. Part A Pure Appl. Chem.*, **48**, pp. 577–582.

217. Pan, B., Xu, G., Zhang, B., Ma, X., Li, H. and Zhang, Y. (2012). Preparation and tribological properties of polyamide 11/graphene coatings, *Polym. Plast. Technol. Eng.*, **51**, pp. 1163–1166.

218. Pang, H., Bao, Y., Lei, J., Tang, J. H., Ji, X., Zhang, W. Q. and Chen, C. (2012). Segregated conductive ultrahigh-molecular-weight polyethylene composites containing high-density polyethylene as carrier polymer of graphene nanosheets, *Polym. Plast. Technol. Eng.*, **51**, pp. 1483–1486.

219. Liang, J., Huang, L., Li, N., Huang, Y., Wu, Y., Fang, S., Oh, J., Kozlov, M., Ma, Y., Li, F., Baughman, R. and Chen, Y. (2012). Electromechanical actuator with controllable motion, fast response rate, and high-frequency resonance based on graphene and polydiacetylene, *ACS Nano*, **6**, pp. 4508–4519.

220. Pan, B., Zhao, J., Zhang, Y. and Zhang, Y. (2012). Wear performance and mechanisms of polyphenylene sulphide/polytetrafluoroethylene wax composite coatings reinforced by graphene, *J. Macromol. Sci. Part B Phys.*, **51**, pp. 1218–1227.

221. Ohno, Y., Maehashi, K., Yamashiro, Y. and Matsumoto, K. (2009). Electrolyte-gated graphene field-effect transistors for detecting ph and protein adsorption, *Nano Lett.*, **9**, pp. 3318–3322.

222. Mohanty, N. and Berry, V. (2008). Graphene-based single-bacterium resolution biodevice and DNA transistor, interfacing graphene derivatives with nanoscale and microscale biocomponents, *Nano Lett.*, **8**, pp. 4469–4476.

223. Liu, J. Q., Tao, L., Yang, W. R., Li, D., Boyer, C., Wuhrer, R., Braet, F. and Davis, T. P. (2010). Synthesis, characterization, and multilayer assembly

of pH-sensitive graphene-polymer nanocomposites, *Langmuir*, **26**, pp. 10068–10075.

224. Chen, L., Chen, G. and Lu, L. (2007). Piezoresistive behaviour study on finger sensing silicone rubber/graphite nanosheet nanocomposites, *Adv. Funct. Mater.*, **17**, pp. 898–904.

225. Georgakilas, V., Otyepka, M., Bourlinos, A. B., Chandra, V., Kim, N., Kemp, K. C., Hobza, P., Zboril, R. and Kim, K. S. (2012). Functionalization of graphene: covalent and non-covalent approaches, derivatives and applications, *Chem. Rev.*, **112**, pp. 6156–6214.

226. Ponnamma, D., Guo, Q., Krupa, I., Al-Maadeed, M. A. S., Varughese, K. T., Thomas, S. and Sadasivuni, K. K. (2015). Graphene and graphitic derivative filled polymer composites as potential sensors, *Phys. Chem. Chem. Phys.*, **17**, pp. 3954–3981.

227. Shi, Y., Wu, J., Sun, Y., Zhang, Y., Wen, Z., Dai, H., Wang, H. and Li, Z. (2012). A graphene oxide based biosensor for microcystins detection by fluorescence resonance energy transfer, *Biosens. Bioelectron.*, **38**, pp. 31–36.

228. Schedin, F., Geim, A. K., Morozov, S. V., Hill, E. W., Blake, P., Katsnelson, M. I. and Novoselov, K. S. (2007). Detection of individual gas molecules adsorbed on graphene, *Nat. Mater.*, **6**, pp. 652–655.

229. Dan, Y., Lu, Y., Kybert, N. J., Luo, Z. and Johnson, A. T. C. (2009). The intrinsic response of graphene vapour sensors, *Nano Lett.*, **9**, pp. 1472–1475.

230. Robinson, J. T., Perkins, F. K., Snow, E. S., Wei, Z. and Sheehan, P. E. (2008). Reduced graphene oxide molecular sensors, *Nano Lett.*, **8**, pp. 3137–3140.

231. (a) Tang, L., Wang, Y., Li, Y., Feng, H., Lu, J. and Li, J. (2009). Preparation, structure, and electrochemical properties of reduced graphene sheet films, *Adv. Funct. Mater.*, **19**, pp. 2782–2789; (b) Yu, M.-R., Wu, R.-J., Suyambrakasam, G., Joly, J. and Chavali, M. (2012). Evaluation of graphene oxide material as formaldehyde gas sensor, *Adv. Sci. Lett.*, **16**, pp. 53–57(5), doi: https://doi.org/10.1166/asl.2012.4264.

232. Sberveglieri, G., ed. (2012). *Gas Sensors: Principles, Operation and Developments* (Springer Science & Business Media).

233. Lei, Y. J. (1993). Sensors for toxic gas detection, *Platinum Met. Rev.*, **37**, pp. 146–150.

234. Hygiene Institute, Chinese Academy of Medical Sciences (1974). Method for detecting poisonous, substance in *Air*, China, **2** (Chinese).

235. Altshuller, A. P. (1968). In *Advances in Chromatography* (Dekker, New York), pp. 229–262.

236. Zlatkis, A. (1976). *Scz. Tech. Arosp. Rep.*, **14**, p. 1

237. Nagashima, K. and Suzuki, S. (1983). The determination of urea by using an enzyme reactor and second-derivative spectrophotometry, *Anal. Chim. Acta*, **151**, pp. 13–18.

238. Bott, B. and Jones, T. A. (1984). A highly sensitive NO2 sensor based on electrical conductivity changes in phthalocyanine films, *Sens. Actuators, B*, **5**, pp. 43–53.

239. Varfolomeev, A. E., Volkov, A. I., Eryshkin, A. V., Malyshev, V. V., Rasumov, A. S. and Yakimov, S. S. (1992). Detection of phosphine and arsine in the air by sensors based on SnO_2 and ZnO, *Sens. Actuators, B*, **7**, pp. 727–729.

240. Torvela, H., Romppainen, P. and Leppävuori, S. (1988). Detection of CO levels in combustion gases bythethick-film SnO_2 sensor, *Sens. Actuators, B*, **14**, pp. 19–25.

241. Kohl, D. (2001). Function and applications of gas sensors, *J. Phys. D: Appl. Phys.*, **34**, p. R125.

242. Piscevic, D., Lawall, R., Veith, M., Liley, M., Okahata, Y. and Knoll, W. (1995). Oligonucleotide hybridization observed by surface plasmon optical techniques, *Appl. Surf. Sci.*, **90**, pp. 425–436.

243. Thiel, A. J., Frutos, A. G., Jordan, C. E., Corn, R. M. and Smith, L. M. (1997). In situ surface plasmon resonance imaging detection of DNA hybridization to oligonucleotide arrays on gold surfaces, *Anal. Chem.*, **69**, pp. 4948–4956.

244. Alizadeh, T. and Soltani, L. H. (2013). Graphene/poly (methyl methacrylate) chemiresistor sensor for formaldehyde odour sensing, *J. Hazard. Mater.*, **248**, pp. 401–406.

245. Dunst, K. J., Scheibe, B., Nowaczyk, G., Jurga, S. and Jasiński, P. (2017). Graphene oxide, reduced graphene oxide and composite thin films NO_2 sensing properties, *Meas. Sci. Technol.*, **28**, p. 054005.

246. Gaikwad, G., Patil, P., Patil, D. and Naik, J. (2017). Synthesis and evaluation of gas sensing properties of PANI based graphene oxide nanocomposites, *Mater. Sci. Eng., B*, **218**, pp. 14–22.

247. Patil, P., Gaikwad, G., Patil, D. R. and Naik, J. (2016). Gas sensitivity study of polypyrrole decorated graphene oxide thick film, *J. Inst. Eng. India Ser. D*, **97**, pp. 47–53.

248. Huang, X., Hu, N., Gao, R., Yu, Y., Wang, Y., Yang, Z., Kong, E. S. W., Wei, H. and Zhang, Y. (2012). Reduced graphene oxide–polyaniline hybrid:

preparation, characterization and its applications for ammonia gas sensing, *J. Mater. Chem.*, **22**, pp. 22488–22495.

249. Ye, Z., Jiang, Y., Tai, H., Guo, N., Xie, G. and Yuan, Z. (2015). The investigation of reduced graphene oxide@ SnO_2–polyaniline composite thin films for ammonia detection at room temperature, *J. Mater. Sci.: Mater. Electron.*, **26**, pp. 833–841.

250. Zhang, Z., Zou, R., Song, G., Yu, L., Chen, Z. and Hu, J. (2011). Highly aligned SnO_2 nanorods on graphene sheets for gas sensors, *J. Mater. Chem.*, **21**, pp. 17360–17365.

251. Deng, S., Tjoa, V., Fan, H. M., Tan, H. R., Sayle, D. C., Olivo, M., Mhaisalkar, S., Wei, J. and Sow, C. H. (2012). Reduced graphene oxide conjugated Cu_2O nanowire mesocrystals for the high-performance NO_2 gas sensor, *J. Am. Chem. Soc.*, **134**, pp. 4905–4917.

252. Cuong, T. V., Pham, V. H., Chung, J. S., Shin, E. W., Yoo, D. H., Hahn, S. H., Huh, J. S., Rue, G. H., Kim, E. J., Hur, S. H. and Kohl, P. A. (2010). Solution-processed ZnO-chemically converted graphene gas sensor, *Mater. Lett.*, **64**, pp. 2479–2482.

253. Yuan, W. and Shi, G. (2013). Graphene-based gas sensors, *J. Mater. Chem. A*, **1**, pp. 10078–10091.

254. Eswaraiah, V., Balasubramaniam, K. and Ramaprabhu, S. (2011). Functionalized graphene reinforced thermoplastic nanocomposites as strain sensors in structural health monitoring, *J. Mater. Chem.*, **21**, pp. 12626–12628.

255. Xia, H., Hong, C., Li, B., Zhao, B., Lin, Z., Zheng, M., Savilov, S. V. and Aldoshin, S. M. (2015). Facile synthesis of hematite quantum-dot/functionalized graphene-sheet composites as advanced anode materials for asymmetric supercapacitors, *Adv. Funct. Mater.*, **25**, pp. 627–635.

256. Han, D., Han, T., Shan, C., Ivaska, A. and Niu, L. (2010). Simultaneous determination of ascorbic acid, dopamine and uric acid with chitosan-graphene modified electrode, *Electroanalysis*, **22**, pp. 2001–2008.

257. Ameen, S., Akhtar, M. S. and Shin, H. S. (2012). Hydrazine chemical sensing by modified electrode based on in situ electrochemically synthesized polyaniline/graphene composite thin film, *Sens. Actuators, B*, **173**, pp. 177–183.

258. Liu, Q., Zhu, X., Huo, Z., He, X., Liang, Y. and Xu, M. (2012). Electrochemical detection of dopamine in the presence of ascorbic acid using PVP/graphene modified electrodes, *Talanta*, **97**, pp. 557–562.

259. Liu, Y., Zhu, L., Luo, Z. and Tang, H. (2013). Fabrication of molecularly imprinted polymer sensor for chlortetracycline based onthecontrolled electrochemical reduction of graphene oxide, *Sens. Actuators, B*, **185**, pp. 438–444.

260. Xu, C., Wang, X., Wang, J., Hu, H. and Wan, L. (2010). Synthesis and photoelectrical properties of β-cyclodextrin functionalized graphene materials with high bio-recognition capability, *Chem Phys Lett.*, **498**, pp. 162–167.

261. Shan, C., Yang, H., Song, J., Han, D., Ivaska, A. and Niu, L. (2009). Direct electrochemistry of glucose oxidase and biosensing for glucose based on graphene, *Anal. Chem.*, **81**, pp. 2378–2382.

262. Kang, X., Wang, J., Wu, H., Aksay, I. A., Liu, J. and Lin, Y. (2009). Glucose oxidase–graphene–chitosan modified electrode for direct electrochemistry and glucose sensing, *Biosens. Bioelectron.*, **25**, pp. 901–905.

263. Wu, H., Wang, J., Kang, X., Wang, C., Wang, D., Liu, J., Aksay, I. A. and Lin, Y. (2009). Glucose biosensor based on immobilization of glucose oxidase in platinum nanoparticles/graphene/chitosan nanocomposite film, *Talanta*, **80**, pp. 403–406.

264. (a) Baby, T. T., Aravind, S. J., Arockiadoss, T., Rakhi, R. B. and Ramaprabhu, S. (2010). Metal decorated graphene nanosheets as immobilization matrix for amperometric glucose biosensor, *Sens. Actuators, B*, **145**, pp. 71–77; (b) Lu, C. H., Yang, H. H., Zhu, C. L., Chen, X. and Chen, G. N. (2009). A graphene platform for sensing biomolecules, *Angew. Chem. Int. Ed.*, **48**, pp. 4785–4787; (c) Jang, H., Kim, Y. K., Kwon, H. M., Yeo, W. S., Kim, D. E. and Min, D. H. (1999). A graphene-based platform for the assay of dup lex-DNA unwinding by helicase, *Angew. Chem. Int. Ed.*, **49**, pp. 5703–5707.

265. Zhou, M., Zhai, Y. and Dong, S. (2009). Electrochemical sensing and biosensing platform based on chemically reduced graphene oxide, *Anal. Chem.*, **81**, pp. 5603–5613.

266. (a) Shan, C., Yang, H., Song, J., Han, D., Ivaska, A. and Niu, L. (2009). Direct electrochemistry of glucose oxidase and biosensing for glucose based on graphene, *Anal. Chem.*, **81**, pp. 2378–2382; (b) Song, Y., Wei, W. and Qu, X. (2011) Colorimetric biosensing using smart materials, *Adv. Mater.*, **23**, p. 4215; (c) Ho, H. A., Najari, A. and Leclerc, M. (2008). Optical detection of DNA and proteins with cationic polythiophenes, *Acc. Chem. Res.*, **41**, p. 168.

267. Yin, H., Zhou, Y., Ma, Q., Ai, S., Ju, P., Zhu, L. and Lu, L. (2010). Electrochemical oxidation behaviour of guanine and adenine on

graphene–nafion composite film modified glassy carbon electrode and the simultaneous determination, *Process Biochem.*, **45**, pp. 1707–1712.

268. Lian, H., Sun, Z., Sun, X. and Liu, B. (2012). Graphene doped molecularly imprinted electrochemical sensor for uric acid, *Anal. Lett.*, **45**, pp. 2717–2727.

269. Yang, Y. C., Dong, S. W., Shen, T., Jian, C. X., Chang, H. J., Li, Y., He, F. T. and Zhou, J. X. (2012). A label-free amperometric immunoassay for thrombomodulin using graphene/silver-silver oxide nanoparticles as an immobilization matrix, *Anal. Lett.*, **45**, pp. 724–734.

270. Xue, R., Kang, T. F., Lu, L. P. and Cheng, S. Y. (2013). Electrochemical sensor based on the graphene-nafion matrix for sensitive determination of organophosphorus pesticides, *Anal. Lett.*, **46**, pp. 131–141.

271. Barkade, S. S., Pinjari, D. V., Singh, A. K., Gogate, P. R., Naik, J. B., Sonawane, S. H., Ashokkumar, M. and Pandit, A. B. (2013). Ultrasound-assisted mini-emulsion polymerization for the preparation of polypyrrole–zinc oxide (PPy/ZnO) functional latex for liquefied petroleum gas sensing, *Ind. Eng. Chem. Res.*, **52**, pp. 7704–7712.

272. Zhang, H., Li, Q., Huang, J., Du, Y. and Ruan, S. C. (2016). Reduced graphene oxide/Au nanocomposite for NO2 sensing at low operating temperature, *Sensors*, **16**, pp. 1152–1160.

273. Chen, W., Li, F., Ooi, P. C., Ye, Y., Kim, T. W. and Guo, T. (2016). Room temperature pH-dependent ammonia gas sensors using graphene quantum dots, *Sens. Actuators, B*, **222**, pp. 763–768.

274. Patil, P., Gaikwad, G., Patil, D. R. and Naik, J. (2016). Synthesis of 1-D ZnO nanorods and polypyrrole/1-D ZnO nanocomposites for photocatalysis and gas sensor applications, *Bull. Mater. Sci.*, **39**, pp. 655–665.

275. Shukla, S. K., Singh, N. B. and Rastogi, R. P. (2013). Efficient ammonia sensing over zinc oxide/polyaniline nanocomposite, *Indian J. Eng. Mater. Sci.*, **20**, pp. 319–324.

276. Patil, U. V., Ramgir, N. S., Karmakar, N., Bhogale, A., Debnath, A. K., Aswal, D. K., Gupta, S. K. and Kothari, D. C. (2015). Room temperature ammonia sensor based on copper nanoparticle intercalated polyaniline nanocomposite thin films, *Appl. Surf. Sci.*, **339**, pp. 69–74.

277. Alfano, B., Polichetti, T., Miglietta, M. L., Massera, E., Schiattarella, C., Ricciardella, F. and Di Francia, G. (2017). Fully eco-friendly H2 sensing device based on Pd-decorated graphene, *Sens. Actuators, B*, **239**, pp. 1144–1152.

278. Anand, K., Singh, O., Singh, M. P., Kaur, J. and Singh, R. C. (2014). Hydrogen sensor based on graphene/ZnO nanocomposite, *Sens. Actuators, B*, **195**, pp. 409–415.

279. Boland, C. S., Khan, U., Backes, C., O'Neill, A., McCauley, J., Duane, S., Shanker, R., Liu, Y., Jurewicz, I., Dalton, A. B. and Coleman, J. N. (2014). Sensitive, high-strain, high-rate bodily motion sensors based on graphene–rubber composites, *ACS Nano*, **8**, pp. 8819–8830.

280. Pandikumar, A., How, G. T. S., See, T. P., Omar, F. S., Jayabal, S., Kamali, K. Z., Yusoff, N., Jamil, A., Ramaraj, R., John, S. A. and Lim, H. N. (2014). Graphene and its nanocomposite material based electrochemical sensor platform for dopamine, *RSC Adv.*, **4**, pp. 63296–63323.

281. Guo, S., Wen, D., Zhai, Y., Dong, S. and Wang, E. (2010). Platinum nanoparticle ensemble-on-graphene hybrid nanosheet: one-pot, rapid synthesis, and used as new electrode material for electrochemical sensing, *ACS Nano*, **4**, pp. 3959–3968.

282. Lian, W., Liu, S., Yu, J., Xing, X., Li, J., Cui, M. and Huang, J. (2012). Electrochemical sensor based on gold nanoparticles fabricated molecularly imprinted polymer film at chitosan–platinum nanoparticles/graphene–gold nanoparticles double nanocomposites modified electrode for detection of erythromycin, *Biosens. Bioelectron.*, **38**, pp. 163–169.

283. Liu, Y., Zhu, L., Luo, Z. and Tang, H. (2013). Fabrication of molecularly imprinted polymer sensor for chlortetracycline based onthecontrolled electrochemical reduction of graphene oxide, *Sens. Actuators, B*, **185**, pp. 438–444.

284. Mao, Y., Bao, Y., Gan, S., Li, F. and Niu, L. (2011). An electrochemical sensor for dopamine based on a novel graphene-molecular imprinted polymers composite recognition element, *Biosens. Bioelectron.*, **28**, pp. 291–297.

285. Pakapongpan, S., Mensing, J. P., Phokharatkul, D., Lomas, T. and Tuantranont, A. (2014). Highly selective electrochemical sensor for ascorbic acid based on a novel hybrid graphene-copper phthalocyanine-polyaniline nanocomposites, *Electrochim. Acta*, **133**, pp. 294–301.

286. Wu, Z., Chen, X., Zhu, S., Zhou, Z., Yao, Y., Quan, W. and Liu, B. (2013). Enhanced sensitivity of ammonia sensor using graphene/polyaniline nanocomposite, *Sens. Actuators, B*, **178**, pp. 485–493.

287. Zhou, Y., Qu, Z. B., Zeng, Y., Zhou, T. and Shi, G. (2014). A novel composite of graphene quantum dots and molecularly imprinted polymer for fluorescent detection of paranitrophenol, *Biosens. Bioelectron.*, **52**, pp. 317–323.

288. Zhang, D., Tong, J., Xia, B. and Xue, Q. (2014). Ultrahigh-performance humidity sensor based on layer-by-layer self-assembly of graphene oxide/polyelectrolyte nanocomposite film, *Sens. Actuators, B*, **203**, pp. 263–270.

289. Zhang, D., Liu, J., Jiang, C., Liu, A. and Xia, B. (2017). Quantitative detection of formaldehyde and ammonia gas via metal oxide-modified graphene-based sensor array combining with neural network model, *Sens. Actuators, B*, **240**, pp. 55–65.

290. Zhang, D., Tong, J. and Xia, B. (2014). Humidity-sensing properties of chemically reduced graphene oxide/polymer nanocomposite film sensor based on layer-by-layer nano self-assembly, *Sens. Actuators, B*, **197**, pp. 66–72.

291. Zeng, Y., Zhou, Y., Zhou, T. and Shi, G. (2014). A novel composite of reduced graphene oxide and molecularly imprinted polymer for electrochemical sensing 4-nitrophenol, *Electrochim. Acta*, **130**, pp. 504–511.

292. Zeng, Y., Zhou, Y., Kong, L., Zhou, T. and Shi, G. (2013). A novel composite of SiO2-coated graphene oxide and molecularly imprinted polymers for electrochemical sensing dopamine, *Biosens. Bioelectron.*, **45**, pp. 25–33.

293. Yuan, W., Huang, L., Zhou, Q. and Shi, G. (2014). Ultrasensitive and selective nitrogen dioxide sensor based on self-assembled graphene/polymer composite nanofibers, *ACS Appl. Mater. Interfaces*, **6**, pp. 17003–17008.

294. Su, P. G. and Shieh, H. C. (2014). Flexible NO2 sensors fabricated by layer-by-layer covalent anchoring and in situ reductions of graphene oxide, *Sens. Actuators, B*, **190**, pp. 865–872.

295. Mao, S., Cui, S., Lu, G., Yu, K., Wen, Z. and Chen, J. (2012). Tuning gas-sensing properties of reduced graphene oxide using tin oxide nanocrystals, *J. Mater. Chem.*, **22**, pp. 11009–11013.

296. Tjoa, V., Jun, W., Dravid, V., Mhaisalkar, S. and Mathews, N. (2011). Hybrid graphene–metal nanoparticle systems: electronic properties and gas interaction, *J. Mater. Chem.*, **21**, pp. 15593–15599.

297. Yun, Y. J., Hong, W. G., Choi, N. J., Park, H. J., Moon, S. E., Kim, B. H., Song, K. B., Jun, Y. and Lee, H. K. (2014). A 3D scaffold for ultra-sensitive reduced graphene oxide gas sensors, *Nanoscale*, **6**, pp. 6511–6514.

298. Yang, Y., Li, S., Yang, W., Yuan, W., Xu, J. and Jiang, Y. (2014). In situ polymerization deposition of porous conducting polymer on reduced graphene oxide for gas sensor, *ACS Appl. Mater. Interfaces*, **6**, pp. 13807–13814.

299. Xue, C., Wang, X., Zhu, W., Han, Q., Zhu, C., Hong, J., Zhou, X. and Jiang, H. (2014). Electrochemical serotonin sensing interface based on thedouble-layered membrane of reduced graphene oxide/polyaniline nanocomposites and molecularly imprinted polymers embedded with gold Nanoparticles, *Sens. Actuators, B*, **196**, pp. 57–63.

300. Xing, X. J., Liu, X. G., He, Y., Lin, Y., Zhang, C. L., Tang, H. W. and Pang, D. W. (2012). Amplified fluorescent sensing of DNA using graphene oxide and a conjugated cationic polymer, *Biomacromolecules*, **14**, pp. 117–123.

301. Wu, Z., Chen, X., Zhu, S., Zhou, Z., Yao, Y., Quan, W. and Liu, B. (2013). Room temperature methane sensor based on graphene nanosheets/polyaniline nanocomposite thin film, *IEEE Sens. J.*, **13**, pp. 777–782.

302. Wang, X., Dong, J., Ming, H. and Ai, S. (2013). Sensing of glycoprotein via a biomimetic sensor based on molecularly imprinted polymers and graphene–Au nanoparticles, *Analyst*, **138**, pp. 1219–1225.

303. Wang, X., Sun, X., Hu, P. A., Zhang, J., Wang, L., Feng, W., Lei, S., Yang, B. and Cao, W. (2013). Colorimetric sensor based on self-assembled polydiacetylene/graphene-stacked composite film for vapour-phase volatile organic compounds, *Adv. Funct. Mater.*, **23**, pp. 6044–6050.

304. Xiang, C., Jiang, D., Zou, Y., Chu, H., Qiu, S., Zhang, H., Xu, F., Sun, L. and Zheng, L. (2015). Ammonia sensor based on polypyrrole–graphene nanocomposite decorated with titania nanoparticles, *Ceram. Int.*, **41**, pp. 6432–6438.

305. Tiwari, D. C., Atri, P. and Sharma, R. (2015). Sensitive detection of ammonia by reduced graphene oxide/polypyrrole nanocomposites, *Synth. Met.*, **203**, pp. 228–234.

306. Si, W., Lei, W., Zhang, Y., Xia, M., Wang, F. and Hao, Q. (2012). Electrodeposition of graphene oxide doped poly (3, 4-ethylenedioxythiophene) film and its electrochemical sensing of catechol and hydroquinone, *Electrochim. Acta*, **85**, pp. 295–301.

307. Parmar, M., Balamurugan, C. and Lee, D. W. (2013). PANI and graphene/PANI nanocomposite films: comparative toluene gas sensing behaviour, *Sensors*, **13**, pp. 16611–16624.

308. Seekaew, Y., Lokavee, S., Phokharatkul, D., Wisitsoraat, A., Kerdcharoen, T. and Wongchoosuk, C. (2014). Low-cost and flexible printed graphene–PEDOT: PSS gas sensor for ammonia detection, *Org. Electron.*, **15**, pp. 2971–2981.

309. Qiu, H., Luo, C., Sun, M., Lu, F., Fan, L. and Li, X. (2012). A chemiluminescence sensor for determination of epinephrine using

graphene oxide–magnetite-molecularly imprinted polymers, *Carbon*, **50**, pp. 4052–4060.

310. Qiu, J. D., Shi, L., Liang, R. P., Wang, G. C. and Xia, X. H. (2012). Controllable deposition of a platinum nanoparticle ensemble on a polyaniline/graphene hybrid as a novel electrode material for electrochemical sensing, *Chem. Eur. J.*, **18**, pp. 7950–7959.

311. Qin, Y., Peng, Q., Ding, Y., Lin, Z., Wang, C., Li, Y., Xu, F., Li, J., Yuan, Y., He, X. and Li, Y. (2015). Lightweight, superelastic, and mechanically flexible graphene/polyimide nanocomposite foam for strain sensor application, *ACS Nano*, **9**, pp. 8933–8941.

312. Promphet, N., Rattanarat, P., Rangkupan, R., Chailapakul, O. and Rodthongkum, N. (2015). An electrochemical sensor based on graphene/polyaniline/polystyrene nanoporous fibres modified electrode for simultaneous determination of lead and cadmium, *Sens. Actuators, B*, **207**, pp. 526–534.

313. Paek, K., Yang, H., Lee, J., Park, J. and Kim, B. J. (2014). Efficient colourimetric pH sensor based on responsive polymer-quantum dot integrated graphene oxide, *ACS Nano*, **8**, pp. 2848–2856.

314. Segal, M. (2009). Selling graphene by the ton, *Nat. Nanotechnol.*, **4**, pp. 612–614.

315. Potts, J. R., Dreyer, D. R., Bielawski, C. W. and Ruoff, R. S. (2011). Graphene-based polymer nanocomposites, *Polymer*, **52**, pp. 5–25.

Chapter 2

Facile Synthesis and Applications of Polyaniline/TiO_2 Hybrid Nanocomposites

Hafeez Anwar, Yasir Javed, Iram Arif, and Uswa Javeed

Department of Physics, University of Agriculture, Faisalabad 38040, Pakistan

hafeez.anwar@gmail.com

The organic/inorganic hybrid nanocomposites, in particular, organic conducting polymer (OCP)-based hybrid nanocomposites, constitute emerging advanced materials combining the unique features of inorganic and organic components. This chapter explores the state-of-the-art nanocomposites based on OCPs, mainly polyaniline (PANI) and titanium dioxide (TiO_2) nanoparticles, and the appropriate methodology to develop new nanocomposites with improved properties. The preparation of these hybrid nanocomposites with fascinating properties has emerged as an attractive alternative in a wide number of applications, especially in photocatalysis, sensors, and energy storage devices, including fuel cells and dye-sensitized solar cells (DSSCs). An overview investigation of the various synthesis methods to prepare these hybrid nanocomposites is presented.

Hybrid Nanocomposites: Fundamentals, Synthesis, and Applications
Edited by Kaushik Pal

ISBN 978-981-4800-34-1 (Hardcover), 978-0-429-00096-6 (eBook)
www.panstanford.com

This chapter also covers a discussion of various properties of the nanocomposites that are significantly different from the individual components and summarizes the recent progress in the use of these advanced hybrid nanocomposites.

2.1 Introduction

The utilization of nanotechnology is widespread, including industry, medicine, sensors, solar energy, display technologies, batteries, optoelectronic devices, and catalysis [1, 2]. The potential of nanomaterials has been probed for different architectures where one type is hybrid structures based on inorganic/organic nanostructures. These hybrid nanostructures have fascinated scientists for almost 20 years for their improved efficiency for applications [3–12]. Although materials scientists deal with nanocomposites, which result in an effective single material based on two or more types of inorganic materials [13], inorganic/organic hybrids have developed a new kind of breakthrough for nanocomposite materials with remarkable characteristics. Such types of nanomaterials carry the characteristics of both organic-inorganic parent constituents [14]. Polymeric nanocomposites have already been explored by chemists for their unique properties and advanced applications [15]. These nanocomposites are a combination of conventional polymers.

The combination of a polymer matrix with inorganic nanoparticles has paved the way for high-performance materials that provide exclusive applications in many fields due to upgraded magnetic, optoelectronic, mechanical, and optical properties. From inorganic/organic nanocomposites, the metal oxides and conducting polymers have shown an important class of hybrid nanocomposites. These hybrids are good for photocopying toners, rechargeable batteries, drug delivery, conductive paints, smart windows, etc. [16, 17]. Many facile synthesis techniques for preparing organic/inorganic nanocomposites have been developed [18], where both materials are mashed up through a mixing or blending process by taking the melted or solution form of the polymer [19]. On the basis of matrix materials, nanocomposites can be divided into three main categories;

- Ceramic matrix nanocomposites
- Polymer matrix nanocomposites
- Metal matrix nanocomposites

The hybrid materials of organic polymers and inorganic nanocomposites show enhanced magnetic, optoelectronic, mechanical, and optical properties. Due to such enhanced properties, these composites have been extensively applicable in many fields, such as optical devices, aerospace, electronics, automotive, military equipment, protective garments, and safety [13].

In addition to this, the combination of metal oxides and conducting polymers introduced a new emerging field, that is, hybrid nanocomposites, for investigation and applications. These hybrid materials are good for photocopying toners, rechargeable batteries, drug delivery, conductive paints, and smart windows [16, 17]. Synthesis of polymer nanocomposites of core–shell inorganic nanoparticles has gained great attention of scientists in recent years due to their excellent properties [20, 21]. The electrical properties of the conducting polymer and magnetic, electrical, optical, or catalytic properties of the metal oxide merge together and emerge into outstanding properties that are extensively employed in the fields of optics, catalysis, and electronics [22].

Our main concern and focus is on conducting polymers and their hybrids with metal oxides. The nanocomposites of conducting polymers with metal oxides make a strong bridge between both of these worlds. In this regard, nanocomposites are divided into two classes:

- Inorganic particles embed in an organic matrix
- Organic polymers embed in an inorganic matrix

In both classes, the preparation of nanocomposites needs some type of encapsulation, in addition to simple mixing or blending (Fig. 2.1).

Figure 2.1 Production of nanocomposites from parent constituents.

2.1.1 Conducting Polymers

In 1958, organic conducting polymers (OCPs) were discovered [23]. This discovery opened the door of an interesting field of research because the OCPs have attractive and fascinating properties as well as a number of applications. At that time, it was considered that the OCPs would show their applicability in many disciplines of science, such as sensors, electrorheological science, membranes, chemicals, electrical science, thermoelectrics, electronics, electrochemical science, electromagnets, and electroluminescence [24–27]. There are many other uses and applications of these OCP hybrid nanomaterials, which are still needed to be explored, but there are some issues regarding their synthesis, which needs to be overcome. The conductivity for such OCPs is not high as that of inorganic semiconductors. The reason for such low conductivity is the low mobility of charge carriers, although the number of charge carriers is considerably large. Increasing the value of conductivity depends upon the orientation and defect-free structure of polymers. The most well-known OCPs are:

- Polypyrrole
- Polyaniline (PANI)
- Polythiophene

OCPs give a wide scope to change their electrical state from semiconducting to metallic by using doping techniques. These materials are organic electrochromic materials accompanied by a chemically functional surface.

2.1.2 Nanocomposites of Conducting Polymers

The nanocomposites of OCPs have been divided into two categories, that is, nanocomposites with organic materials and nanocomposites with inorganic materials, and both have a different synthesis process.

Electrochemical synthesis–based organic–inorganic nanocomposites contain SnO_2, PB, SiO_2, MnO_2, CB, WO_3, etc., and exhibit electrochromic, charge storage, and optical activities. Nanocomposites containing Pt, Cu, Pd, etc., display catalytic activities, and nanocomposites containing γ-Fe_2O_3 and magnetic macroanions show magnetic susceptibility. Chemical synthesis–

based inorganic–organic nanocomposites contain SiO_2, SnO_2, $BaSO_4$, etc., are materials of core and display colloidal stability; nanocomposites containing Fe_2O_3, ZrO_2, TiO_2, etc., show enhanced mechanical and physical properties; nanocomposites containing Fe_3O_4, γ-Fe_2 O_3, etc., as magnetic particles display magnetic susceptibility; nanocomposites containing Pt, PtO_2, TiO_2, Pd, POM, etc., exhibit piezoresistive, dielectric, energy storage, and catalytic activities; and nanocomposites containing –NH_2/–COOH functional groups on the surface and colloidal silica as the core display surface activeness [19].

There are two different approaches that can be applied for the development of composite materials:

- Preparation of nanocomposite materials in which known active building blocks are employed that could proceed with each other and give the final material in which the parent constituents maintain at least their partial original integrity
- Preparation of nanocomposite materials in which the parent constituents are completely transformed into a new material

2.1.2.1 Building block approach

In this approach, building blocks maintain partially their original molecular integrity throughout the composite formation. This shows that some source structure will also be present in the product structure. The novel material does not exhibit the typical properties of the source material, but during the matrix transformation, it does not affect the typical properties of the source material. Figure 2.2 shows an example of the building blocks of nanoparticles with the reactive organic group.

2.1.2.2 In situ approach

The opposite of the building block approach is the in situ approach. It is a chemical transformation–based process. In this approach known discrete molecules are converted into multidimensional systems that are quite different from the source material. There are three basic types of in situ processes:

- In situ formation of inorganic-based materials
- Formation of organic polymers in the presence of preformed inorganic materials,

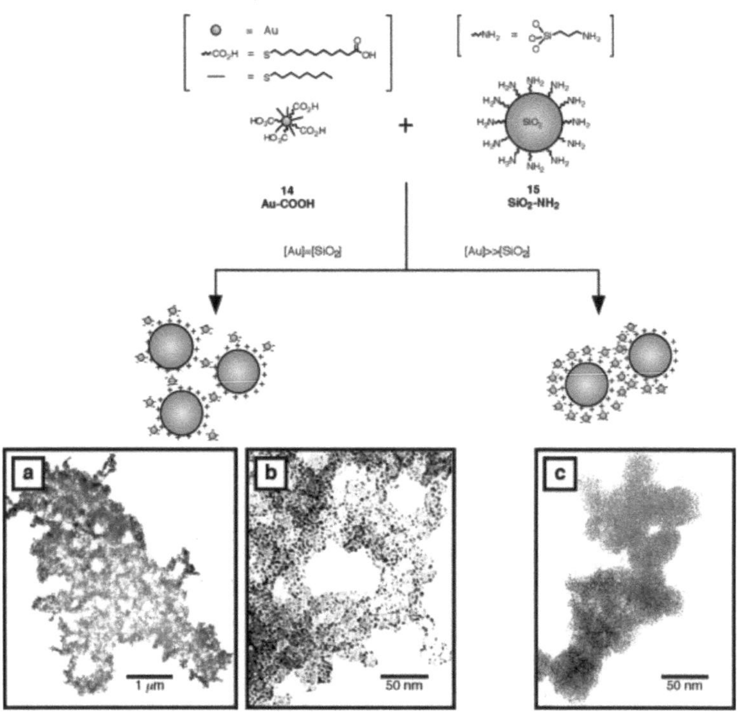

Figure 2.2 Nanocomposites assembled by the building block approach. (a, b) Large nanoparticle assembly formed by 1:1 (w/w) AuCOOH/SiO_2-NH_2 mixture. (c) Smaller-scale aggregate formed from 100:1 (w/w) Au-COOH/SiO_2-NH_2. Reprinted with permission from Ref. [29]. Copyright (2003) American Chemical Society.

- Hybrid materials through the formation of both organic and inorganic components simultaneously

In the next section, the development of OCPs in the presence of preformed inorganic materials will be discussed.

2.1.2.2.1 *Preparation of organic polymers in the presence of preformed inorganic materials*

For the development of hybrid materials from OCPs in the presence of metal oxides, one must know about the solutions of several possibilities to get a successful output from these two materials with different natures. It might be possible that the metal oxide has an exposed surface but might be not be able to provide an optimum functional surface. Therefore, to cope with this issue, there are two possibilities:

- Either it could be modified with organic groups that are nonreactive (e.g., alkyl chains)
- Either it could be modified with organic groups that have reactive surface groups [28, 29]

The compound can be treated chemically depending on such preconditions. The pure inorganic surface can be treated in two ways, either with silane coupling agents or with surfactants to make it suitable with functional monomers or organic monomers. The inorganic components are linked with nonreactive organic components through the surface. These inorganic components can be dissolved into a monomer that is eventually polymerized. The product after the organic polymerization is a blend, and in such circumstances inorganic components interact weakly with organic compounds [13].

A porous 3D inorganic system is employed as an inorganic component for the fabrication of composite materials. A unique approach must be applied that depends upon functionalization of surface, size of pores, and stiffness of the inorganic network. Sometimes the intercalation of organic material in the pores is complicated and hard due to limited diffusion. The obtained composites could be considered as a host–guest composite material. There are two possible approaches for such types of nanocomposite components.

- For melting and soluble polymers, the host channels are applied to the preformed polymer of direct threading.
- For the hosts channels and pore polymerization, the in situ process is applied [30].

In this chapter, we will discuss hybrid nanocomposites of PANI and TiO_2 in detail, along with facile synthesis, characterization, and applications.

2.1.3 Polyaniline

In the family of OCPs, the most favorable type is PANI, due to its good stability, easy fabrication, low-price monomer, and tunable characteristic. PANI has unique properties that are not found commonly in other OCPs. It has many types that are different in the sense of oxidation degree and extent of protonation or both [31].

2.1.3.1 Structure of polyaniline

The π-levels at the boundaries of the conduction and valence bands make the OCPs more enchanting chemically and physically. To investigate the π-levels of the highest valence band spectroscopic techniques have been used. The range of ultraviolet and X-ray wavelengths apply for such studies through photoelectron spectroscopy [32].

PANI is basically a phenylene-based material having an –NH group in the polymer chain, which is bounded to either edge of a ring of phenylene. The fundamental scheme of PANI is shown in Fig. 2.3. Due to the combination of oxidized {–N=Q=N–} and reduced {–NH–B–NH–} PANI can be considered a polymer having a state of mixed oxidization. Also, the oxidized and reduced units are repeated, where =Q= and –B– represent a quinoid and a benzenoid unit, respectively, fabricating a chain of the polymer and a general state of oxidation is represented as $(1 - y)$.

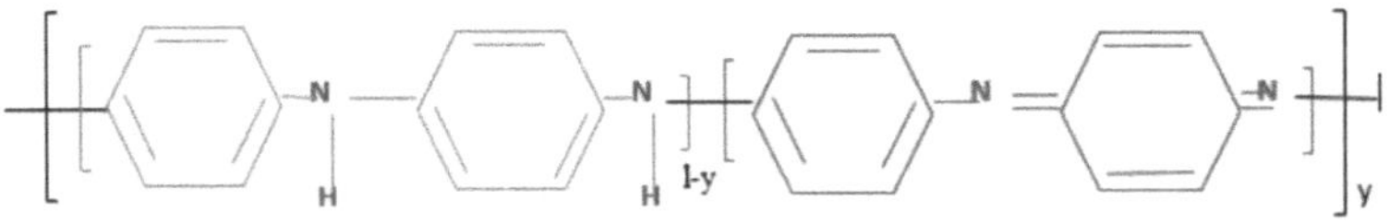

Figure 2.3 Structure of polyaniline.

There are two types of states: one is a completely pernugraniline-based oxidized state in which $1 - y = 1$ [33], and the second is a fully leucoemeraldine-based reduced state for which $y - 1 = 0$. PANI can have many structures in any oxidation state, depending upon the oxidation state of nitrogen atoms that are present as imine/amine configuration. The range of the adopting of structures is between $1 - y = 1$ and $y - 1 = 0$ states. $1 - y = 0.5$ is called the half-oxidized emeraldine state, which is a semiconductor in behavior and is made up of two benzenoid units and one quinoid unit, which form an alternating sequence. An emeraldine conducting salt is a form of protonating [34]. The insulating forms like leuocemeraldine base (LB), emeraldine base (EB), and pernigraniline base (PB) are not only different in excitations but electronic structures are also contrasted. For the fabrication of an emeraldine salt (ES) system the PB system can be doped (n-doped) by the redox reaction, the LB system can be oxidatively doped (p-doped), and the EB system can be doped with protonic acid. The LB can be converted into a conductive ES, a state of PANI, by doping the nonredox intermediate state of PANI into acids. It could be the reason of conductivity when the imine nitrogen is doped with protons, normally producing radical cations on its sides. This is also called a self-doped polymer because when proton doping occurs a counterion is introduced and this counterion is attached to the parent polymer through sulfonating benzene rings partially. Both organic and inorganic acids are effective because both lead to solubility in the wide range of organic solvents with organic sulfonic-based acids [32].

The three ideal oxidation states of PANI are given below:

- Leucoemeraldine, clear or white and colorless $[C_6H_4NH]_n$
- Emeraldine: for ES it's green; for EB it's blue $\{[C_6H_4NH]_2[C_6H_4N]_2\}_n$
- (Per)Nigraniline violet or blue $[C_6H_4N]$

Different forms of PANI have been shown in Fig. 2.4. This kind of polymer is the first one that achieved worldwide availability at the commercial level [35].

PANI is also famous for its electrochromic property. Color changes occur during electrochemical reactions, depending upon the potential range, generally from −0.2 to −0.8 V against a calomel electrode, which is standard. The color changes at different potentials in PANI are due to the formation of different structures in polymers [36]. For example, in 0.1 M H_2SO_4 aqueous solution, the cyclic voltammetry graph of the PANI film is largely due to the capacitive current background along with two anodic peaks having values of 0.2 and 0.7 V. The transparent yellow color turns into green at 0.2 V and changes to dark blue at potentially more than 0.3 V. The green color appears due to the fabrication of Wurster-type cations and turns to dark blue due to the fabrication of domain format and doped states with $(SO_4)_2$. At 0.7 V, deterioration of electrochromic behavior is observed due to hydrolysis of the domain structure [37].

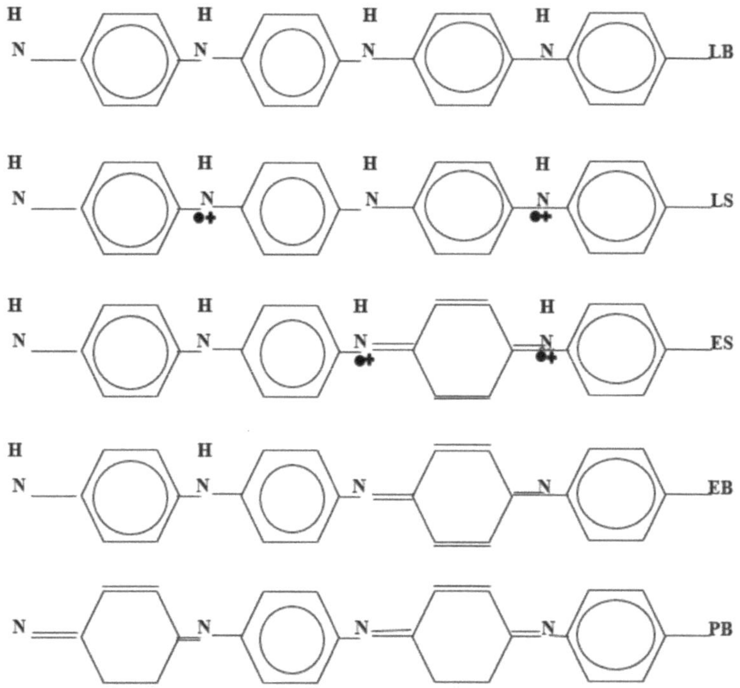

Figure 2.4 Different types of PANI: leuocemeraldine base (LB), leucoemeraldine salt (LS), emeraldine salt (ES), emeraldine base (EB), and pernugraniline base (PB).

2.1.3.2 Synthesis of polyaniline

The synthesis of PANI can be achieved by different procedures that are usually electrochemical and chemical routes. To get PANI, aniline gets oxidized chemically with ammonium peroxodisulfate (APS)-type oxidizing agents. APS is preferable because it has some dominant properties such as nonmetallic oxidizing agent, no ion interference, and maximum redox potential. The electrochemical method of fabrication of PANI is through potential sweeping [38, 39], potentiostatic [40], and galvanostatic [41] methods. PANI has a wide range of applications that are not only chemical but also electrochemical, such as membrane technology, microelectronics, sensors, electrochromic devices, conductive textiles, and lithium batteries [42]. Redox doping and protonation are the techniques used to control the electronic properties of PANI reversibly.

The most popular applications of PANI are enzyme immobilization [43, 44], light-emitting diodes [45], electrochemical-based light conversion into electricity [46, 47], batteries [48], and electrochromical devices [49].

2.1.4 Titanium Dioxide

Titania is another famous name of titanium dioxide. It is not found naturally in pure form. It is extracted from leuxocene ores or ilmenite. Also, rutile beach sand is one of its easily mined purest forms. TiO_2 exhibits many prominent characteristics such as chemical and electrical properties. These properties make it attractive and suitable for many applications. Many techniques have been introduced to prepare titania at laboratory and industrial levels, such as sol–gel method [50, 51] and gas condensation method [52]. To prepare nanoparticles of titania, microemulsion is an advanced technique [53].

2.1.4.1 Structure of TiO_2

TiO_2 has a closely packed hexagonal crystalline structure with a space group of P_{63}mmc and a space group number 194 [54]. Basically, titania is a polymorphic material. On the basis of symmetry, it is categorized into three polymorphous systems:

- Rutile
- Anatase
- Brookite

The three structures show stability according to their nanosizes [55]. Rutile and anatase exhibit tetragonal symmetry with a space group of P_{42}/mnm and I_{41}/amd [56], respectively, while brookite exhibits orthorhombic symmetry having a space group of Pbca [57], as shown in Fig. 2.5.

In these structures, every titanium atom is shared with six oxygen atoms at same distances and every oxygen atom is shared with three atoms of titanium. On the bases of relative spacing, the octahedral structure is different among three types. There are three faces of rutile: (110) and (100) are extremely low in energy but these two are very effective and polycrystalline. (110) shows good stability at high temperatures. The titania atoms that are exposed show very low electronic density. (001) does not show good stability at high temperatures. This type of titania could not show stability just above 475°C. In (001) two rows of O_2 have changed with a single row of titanium-exposed atoms. These rows are not axial but equatorial type.

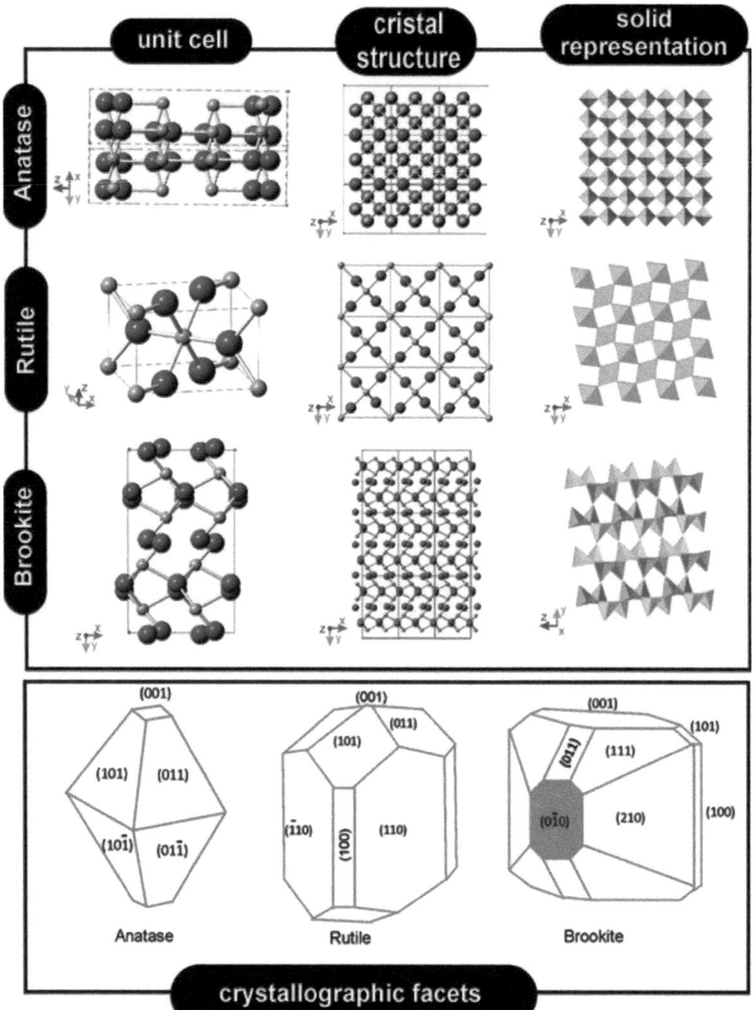

Figure 2.5 Crystallographic structures of different TiO_2 polymorphs: anatase, rutile, and brookite [58].

It is commonly and frequently employed in paints as a pigment, sunscreens, toothpaste, ointments, and coatings. It is assumed as a quality-of-life compound that is used not only for domestic purposes but also at the industrial level. It's a very important white pigment because it is the only market product that can provide strong binding due to its high rating of refractive index [59]. Also, it is applicable as gas sensors [60], humidity sensors [61, 62], ceramic membranes [63], and photocatalysts [64, 65].

2.2 PANI/TiO_2 Hybrid Nanocomposites

The new field of materials science and modern technology is established by polymer semiconductors [66]. The most common composites of titania are with poly(methyl methacrylate) (PMMA) [67], poly(phenylene vinylene) (PPV) [68], and conducting PANI [69–71]. These composites exhibit not only prominent features but also show versatile applications in many fields of science and technology. The composite of PANI/TiO_2 is very attractive due to its improved electrical, optical, and catalytic properties. For the past few years, these composites are under investigation. The researchers are mainly concerned with the optical characteristics of the nanocomposites of polymer surface-modified TiO_2. This material also has a high dielectric constant and the composites having the maximum dielectric constant are extensively employed in integrated electronic-based circuits. These dielectrics are mostly used in capacitors [72]. One of the great aspects of such composites is to minimize the current leakage and voltage breakdown. For the manufacturing of metal oxide-based devices at a small scale, the submicron-width gate materials are needed.

2.2.1 Different Structures of PANI/TiO_2 Hybrid Nanocomposites

In this section, different nanostructures of PANI/TiO_2 are discussed.

Figure 2.6 shows scanning electron microscopy (SEM) images of TiO_2, PANI, and PANI/TiO_2 nanocomposites. In Fig. 2.6a, it is clear that the nanoparticles of TiO_2 were dispersed uniformly. Figure 2.6b shows that PANI is made up in the form of nanorods. After grinding

TiO_2 nanoparticles with PANI, no PANI nanorods were observed, as shown in Fig. 2.6c, which may be because PANI nanorods were crashed and covered with TiO_2 nanoparticles.

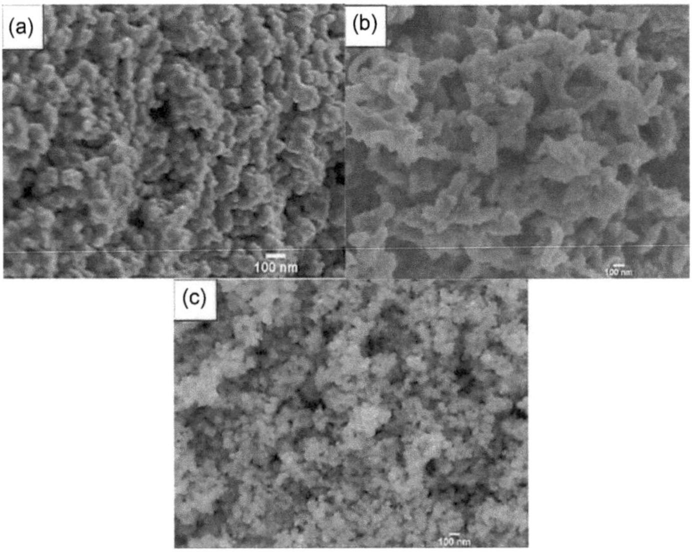

Figure 2.6 SEM images of (a) TiO_2, (b) PANI, and (c) TiO_2/ PANI [73].

Figure 2.7 shows that the TiO_2/PANI core–shell nanocomposites were prepared successfully. Figures 2.7a and 2.7b showed that the PANI is decorated completely on the surface of TiO_2 nanoparticles that creates repulsion between them and prevents agglomeration. When 20 wt% and 40 wt% of TiO_2/PANI core–shell nanocomposites were added in poly(vinyl alcohol) (PVA), it provides a homogeneous fibrous structure due to crosslinking between PVA and PANI, as shown in Fig. 2.7c,d [74].

Figure 2.8 shows in transmission electron microscopy (TEM) images that the prepared PANI/TiO_2 nanocomposites have a tubular morphology and crystalline structure and the ratio of TiO_2 nanowires in the PANI/TiO_2 nanocomposites varied from 5 to 15 wt% [75].

Figure 2.9 shows the morphology of TiO_2/PANI electrodes in which TiO_2 was immersed in PANI for different times, that is, 30 min., 60 min., 90 min., and 120 min.

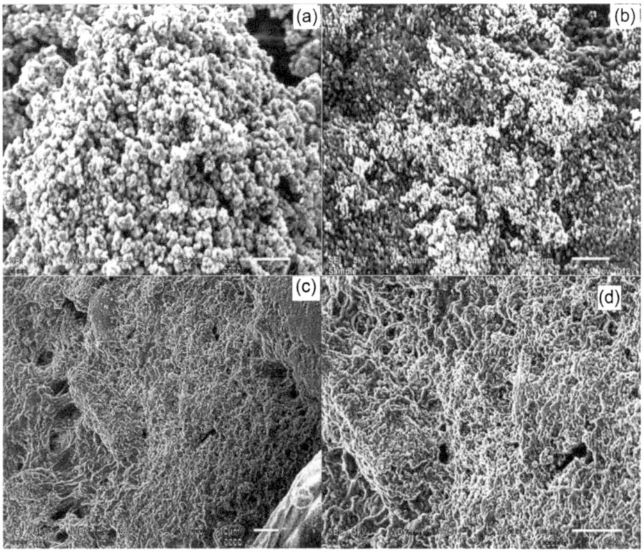

Figure 2.7 SEM images of (a , b) TiO_2/PANI core–shell nanocomposites at two magnifcations; (c, d) TiO_2/PANI core–shell nanocomposites loaded in a PVA stabilizer [74].

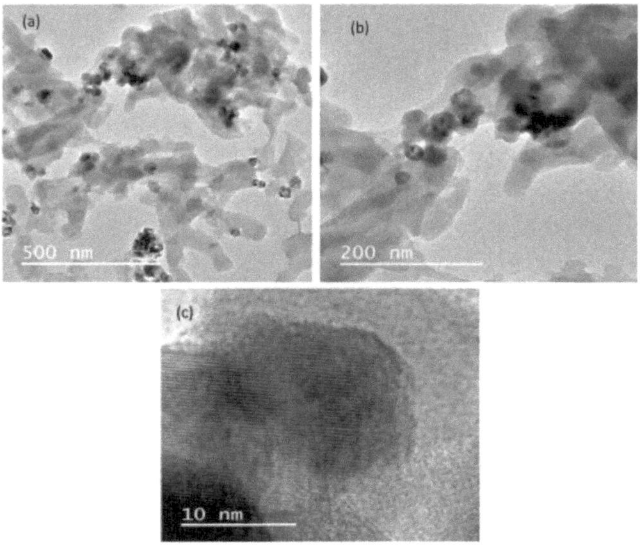

Figure 2.8 TEM image of PANI–TiO_2 nanocomposite. (a, b) Hexagonal TiO_2 nanoparticles were observed on the edges of the PANI tubule, and (c) highly crystalline TiO_2 nanoparticle on the edges on PANI. Reproduced from Ref. [86]. Copyright © 2017, Springer-Verlag GmbH Germany, part of Springer Nature.

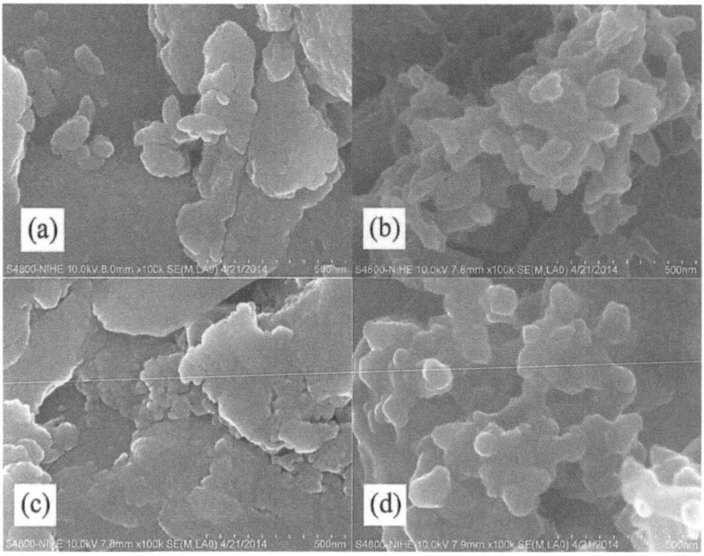

Figure 2.9 SEM image of TiO_2/PANI electrodes prepared under different condition: TiO_2 immersed in a PANI solution for (a) 30 min., (b) 60 min., (c) 90 min., and (d) 120 min. [76].

2.2.2 Synthesis of PANI/TiO_2 Hybrid Nanocomposites

PANI/TiO_2-based nanocomposites have high dielectric constants that are of great importance because this property makes them potential candidates for the development of microelectromechanical systems (MEMS) and multiple access memories. Only materials having a high dielectric constant can fulfill this requirement. PANI/TiO_2 yields a great dielectric constant because Titania itself has a high constant and wide bandgap; on the other hand, PANI is stable at high temperatures. Although OCPs having nanosized semiconductors exhibit a dielectric constant and transportability, they are not studied and investigated extensively [77].

There are a number of methods that are adapted for the synthesis of composites of PANI and titania.

2.2.2.1 Chemical methods

The chemical methods that are commonly used for the fabrication of PANI/TiO_2 nanocomposites are:

- Chemical oxidation method
- Hydrothermal method
- Sol–gel method

2.2.2.1.1 *Chemical oxidation*

A wide research on PANI/TiO_2 nanocomposites has been done. The main concern of the researchers is to investigate the morphology characterization and technique of preparation. The morphology characterization is much important in regard to the shape and size of oxide small particles, the interface between inorganic and organic phases, the kind of interaction, and the degree of dispersion.

Chemical oxidation was used, at room temperature, for the synthesis of PANI/TiO_2 nanocomposites with different ratios of TiO_2. For the characterization of such samples a UV-visible spectrometer, SEM, Fourier transform infrared (FTIR), X-ray diffraction (XRD), energy dispersive X-ray analysis (EDAX), and conductivity measurements were used. Due to the interlink chains of conjugated polymer and nanoparticles of TiO_2 a red shift occurred due to the incorporation of titania nanoparticles at 310 nm with p-n-p transitions. The molecular structure of titania was affirmed by FTIR. The additional bands at 1105 cm^{-1} and 1623 cm^{-1} in FTIR spectra are due to the presence of Ti–O–C and Ti–O in PANI/TiO_2 nanocomposites in stretching mode. It could be considered that Ti compounds are produced along with a straight-line arrangement system of titania particles. Patterns of XRD showed that the amorphous structure of titania reduced and nanocomposites grew strongly along the (110) direction. The new system of nanocrystalline titania was tetragonal. The 200 nm size for 50% PANI/TiO_2 nanocomposites and uniform granular morphology was confirmed by SEM [78].

Anatase titania with PANI powder in an emeraldine system was synthesized through chemical oxidation of aniline along with APS as an oxidant. Aniline and APS were bought from Emerck and applied as received. Around 1 g nanopowder of anatase titania was taken in a round-shaped flask. The titania was bought from Sisco Research Laboratories. In t, 50 mL of 1 M HCl was poured as a catalyst, which was purchased from Emerck. After stirring titania with HCl for 2–3 h, 2 mL aniline was added slowly dropwise. After this addition, for polymerization, 4.99 g of APS was poured dropwise, which was mixed in 50 mL HCl. The solution was stirred 4–5 h until the green color of ES appeared, which meant the full polymerization component. HCl was used to wash the precipitate, followed by distilled water [79].

2.2.2.1.2 *Hydrothermal method*

The hydrothermal method is an effective method for the fabrication of PANI/TiO_2 nanocomposites. Different techniques were used to investigate the structure, optical, thermal, and electrical properties by XRD and TEM. XRD displayed two individual peaks of PANI and titania for the less than 20% of titania composites. TEM results described that the tubular structure of titania nanoparticles had a diameter of 8–12 nm and a length of 80–150 nm. Conductivity for direct current (DC) fell down by adding titania in PANI/TiO_2 nanocomposites. During the mother phase of titania, the photoluminescence properties were found to be controllable with the fraction of weight [80].

A core-shell array of PANI/TiO_2 nanorods was prepared by using both electropolymerization and hydrothermal methods. For PANI/TiO_2 nanorods better cycling performance of 62.1% after 1000 cycles, high coloration efficiency, and optical modulation up to 57.6% at 700 nm were achieved (Fig. 2.10) [81].

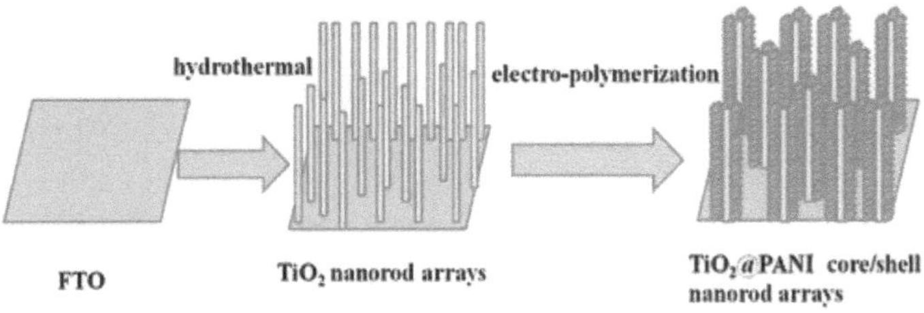

Figure 2.10 Core–shell array of PANI/TiO_2 nanorods. Reprinted with permission from Ref. [81]. Copyright (2013) American Chemical Society.

2.2.2.1.3 *Sol–gel method*

PANI/TiO_2 hybrid nanocomposites were prepared by using sol–gel chemical synthesis. TiO_2 nanoparticles with a diameter of 3–5 nm were added to aniline that was chemically polymerized. The addition ratio of TiO_2 to PANI changed the morphology of hybrids from plate-like small grains to aggregates. The results of cyclic voltammetry showed that the structure of plate-like grains was more suitable for electrochemical stability. TiO_2 and PANI established a chemical bond relation. Danielle C. Schnitzler and Aldo J. G. Zarbin [82] prepared PANI/TiO_2 nanocomposites via sol–gel methodology and followed two different routes in terms of adding PANI before and after the sol process. They also investigated the effect of PANI by varying its amounts and reported that different routes did not affect the final product much except the different amounts of PANI in the final product.

2.2.2.2 In situ polymerization

In situ polymerization is the major synthesis technique used for the synthesis of PANI/TiO_2 nanocomposites. In the past few years, different polymerization techniques were used to grow polymer nanocomposites. The mechanism of in situ polymerization is given in Fig. 2.11. Atomic transfer radical polymerization (ATRP) is one of the recently reported methods that allows better control of the distribution of polymers over nanomaterial surfaces [83, 84]. In most cases, aniline is polymerized chemically [77, 79, 85–89] in the presence of TiO_2 and APS is used as a precursor. In this typical procedure, the assessed amount of TiO_2 is sonicated in acid, usually HCl, for better dispersion. To avoid agglomeration, vigorous mechanical stirring is carried out, then an oxidizing agent is added, and the temperature is usually kept low. The nanocomposites are obtained in the form of powder.

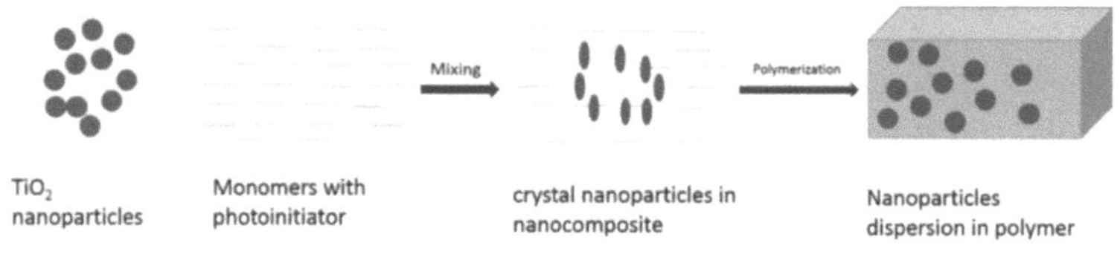

Figure 2.11 In situ polymerization in combination with chemical oxidation of aniline from the surfaces of TiO_2 nanoparticles.

2.2.2.3 The electrochemical method

An advantage of this synthesis technique is that polymerization in an electrolyte produces flexible films. In this technique, electrolyte films are produced by oxidative coupling. This technique is like the electrochemical deposition of metals. Letheby in 1862 reported the first synthesis of polyemarldine salt by an electrochemical technique. Mohiner has defined some mechanistic features of synthesis of PANI by electrochemical synthesis. An electrochemical technique is a most adaptable method for synthesis of better-ordered and cleaned polymer films [90]. The electrochemical polymerization method is much faster and environmentally friendly than chemical polymerization due to the absence of oxidants [91]. In 1993, for the first time, Kuwabata et al. [92] purposed the electrochemical method for the synthesis of PANI/TiO_2 nanofilms. Later, Luo et al. [93] used this technique and formed two types of multilayered electrodes, TiO_2/PANI/PATP and PANI/PATP, for solar cell applications and reported that the response of TiO_2/PANI/PATP-based electrodes is much better than the other one and it covers violet and red-light regions. Kuwabata et al. also used the electrochemical method for the synthesis of PANI/TiO_2 nanocomposites by electropolymerization of aniline [92]. Mickova et al. [94] also prepared PANI/TiO_2 nanocomposites by using the electrochemical method, characterized its photoelectrochemical and electrochemical properties, and also investigated the surface morphology [95]. Pavol Kunzo et al. also synthesized PANI/TiO_2 nanocomposites by using an electrochemical method and used them as gas sensors [96]. Murat Ates and ErhanTopkayaalso used this technique to form PANI films with TiO_2, Ag, and Zn and characterized their corrosion-resisting properties [97].

2.2.2.4 Enzymatic synthesis

The enzymatic polymerization method is also used for the synthesis of PANI/TiO_2 nanocomposites due to easy experimental conditions as compared to other polymerization methods. Different oxidation-reduction enzymes like soybean peroxides and horseradish peroxidase (HRP) can oxidize polymers [98]. This method is environmentally friendly because the oxidative agents are mostly derived from renewable resources. Most of the research related to enzymatic polymerization of aniline was carried out using polyelectrolyte templates such as sulfonated polystyrene (SPS) and poly(vinyl phosphonic acid). In an enzymatic route for the synthesis of PANI/TiO_2 reported by Nabid et al. [99], HRP was used to catalyze the polymerization and H_2O_2 as oxidant. The presence of

SPS affected the polymerization reaction. PANI was deposited on the surface of TiO_2, forming a core–shell structure, as shown in Fig. 2.12. These nanoparticles showed a great effect on the electron exchange assistance because the reversibility is better for the PANI/TiO_2/Pt electrode with an *E* value of 39 mV in comparison to that of the PANI/Pt electrode with an *E* value of 281 mV [99].

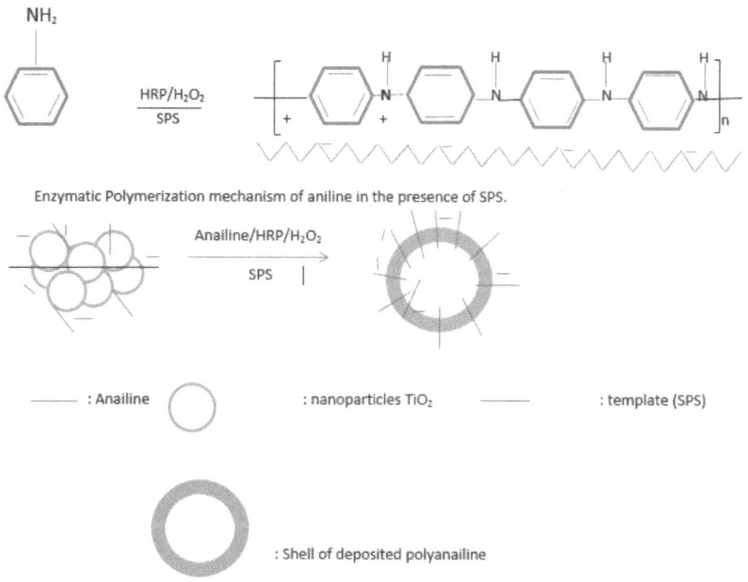

Figure 2.12 Enzymatic polymerization mechanism of aniline in the presence of SPS.

2.2.2.5 The self-assembly method

In this method, nanostructures assemble themselves in larger structures due to different forces acting among them, such as intermolecular forces [100] and hermitian interactions [101, 102]. Lijuan Zhang and Meixiang Wan prepared PANI/TiO_2 nanotubes by using the self-assembly technique and β-naphthalenesulfonic acid (β-NSA) was used as the dopant. Characterization was carried out using XRD and Raman spectroscopy and both did not interact chemically [103]. Cui et al. used the self-assembly method in combination with sol–gel and electrostatic layer-by-layer techniques to form PANI/TiO_2 nanocomposite-based films and fabricate a gas sensor that showed excellent sensitivity toward NH_3 gas [104]. Li et al. synthesized a PANI/TiO_2 nanocomposite with the help of self-assembly and graft polymerization methods. In this work, a layer of PANI was chemically grafted on the monolayer of TiO_2 nanoparticles and γ-aminopropyl triethoxysilane was used as a coupling agent. It showed better photocatalytic activity as compared to PANI [105]. Xie et al. also used both techniques, that is, self-assembly and graft polymerization, for the synthesis of PANI/TiO_2 nanotubes and characterized them by FTIR, field emission scanning electron microscopy (FESEM), and cyclic voltammetry [106]. PANI/TiO_2 composite nanofibres were also prepared by self-assembly and reported high yield as compared to other techniques [107].

2.2.2.6 Template polymerization

Template polymerization is also known as replica polymerization/ matrix polymerization. The word "replica" was first used in 1954 [108] but later changed to "template polymerization" [109]. It is a technique in which a monomer having low molecules is converted into a large polymer by a chain reaction in the presence of a template polymer [110]. There are two types of templates, soft template and hard template, like micelles and surfactants, and are usually used in the synthesis of PANI/TiO_2 nanocomposites. Xiong et al. (2004) used this method for the synthesis of PANI/TiO_2 bilayer microtubes [111]. Li et al. also used this method for the synthesis of PANI/TiO_2 nanobelts in the presence of a cotton template and investigated different properties and effects of the molar ratio on microwave and photocatalytic properties [112]. Lusheng Su Yong also used this method to prepare PANI/TiO_2 nanocomposites; photosensitive and thermoelectric properties were studied by SEM, Raman spectroscopy, and FTIR [113]. Sun et al. also synthesized PANI/ TiO_2 nanocomposites by varying the amount of TiO_2 nanoparticles by using cetyltrimethylammonium bromide (CTAB) as a template [114].

2.2.2.7 Gamma irradiation

The γ-irradiation technique is one of the most frequently used techniques for the synthesis of nanocomposites at normal pressure and room temperature. It is easy to control and adapt, and the product nanocomposites also have fewer impurities [115]. El-

Arnaout et al. synthesized PANI/TiO_2 nanocomposites by using the γ-irradiation technique, in which aniline radicals that behave as cations were adsorbed on the surface-growing TiO_2 nanoparticles in the presence of γ-radiation and their electroresponsive properties were characterized [115]. Karim et al. also prepared PANI/TiO_2 nanocomposites by using the γ-irradiation method in which an aqueous mixture of radical-free TiO_2 nanoparticles and aniline was irradiated by γ-rays and characterization showed that the degradation temperature increased as compared to PANI [116]. Safify et al. also used γ-irradiation to synthesize PANI/TiO_2 nanocomposites and UV-visible spectroscopy was used for the confirmation and characterization of the nanocomposites. TEM and XRD were also used for the analysis, and they showed that doping with TiO_2 or an increase in γ-rays can affect the dielectric constant of the composites [117].

2.2.2.8 The microemulsion method

A microemulsion, which is defined as a thermodynamically stable and isotropic transparent solution of two immiscible liquids basically consisting of oil, water, and surfactant molecules, has been employed as a polymerization medium to obtain spherical latex particles [118]. The microemulsion method for the synthesis of nanostructures has gained the interest of researchers because it provides control over various morphological properties such as shape, geometry, surface area, etc., of nanostructures [119]. Li et al. prepared PANI/TiO_2 nanocomposites by using facile emulsion polymerization homogeneously in the presence of a peroxotitanium complex (PTC), which is used both as an oxidant agent and as a precursor [120]. Wei et al. also used this technique for the synthesis and studied the photoinduced charge transfer efficiency of the material [121]. PANI/TiO_2 nanocomposites in which TiO_2 was decorated on the core of PANI were synthesized in an ionic liquid–water microemulsion in the presence of TiO_2 nanoparticles. TiO_2 nanoparticles were dispersed in n-butanol and OP-10 to minimize the agglomeration [42, 122].

2.2.2.9 The inverse emulsion method

Karim et al. reported PANI/TiO_2 nanocomposites by using the inverse emulsion synthesis technique in which polymerization of an organic solvent ($CHCl_3$, chloroform) in the presence of HCl as a dopant and CTAB as an emulsifier. The PANI/TiO_2 nanocomposites were of a diameter ranging from 50 to 200 nm [123]. Aniline and TiO_2 were polymerized into textiles by the inverted emulsion polymerization method. An aqueous mixture of aniline, a free-radical oxidant, and TiO_2 nanoparticles were utilized to synthesize the hybrid nanocomposites [124].

2.2.2.10 One-pot polymerization

One-pot synthesis of TiO_2 and PANI/TiO_2 was done from various precursors such as tetrabutyl titanate and titanium tetrachloride. Wei et al. reported a synthesis method that produces mesoporous TiO_2 in th eanatase phase in a long reaction needing several hours [125]. Han et al. (2008) also reported the core–shell nanostructure of PANI/TiO_2 by one-pot polymerization using APS as an oxidant agent. Han et al. also reported the one-pot method in which TiO_2 was used with aniline and polymerized it. This technique is comparable to the in situ polymerization technique, with only one different step: the addition of the surfactant, mostly anionic surfactants [126]. He et al. used the one-pot synthesis technique to prepare the hierarchical structure of nanocomposites and characterize their performance in dye-sensitized solar cells (DSSCs) [127].

2.2.3 Effect of Surfactants

The potential incorporation of a surfactant into a conducting polymer is likely to improve the electrical, thermooxidative, and hydrolytic stability due to the introduction of the bulky hydrophobic component. PANI/TiO_2 composites' thermal stability is affected by surfactants [128]. Protonic acid surfactants such as dodecylbenzenesulfonic acid (DBSA) is mostly investigated as a doping agent to enhance the dispersibility of the nanostructure and thermal stability [129]. Han et al. reported that the TiO_2/PANI–cationic surfactant has the highest conductivity as compared to TiO_2/PANI–anionic surfactant. The magnetic susceptibility is used to measure the magnetic properties of the TiO_2/PANI nanocomposites. The magnetic susceptibility of these nanocomposites showed negative values. So these nanocomposites are diamagnetic [130]. The product is higher in the presence of surfactants as compared to the surfactant-free case. The incorporation of bulky surfactant anions into TiO_2/PANI is the cause of the increase in the polymer resultant. This interaction can be an ionic interaction between the polycations and the surfactants during polymerization of PANI/TiO_2 with APS as an oxidant [129, 131]. The addition of a small amount of TiO_2 nanoparticles greatly increased electrical conductivity from 5.89 to 14.2 S/cm. These nanosized powders could transfer into the organic phase. With the increase in the amount of sodium dodecyl sulfate (SDS), the dispersibility into the organic solvent was increased. Consequently, the electrical conductivity of the product was also decreased. The obtained composites showed 14.16 S/cm of conductivity at the maximum, while the value was almost independent of the PANI coating ratio in the range of 100–20 wt%. The conductivity value of composite with 20 wt% PANI was 70,000 times higher than that of raw titania. Modified titania had properties of PANI and titania together. In addition, these composites showed a photoconductive response against UV irradiation, which might show the existence of a p-n junction between titania and PANI [129].

2.3 Properties of Hybrid Composites

PANI/TiO_2 nanocomposites have gained a much higher place in materials science due to their enhanced electrical, optical, and dielectric properties.

2.3.1 Optical/Photocatalytic Properties

Most semiconductors are used as photocatalysts for various applications, as mentioned later. TiO_2 is one of such materials having excellent photocatalytic activity. Its photocatalytic activity depends upon the crystal size, structure, phase, and surface area. The electron and hole recombination rate depends on the crystal size. The adsorption increases in nanostructures due to the large surface area and the rate of reaction is also increased. It has a wide bandgap, that is, 3.2 eV, that allows only the absorption of UV light [132]. The increase in its range for photoresponse has attracted researchers since past decades. Different kinds of doping have been reported for past several years, but PANI/TiO_2 nanocomposites showed an excellent response [133] and their range increased to visible and IR region, showing maximum sensitivity for the 600–700 nm electromagnetic spectra [134]. A photon with suitable energy can be used to excite TiO_2, causing the generation of an electron–hole pair in the material. The transfer of electrons from TiO_2 to PANI increases the reaction rate, which leads toward high catalytic activity. The level of the valence band of PANI is much lower than TiO_2, so holes can be trapped in it. Therefore, it decreases recombination, as shown in Fig. 2.13 [135].

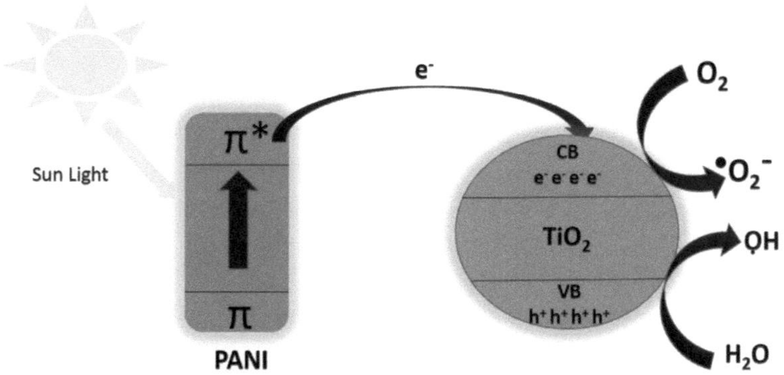

Figure 2.13 Mechanism of charge transfer in PANI/TiO_2 nanocomposites under sunlight.

Different PANI/TiO_2 nanocomposites were synthesized by varying the amount of PANI and their photocatalytic activity characterized, which showed PANI/TiO_2 nanocomposites with 15% of PANI exhibit the highest photocatalytic activity for the degradation of azo dye in wastewater, as shown in Fig. 2.14 [136].

2.3.2 Electrical/Dielectric Properties

Materials having high dielectric constants have major contributions in the fabrication of a new generation of MEMS and dynamic random access memories (DRAMs). Therefore, these materials have gained the attention of researchers. The basic purpose of these materials is to control the breakdown voltage and leakage currents in electronic circuits, and thin layers of the order of microns are needed to design the devices, and large dielectric constants are required to fulfill these criteria. PANI/TiO_2 nanocomposites have large dielectric constants that make them favorable materials for fabrication of electronic devices. The dielectric constant of alternating current (AC) and DC conductivities of PANI/TiO_2 nanocomposites are widely investigated and it is reported that the concentration of PANI does not affect DC conductivity but AC measurements showed the correlated barrier hopping (CBH) conduction process [77].

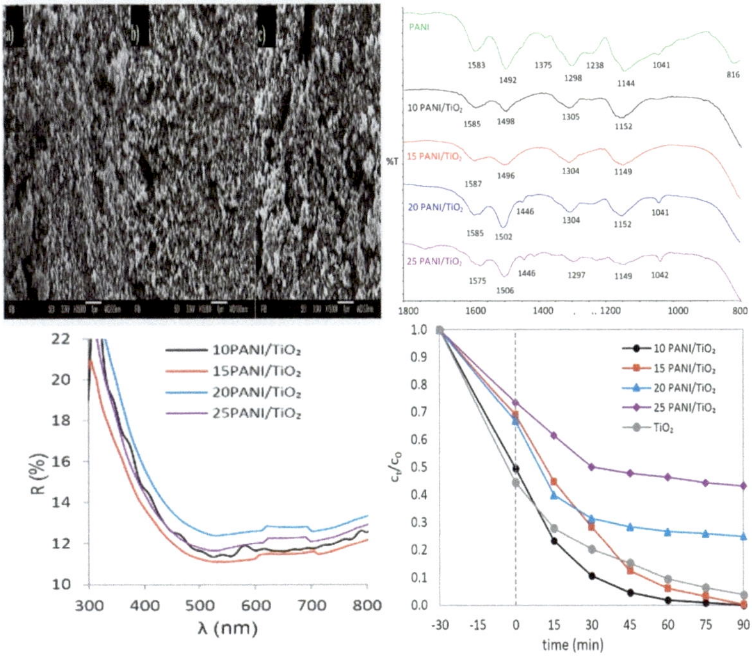

Figure 2.14 (a) SEM micrographs of different PANI/TiO_2 hybrids with different ratios of PANI, (b) FTIR spectra of different PANI/TiO_2 hybrids with different ratios of PANI, (c) UV-visible reflectance spectra for different PANI/TiO_2 hybrids with different ratios of PANI, and (d) change in concentration of dye during photocatalysis of PANI/TiO_2 hybrids with different pH values [136].

AC conductivity and the dielectric constant both depend on the amount of TiO_2 in the nanocomposites. At low frequencies there is no change in the AC conductivity, but at frequencies above 10^5 kHz, it increases steeply, which also shows PANI/TiO_2 nanocomposites belong to disordered materials [137]. Mo et al. (2008) did a detailed study of the dielectric properties of PANI/TiO_2 nanocomposites with varying amounts of TiO_2 and reported that the electrical conductivity increased as the amount of TiO_2 increased [138].

2.4 Applications of PANI/TiO_2 Composites

PANI/TiO_2 composites/hybrids have various applications due to the electric and photocytalic conductivity, which are discussed next.

2.4.1 Photocatalysis

Degradation of organic pollutants is mostly carried out by photocatalysis, and TiO_2 is a promising photocatalyst due to its stability and good photocatalytic activity [139]. It has the only range in the UV region. To enhance its range of photocatalysis, it is doped with various materials and PANI is the best combination. PANI/TiO_2 hybrid nanocomposites are mostly used as photocatalysts due to their better range in UV and visible light, stability, and recycling ability [101]. PANI is not only a good donor but also an acceptor of electrons in photocatalysis. Jinzhang et al. (2007) prepared PANI/TiO_2 nanocomposites via the chemical polymerization technique and found that 67.1% and 83.2% of rhodamine-B could be degraded under sunlight and UV irradiation, respectively, within 120 min, using the PANI/TiO_2 composite film as a photocatalyst [140]. Wei et al. synthesized PANI/TiO_2 nanocomposites by the hydrothermal technique and the photocatalytic properties of the samples were investigated by the photodegradation of gaseous acetone under UV (λ = 254 nm) and visible light irradiation (λ > 400 nm). In fact, the photocatalytic effects exhibited by the composite materials were superior to that of pure TiO_2 and PANI samples [141]. Li et al. (2015) prepared a new type of macroporous PANI/TiO_2 nanocomposite by in situ oxidative polymerization that showed excellent photocatalytic activity and regeneration ability under visible light for the degradation of organic wastewater [142]. Marija et al. (2017) developed a novel photocatalytic system based on carbonized PANI/TiO_2 nanocomposites by using a simple bottom-up approach. The photocatalytic degradation of methylene blue and rhodamine-B was characterized that showed excellent photocatalytic activity [143].

2.4.2 Smart Corrosion-Resistant Coatings

Corrosion-resistant materials have gained the interest of researchers for years as they protect steel from corrosion, and PANI/TiO_2 nanocomposites showed excellent behavior in this regard. This excellent improvement in the behavior of coatings is due to the increase in the diffusion barrier, redox properties of PANI, and large surface area for the liberation of dopants due to nanosized TiO_2 [144]. Radhakrishnan et al. (2009) prepared a coating and applied it to steel plates, in which PANI/TiO_2 nanocomposites were synthesized by in situ polymerization, and the plates showed excellent corrosion resistance in tough environmental conditions [145]. Pagotto et al. (2016) studied the corrosion resistivity of multilayer PANI and PANI/TiO_2 nanocomposites separately and concluded that PANI/TiO_2 nanocomposites show much better resistivity and less porosity as the thickness of the coating increases [146]. PANI/TiO_2 nanocomposites in poly(vinyl acetate) (PVAc) showed the best corrosion resistivity in HCl [147].

2.4.3 Sensors

PANI is known as a good sensing polymer due to gas-sensing ability for various gases such as CO, H_2, NH_3, methanol, hydrazine, H_2S, and NO_2 [148]. It is due to its electrical conductivity at room temperature. Its electrical conductivity can be changed by doping with metal oxides [149]. The sensing process of a gas sensor based on chemical reaction takes place on the surface, and it depends on five factors: surface modification, chemical components, microstructures of sensing layers, humidity, and temperature [150]. TiO_2/PANI nanocomposites show enhanced sensing activity as compared to PANI or TiO_2 individually. PANI is a p-type hole material; on the other hand, TiO_2 is a good electron acceptor, that is, n-type material, and the p-n contact between PANI/TiO_2 enhances the adsorption of gas molecules [151]. Srivsatava et al. reported that the chemiresistor H_2 gas sensor based on TiO_2/PANI nanocomposites showed a higher response as compared to a pure PANI-based sensor [152]. Ansari and Mohammad studied the sensitivity of *p*-toluenesulfonic acid (pTSA)/PANI and TiO_2 for ammonia. Their result showed that the high surface area of TiO_2 is responsible for ammonia sensitivity

[153]. Pawar et al. reported an ammonia gas sensor based on PANi/TiO_2 nanocomposites that were deposited on a glass substrate by the spin-coating technique, and it showed good selectivity for ammonia gas at room temperature [154].

2.4.4 Energy Storage Devices

Conducting polymers have gained the attention of researchers since 1960 for energy storage and sensing devices [149] due to their conductivity and dielectric properties and low cost. Harb et al. fabricated organic field-effect transistors (OFETs) for this purpose [155]. Results were much better compared to inorganic batteries and carbon-based capacitors [156]. Reddy et al. reported that multiwalled carbon nanotube (MWCNT)/PANI nanocomposites coated with TiO_2 by the chemical polymerization method showed a capacitance of 443.57 F/g at a 2 mV/s scan rate [157]. Xie et al. prepared PANI nanowire/TiO_2 nanotube arrays that showed a capacitance of 732 F/g with 1 M HCl. Shao et al. prepared a supercapacitor electrode based on a PANI/TiO_2 nanotube array by a simple ionization method, and it showed a specific capacitance of 897.35 F/g in 0.05 M H_2SO_4 at a scan rate of 1.2 mV/s [158]. Bian et al. synthesized PANI/TiO_2 nanocomposites by one-pot in situ oxidation polymerization, and they showed a maximum specific capacitance of 330 F/g [159].

2.4.5 Fuel Cells

The fuel cell is an important part of energy conversion technology, which converts the chemical energy of fuels into electrical energy without the environmental hazards with excellent efficiency. The issue with this technology is its high cost. Therefore, new materials for fuel cells that are environment friendly have become a major interest of researchers. In most cases, platinum with porous carbon is used for electrochemical catalysis [160]. PANI/TiO_2 nanocomposites were also investigated for direct methanol fuel cells [161]. Qiao et al. fabricated a unique PANI/TiO_2 nanocomposite-based anode for microbial fuel cells that showed that the nanocomposites having 30 wt% of PANI provide the best electrocatalytic activity as compared to previously reported microbial fuel cells [162]. Ganesan et al. reported that PANI/TiO_2 nanocomposites can also be used as catalysts in electrocatalysis of ascorbic acid in fuel cells as they remain stable during the reaction [163].

2.4.6 Dye-Sensitized Solar Cells

The main issue researchers are facing in third-generation solar cells, namely dye-sensitized solar cells (DSSCs), is efficiency [164]. PANI was also used as a counterelectrode of DSSCs with efficiency much higher than the conventional Pt electrode, that is, 7.15% [165]. The dye-absorbed TiO_2/PANI electrode-based DSSCs significantly improved the conversion efficiency and may be attributed to the high charge carrier transportation between the TiO_2 and the PANI layer. Zhang et al. fabricated two different kinds of TiO_2/sulfonated PANI nanocomposites and investigated their photocurrent responses and reported that that the self-doped layer-by-layer nanocomposite film shows a much better response [166]. Ameen et al. prepared dye-absorbed TiO_2/PANI electrodes by plasma-enhanced polymerization for DSSCs and reported that the electrical conductivity of PANI is remarkably improved and the overall conversion efficiency of a fabricated solar cell was 0.68% [69]. Kawata et al. synthesized novel PANI/TiO_2 nanocomposites by chemical oxidation polymerization and applied them on fluorine-doped tin oxide (FTO) by the spin-coating technique to fabricate a counterelectrode of DSSCs and reported that the efficiency of solar cells improved significantly [167].

Al-Daghman et al. (2015) synthesized PANI/TiO_2 nanocomposites by the sol–gel method and fabricated an assembly of indium tin oxide (ITO)/TiO_2/PANI/Ag a in sandwich panel structure [168]. The FTIR and UV-visible spectroscopy characterization showed smooth interaction among the TiO_2 nanoparticles and PANI chains, as shown in Fig. 2.15.

The *I–V* characteristic for DSSCs has a high open-circuit voltage of 0.656 V and a short-circuit current density of 315 mA/cm under simulated solar radiation (50 mW/cm^2), as shown in Fig. 2.16.

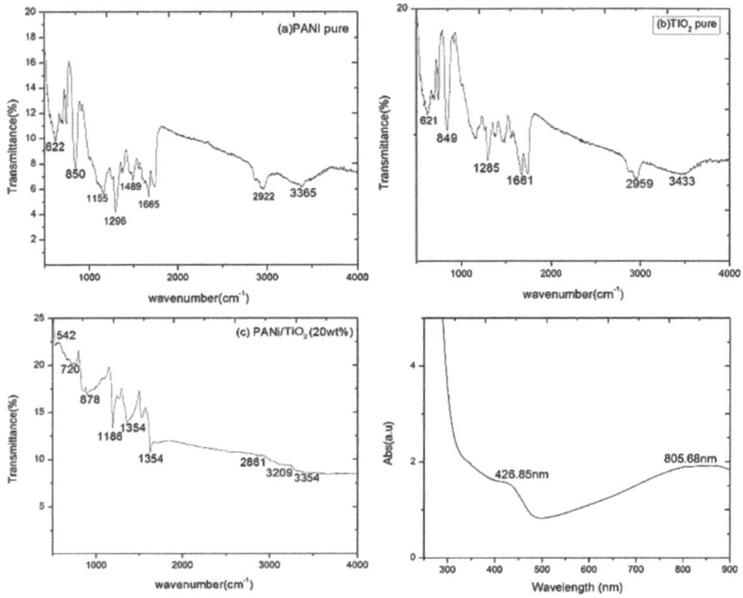

Figure 2.15 (a) FTIR spectra of PANI (EB), (b) FTIR spectra of TiO_2, (c) FTIR spectra of PANI/TiO_2 nanocomposites, and (d) absorption spectrum of PANI (EB) in the visible spectrum [168].

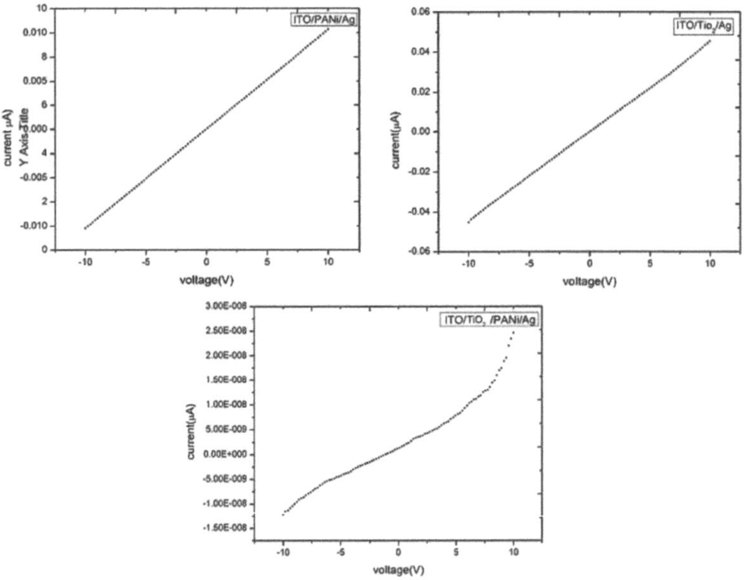

Figure 2.16 *I–V* characteristic of the sandwich-type structure of PANI, TIO_2, and ITO/TiO_2/PANI/Ag [168].

2.5 Conclusion

Conducting polymer/semiconductor hybrid nanocomposites are novel and multifunctional materials with unique physical and chemical properties. These properties lead researchers to employ these materials for various applications in different fields of science and technology. PANI/TiO_2 nanocomposites have great applications in photocatalysis, sensors, energy storage devices (including fuel cells and solar cells), and clean-up of the environment. There are different approaches and methods to synthesize these nanocomposites, and selection of any of the preparation methods affects their physical and chemical properties such as structural, morphological, optical, and electrical/dielectric properties. For applications in optoelectronics and electronics, it is critical to use these nanocomposites with controlled particle size and uniform distribution of semiconductors within the conducting polymers. Therefore, researchers are extensively working on encapsulation strategies to exploit their efficient use in these applications. On a final note, although the field of organic/inorganic nanocomposites is making rapid progress, nevertheless there is a lot of work that needs to be done in terms of their facile synthesis methods to get full control of desired properties and hence their utilization in various applications.

Acknowledgments

Hafeez Anwar acknowledges financial support from the Higher Education Commission (HEC) under the project no. 21-266 SRGP/R&D/HEC/2014.

References

1. Jayasena, B. and Subbiah, S. (2011). A novel mechanical cleavage method for synthesizing few-layer graphenes, *Nanoscale Res. Lett.*, **6**(1), p. 95.
2. Thangavelu, R. M., Gunasekaran, D., Jesse, M. I., SU, M. R., Sundarajan, D. and Krishnan, K. (2018). Nanobiotechnology approach using plant rooting hormone synthesized silver nanoparticle as "nanobullets" for

the dynamic applications in horticulture–an in vitro and ex vitro study, *Arabian J. Chem.*, **11**(1), pp. 48–61.

3. Asefa, T., Yoshina-Ishii, C., MacLachlan, M. J. and Ozin, G. A. (2000). New nanocomposites: putting organic function "inside" the channel walls of periodic mesoporous silica, *J. Mater. Chem.*, **10**(8), pp. 1751–1755.
4. Backov, R., Bonnet, B., Jones, D. J. and Rozière, J. (1997). Assembly of partially oxidized tetrathiafulvalene in layered phosphates. Formation of conducting organic– inorganic hybrids by intercalation, *Chem. Mater.*, **9**(8), pp. 1812–1818.
5. Boury, B. and Corriu, R. (2000). Adjusting the porosity of a silica-based hybrid material, *Adv. Mater.*, **12**(13), pp. 989–992.
6. Choudhury, K. R., Winiarz, J. G., Samoc, M. and Prasad, P. N. (2003). Charge carrier mobility in an organic-inorganic hybrid nanocomposite, *Appl. Phys. Lett.*, **82**(3), pp. 406–408.
7. Giannelis, E. P. (1996). Polymer layered silicate nanocomposites, *Adv. Mater.*, **8**(1), pp. 29–35.
8. Gómez-Romero, P., Chojak, M., Cuentas-Gallegos, K., Asensio, J. A., Kulesza, P. J., Casañ-Pastor, N. and Lira-Cantú, M. (2003). Hybrid organic–inorganic nanocomposite materials for application in solid state electrochemical supercapacitors, *Electrochem. Commun.*, **5**(2), pp. 149–153.
9. Judeinstein, P. and Sanchez, C. (1996). Hybrid organic–inorganic materials: a land of multidisciplinarity, *J. Mater. Chem.*, **6**(4), pp. 511–525.
10. Kimizuka, N. and Kunitake, T. (1996). Organic two-dimensional templates for the fabrication of inorganic nanostructures: organic/inorganic superlattices, *Adv. Mater.*, **8**(1), pp. 89–91.
11. Kryszewski, M. (2000). Nanointercalates—novel class of materials with promising properties, *Synth. Met.*, **109**(1), pp. 47–54.
12. Valkenberg, M. H. and Hölderich, W. F. (2002). Preparation and use of hybrid organic–inorganic catalysts, *Catal. Rev.*, **44**(2), pp. 321–374.
13. Jeon, I. and Baek, J. (2010). Nanocomposites derived from polymers and inorganic nanoparticles, *Materials*, **3**(6), pp. 3654–3674.
14. Nandapure, B., Kondawar, S., Salunkhe M. and Nandapure, A. (2013). Magnetic and transport properties of conducting polyaniline/nickel oxide nanocomposites, *Adv. Mater. Lett.*, **4**(2), pp. 134–140.
15. Gajendran, P. and Saraswathi, R. (2008). Polyaniline-carbon nanotube composites, *Pure Appl. Chem.*, **80**(11), pp. 2377–2395.

16. Butterworth, M., Corradi, R., Johal, J., Lascelles, S., Maeda, S. and Armes, S. (1995). Zeta potential measurements on conducting polymer-inorganic oxide nanocomposite particles, *J. Colloid Interface Sci.*, **174**(2), pp. 510–517.

17. Jarjayes, O., Fries, P. and Bidan, C. (1995). New nanocomposites of polypyrrole including γ-Fe_2O_3 particles: electrical and magnetic characterizations, *Synth. Met.*, **69**(1), pp. 343–344.

18. Braun, P. V. and Stupp, S. I. (1999). CdS mineralization of hexagonal, lamellar, and cubic lyotropic liquid crystals, *Mater. Res. Bull.*, **34**(3), pp. 463–469.

19. Tsai, H.-L., Schindler, J. L. Kannewurf, C. R. and Kanatzidis, M. G. (1997). Plastic superconducting polymer– NbSe2 nanocomposites, *Chem. Mater.*, **9**(4), pp. 875–878.

20. Govindaraj, B., Sastry, N. and Venkataraman, A. (2004). Studies on γ-Fe_2O_3–high-density polyethylene composites and their additives, *J. Appl. Polym. Sci.*, **92**(3), pp. 1527–1533.

21. Khasim, S., Raghavendra, S., Revanasiddappa, M. and Ambika Prasad, M. (2005). Synthesis, characterization and electrical properties of polyaniline/$BaTiO_3$ composites, *Ferroelectrics*, **325**(1), pp. 111–119.

22. Kalyane, S. (2017). Conductivity study of polyaniline-PbO composites, *Int. J. Pure Appl. Phys.*, **13**(1), pp. 109–116.

23. Shinde, S. S. and Kher, J. A. (2014). A review on polyaniline and its noble metal composites, *Int. J. Innovative Res. Sci. Eng. Technol.*, **3**(9), pp. 16570–16576.

24. Angelopoulos, M. (2001). Conducting polymers in microelectronics, *IBM J. Res. Dev.*, **45**(1), pp. 57–75.

25. Molapo, K. M., Ndangili, P. M., Ajayi, R. F. Mbambisa, G., Mailu, S. M., Njomo, N., Masikini, M., Baker, P. and Iwuoha, E. I. (2012). Electronics of conjugated polymers (I): polyaniline, *Int. J. Electrochem. Sci.*, **7**(12), pp. 11859–11875.

26. Ramana, G. V., Kumar, P. S., Srikanth, V. V., Padya, B. and Jain, P. (2015). Electrochemically active polyaniline (PANi) coated carbon nanopipes and PANi nanofibers containing composite, *J. Nanosci. Nanotechnol.*, **15**(2), pp. 1338–1343.

27. Schoch, K. (1994). Update on electrically conductive polymers and their applications, *IEEE Electr. Insul. Mag.*, **10**(3), pp. 29–32.

28. Muller, A. and Gouzerh, P. (2012). From linking of metal-oxide building blocks in a dynamic library to giant clusters with unique properties

and towards adaptive chemistry, *Chem. Soc. Rev.*, **41**(22), pp. 7431–7463.

29. Shenhar, R. and Rotello, V. M. (2003). Nanoparticles: scaffolds and building blocks, *Acc. Chem. Res.*, **36**, pp. 549–561.
30. Dehghani, Z., Nazerdeylami, S., Saievar-Iranizad, E. and MajlesAra, M. H. (2011). Synthesis and investigation of nonlinear optical properties of semiconductor ZnS nanoparticles, *J. Phys. Chem. Solids*, **72**(9), pp. 1008–1010.
31. Feast, W. J., Tsibouklis, J., Pouwer, K. L., Groenendaal, L. and Meijer, E. W. (1996). Synthesis, processing and material properties of conjugated polymers, *Polymer*, **37**(22), pp. 5017–5047.
32. McCall, R. P., Ginder, J. M., Leng, J. M., Ye, H. J., Manohar, S. K., Masters, J. G., Asturias, G. E., MacDiarmid, A. G. and Epstein, A. J. (1990). Spectroscopy and defect states in polyaniline, *Phys. Rev. B*, **41**(8), pp. 5202–5211.
33. MacDiarmid, A. G. and Epstein, A. J. (1989). Polyanilines: a novel class of conducting polymers, *Faraday Discuss. Chem. Soc.*, **88**(1), pp. 317–332.
34. Epstein, A. J., Ginder, J. M., Zuo, F., Bigelow, R. W., Woo, H. S., Tanner, D. B., Richter, A. F., Huang, W. S. and MacDiarmid, A. G. (1987). Insulator-to-metal transition in polyaniline, *Synth. Met.*, **18**(3), pp. 303–309.
35. Boeva, Z. A. and Sergeyev, V. G. (2014). Polyaniline: synthesis, properties, and application, *Polym. Sci.*, **56**(1), pp. 144–153.
36. Chen, W., Hu, C., Hsu, C. and Ho, K. (2009). A study on the electrochromic properties of polyaniline/silica composite films with an enhanced optical contrast, *Electrochim. Acta*, **54**(18), pp. 4408–4415.
37. Watanabe, A., Mori, K., Iwasaki, Y., Nakamura, Y. and Niizuma, S. (1987). Electrochromism of polyaniline film prepared by electrochemical polymerization, *Macromolecules*, **20**(8), pp. 1793–1796.
38. Anderson, R. E., Ostrowski, A. D., Gran, D. E., Fowler, J. D., Hopkins, A. R. and Villahermosa, R. M. (2008). Diameter-controlled synthesis of polyaniline nanofibers, *Polym. Bull.*, **61**(5), pp. 563–568.
39. Mahapatra, S. S., Shekhar, S., Thakur, B. K. and Priyadarshi, H. (2014). Synthesis and characterization of electrodeposited C-PANI-Pd-Ni composite electrocatalyst for methanol oxidation, *Int. J. Electrochem.*, **2014**(1), pp. 1–8.
40. Zhou, H., Wen, J., Ning, X., Fu, C., Chen, J. and Kuang, Y. F. (2007). Electrosynthesis of polyaniline films on titanium by pulse potentiostatic method, *Synth. Met.*, **157**(2), pp. 98–103.

41. Zhou, H. H., Jiao, S. Q., Chen, J. H., Wei, W. Z. and Kuang, Y. F. (2004). Relationship between preparation conditions, morphology and electrochemical properties of polyaniline prepared by pulse galvanostatic method (PGM), *Thin Solid Films*, **450**(2), pp. 233–239.

42. Ghorbani, M., Lashkenari, M. S. and Eisazadeh, H. (2011). Synthesis and thermal stability studies of polyaniline/silver nanocomposite based on reduction of silver ions using polyaniline, *High Performance Polym.*, **23**(7), pp. 513–517

43. Hourquebie, L. O. P. and Buvat, P. (1997). An illustration of dielectric properties of conductive polymers, *SPE/ANTEC 1997 Proceedings*, CRC Press.

44. Yang, Y. and Mu, S. (1997). Bioelectrochemical responses of the polyaniline horseradish peroxidase electrodes, *J. Electroanal. Chem.*, **432**(1–2), pp. 71–78.

45. Bartlett, P. N. and Whitaker, R. G. (1987). Strategies for the development of amperometric enzyme electrodes, *Biosensors*, **3**(6), pp. 359–379.

46. Acevedo, D. A., Lasagni, A. F., Barbero, C. A. and Mücklich, F. (2007). Simple fabrication method of conductive polymeric arrays by using direct laser interference micro-/nanopatterning, *Adv. Mater.*, **19**(9), pp. 1272–1275.

47. Ayad, M. and Rehab, A. (2008). Study the effect of inorganic salts on the chemically polymerized aniline films using quartz crystal microbalance, *Polym. Adv. Technol.*, **19**(5), pp. 414–418

48. Kobayashi, T., Yoneyama, H. and Tamura, H. (1984). Electrochemical reactions concerned with electrochromism of polyaniline film-coated electrodes, *J. Electroanal. Chem. Interfacial Electrochem.*, **177**(1–2), pp. 281–291.

49. Batich, C., Laitinen, H. and Zhou, H. (1990). Chromatic changes in polyaniline films, *J. Electrochem. Soc.*, **137**(3), pp. 883–885.

50. Antonelli, D. and Ying, J. Y. (1995). Synthesis of hexagonally packed mesoporous TiO_2 by a modified sol–gel method, *Angew. Chem. Int. Ed.*, **34**(18), pp. 2014–2017.

51. Lusvardi, G., Barani, C., Giubertoni, F. and Paganelli, G. (2017). Synthesis and characterization of TiO_2 nanoparticles for the reduction of water pollutants, *Materials*, **1208**(10), pp. 1–11.

52. Nagaveni, K., Hegde, M. S., Ravishankar, N., Subbanna, G. N. and Madras, G. (2004). Synthesis and structure of nanocrystalline TiO_2 with lower band gap showing high photocatalytic activity, *Langmuir*, **20**(7), pp. 2900–2907.

53. Malgras, V., Jood, P., Sun, Z., Dou, S. X., Yamauchi, Y. and Kim, J. H. (2014). Channelled porous TiO_2 synthesized with a water-in-oil microemulsion, *Chem. Eur. J.*, **20**(33), pp. 10451–10455.

54. Pawar, R. R. and Deshpande, V. T. (1968). The anisotropy of the thermal expansion of α-titanium, *Acta Crystallogr. Sect. A*, **24**(2), pp. 316–317.

55. Fujishima, A., Zhang, X. and Tryk, D. A. (2008). TiO_2 photocatalysis and related surface phenomena, *Surf. Sci. Rep.*, **63**(12), pp. 515–582.

56. Howard, C. J., Sabine, T. M. and Dickson, F. (1991). Structural and thermal parameters for rutile and anatase, *Acta Crystallogr. Sect. A*, **47**(4), pp. 462–468.

57. Paola, A., Bellardita, M. and Palmisano, L. (2013). Brookite, the least known TiO_2 photocatalyst, *Catalysts*, **3**(1), pp. 36–73.

58. Pimentel, A., Nunes, D., Pereira, S., Martins, R. and Fortunato, E. (2016). Photocatalytic activity of TiO_2 nanostructured arrays prepared by microwave-assisted solvothermal method, in *Semiconductor Photocatalysis: Materials, Mechanisms and Applications*, Cao, W., ed. (InTech, Rijeka, Croatia), pp. 81–103, doi:10.5772/63237.

59. Foger, K. (1984). Dispersed metal catalysts, in *Catalysis: Science and Technology*, Anderson, J. R. and Boudart, M., eds. (Springer-Verlag, Berlin, Heidelberg), pp. 227–305.

60. Azad, A. M., Younkman, L. B., Akbar, S. A. and Alim, M. A. (1994). Characterization of TiO_2-based sensor materials using immittafice spectroscopy, *J. Am. Ceram. Soc.*, **77**(2), pp. 481–486.

61. Anbia, M. and Fard, S. M. (2011). A humidity sensor based on Nb-doped nanoporous TiO_2 thin film, *Sens. Transducers*, **134**(11), pp. 56–64.

62. Yeh, Y. C., Tseng, T. Y. and Chang, D. A. (1990). Electrical properties of TiO_2-$K_2Ti_6O_{13}$ porous ceramic humidity sensor, *J. Am. Ceram. Soc.*, **73**(7), pp. 1992–1998.

63. Xu, Q. and Anderson, M. A. (1994). Sol–gel route to synthesis of microporous ceramic membranes: preparation and characterization of microporous TiO_2 and ZrO_2 xerogels, *J. Am. Ceram. Soc.*, **77**(7), pp. 1939–1945.

64. Cho, S. and Choi, W. (2001). Solid-phase photocatalytic degradation of PVC–TiO_2 polymer composites, *J. Photochem. Photobiol., A*, **143**(2), pp. 221–228.

65. Hoffmann, M. R., Martin, S. T., Choi, W. and Bahnemann, D. W. (1995). Environmental applications of semiconductor photocatalysis, *Chem. Rev.*, **95**(1), pp. 69–96.

66. Carlos, L., Ferreira, R. S. and de Zea Bermudez, V. (2003). Light emission from organic-inorganic hybrids lacking activator centers, in *Handbook of Organic-Inorganic Hybrid Materials and Nanocomposites*, Nalwa, H. S., ed. (American Scientific, USA), pp. 353–380.

67. Zhang, J., Ju, X., Wang, B.-J., Li, Q.-S., Liu, T. and Hu, T.-D. (2001). Study on the optical properties of PPV/TiO_2 nanocomposites, *Synth. Met.*, **118**(1), pp. 181–185.

68. Wang, D., Wang, Y., Li, X., Luo, Q., An, J. and Yue, J. (2008). Sunlight photocatalytic activity of polypyrrole–TiO_2 nanocomposites prepared by 'in situ'method, *Catal. Commun.*, **9**(6), pp. 1162–1166.

69. Ameen, S., Akhtar, M. S., Kim, G.-S., Kim, Y. S., Yang, O.-B. and Shin, H.-S. (2009). Plasma-enhanced polymerized aniline/TiO_2 dye-sensitized solar cells, *J. Alloys Compd.*, **487**(1), pp. 382–386.

70. Sung, J., Lee, I., Choi, H. and Choi, S. (2005). Electrorheological response of polyaniline-TiO_2 composite suspensions, *Int. J. Mod. Phys. B*, **19**(07n09), pp. 1128–1134.

71. Xu, J.-C., Liu, W.-M. and Li, H.-L. (2005). Titanium dioxide doped polyaniline, *Mater. Sci. Eng., C*, **25**(4), pp. 444–447.

72. Singu, B., Male, U., Srinivasan, P. and Pabba, S. (2014). Use of surfactant in aniline polymerization with TiO_2 to PANI-TiO_2 for supercapacitor performance, *J. Solid State Electrochem.*, **18**(7), pp. 1995–2003.

73. Zheng, H., Ncube, N., Raju, K., Mphahlele, N. and Mathe, M. (2016). The effect of polyaniline on TiO_2 nanoparticles as anode materials for lithium ion batteries, *SpringerPlus*, **5**, p. 630.

74. Arora, R., Srivastav, A., Mandal, U. K. and Sharma, P. (2016). TiO_2/PANI nanocomposite loaded in PVA for anticorrosive applications, *Mater. Sci.-Poland*, **34**(4), pp. 721–725.

75. Youssef, M. A. (2014). Morphological studies of polyaniline nanocomposite based mesostructured TiO_2 nanowires as conductive packaging materials, *R. Soc. Chem. Adv.*, **4**(1), pp. 6811–6820.

76. Pham, T. T., Nguyen, T. D., Xuan, T. M., Mai, T. T. T., Tran, H. Y. and Phan, T. B. (2015). Influence of polyaniline on photoelectrochemical characterization of TiO_2-PANI layers, *Adv. Nat. Sci.: Nanosci. Nanotechnol.*, **6**(1), pp. 25008–25012.

77. Ashis, D., Sukanta, D., Amitabha, D. and De, S. K. (2004). Characterization and dielectric properties of polyaniline–TiO_2 nanocomposites, *Nanotechnology*, **15**(9), p. 1277.

78. Deivanayaki, S., Ponnuswamy, V., Ashokan, S., Jayamurugan, P. and Mariappan, R. (2013). Synthesis and characterization of TiO_2-doped

Polyaniline nanocomposites by chemical oxidation method, *Mater. Sci. Semicond. Process.*, **3**(16), pp. 554–559.

79. Arora, R., Mandal, U. K., Sharma, P. and Srivastav, A. (2015). Synthesis and thermal properties of polyaniline-TiO_2 nanocomposites PVA based film, *Mater. Today: Proc.*, **2**(4–5), pp. 2215–2225.

80. Mostafa, Y. N., Mohamed, M. B., Imam, N. G., Alhamyani, M. and Heiba, Z. K. (2016). Electrical and optical properties of hydrogen titanate nanotube/PANI hybrid nanocomposites, *Colloid Polym. Sci.*, **294**, pp. 215–224.

81. Cai, G., Tu, J., Zhou, D., Zhang, J., Xiong, Q., Zhao, X., Wang, X. and Gu, C. (2013). Multicolor electrochromic film based on TiO_2@polyaniline core/shell nanorod array, *J. Phys. Chem. C*, **117**(31), pp. 15967–15975.

82. Schnitzler, D. and Zarbin, A. J. G. (2004). Organic/inorganic hybrid materials formed from TiO_2 nanoparticles and polyaniline, *J. Braz. Chem. Soc.*, **15**(3), pp. 378–384.

83. Li, C., Wang, J., Guo, H. and Ding, S. (2015). Low temperature synthesis of polyaniline–crystalline TiO_2–halloysite composite nanotubes with enhanced visible light photocatalytic activity, *J. Colloid Interface Sci.*, **458**, pp. 1–13.

84. Li, C., Zhou, T., Zhu, T. and Li, X. (2015). Enhanced visible light photocatalytic activity of polyaniline–crystalline TiO_2–halloysite composite nanotubes by tuning the acid dopant in the preparation, *RSC Adv.*, **5**(119), pp. 98482–98491.

85. Guo, N., Liang, Y., Lan, S., Liu, L., Zhang, J., Ji, G. and Gan, S. (2014). Microscale hierarchical three-dimensional flowerlike TiO_2/PANI composite: synthesis, characterization, and its remarkable photocatalytic activity on organic dyes under UV-light and sunlight irradiation, *J. Phys. Chem. C*, **118**(32), pp. 18343–18355.

86. Kalikeri, S., Kamath, N., Gadgil, D. J. and Kodialbail, V. S. (2017). Visible light-induced photocatalytic degradation of Reactive Blue-19 over highly efficient polyaniline-TiO_2 nanocomposite: a comparative study with solar and UV photocatalysis, *Environ. Sci. Pollut. Res.*, pp. 1–14.

87. Kim, B.-S., Lee, K.-T., Huh, P.-H., Lee, D.-H., Jo, N.-J. and Lee, J.-O. (2009). In situ template polymerization of aniline on the surface of negatively charged TiO_2 nanoparticles, *Synth. Met.*, **159**(13), pp. 1369–1372.

88. Reddy, K. R., Karthik, K., Prasad, S. B., Soni, S. K., Jeong, H. M. and Raghu, A. V. (2016). Enhanced photocatalytic activity of nanostructured titanium dioxide/polyaniline hybrid photocatalysts, *Polyhedron*, **120**, pp. 169–174.

89. Tai, H., Jiang, Y., Xie, G., Yu, J. and Chen, X. (2007). Fabrication and gas sensitivity of polyaniline–titanium dioxide nanocomposite thin film, *Sens. Actuators, B*, **125**(2), pp. 644–650.

90. MacDiarmid, A. G. (2001). "Synthetic metals": a novel role for organic polymers (Nobel lecture), *Angew. Chem. Int. Ed.*, **40**(14), pp. 2581–2590.

91. Wang, H., Lin, J. and Shen, Z. X. (2016). Polyaniline (PANi) based electrode materials for energy storage and conversion, *J. Sci.: Adv. Mater. Devices*, **1**(3), pp. 225–255.

92. Kuwabata, S., Takahashi, N., Hirao, S. and Yoneyama, H. (1993). Light image formations on deprotonated polyaniline films containing titania particles, *Chem. Mater.*, **5**(4), pp. 437–441.

93. Luo, J., Huang, H., Lin, Z. and Hepel, M. (2003). Photoelectrochemical behavior of p-ATP/PANI film and nanoparticulate p-ATP/PANI/TiO_2 film on Au electrodes, in *Conducting Polymers and Polymer Electrolytes* (ACS Publications), pp. 113–127.

94. Mickova, I., Prusi, A., Grcev, T. and Arsov, L. (2006). Electrochemical polymerization of aniline in presence of TiO_2 nanoparticles, *Bull. Chem. Technol. Maced.*, **25**(1), pp. 45–50.

95. Savitha, K. and Prabu, H. G. (2011). One-pot synthesis of PANI–TiO2 (anatase) hybrid of low electrical resistance using TiCl4 as precursor, *Mater. Chem. Phys.*, **130**(1), pp. 275–279.

96. Kunzo, P., Lobotka, P., Kovacova, E., Chrissopoulou, K., Papoutsakis, L., Anastasiadis, S. H., Krizanova, Z. and Vavra, I. (2013). Nanocomposites of polyaniline and titania nanoparticles for gas sensors, *Phys. Status Solidi A*, **210**(11), pp. 2341–2347.

97. Ates, M. and Topkaya, E. (2015). Nanocomposite film formations of polyaniline via TiO_2, Ag, and Zn, and their corrosion protection properties, *Prog. Org. Coat.*, **82**, pp. 33–40.

98. Nagarajan, R., Tripathy, S., Kumar, J., Bruno, F. F. and Samuelson, L. (2000). An enzymatically synthesized conducting molecular complex of polyaniline and poly (vinylphosphonic acid), *Macromolecules*, **33**(26), pp. 9542–9547.

99. Nabid, M., Golbabaee, M., Moghaddam, A., Dinarvan, R. and Sedghi, R. (2008). Polyaniline/TiO_2 nanocomposite: enzymatic synthesis and electrochemical properties, *Int. J. Electrochem. Sci.*, **3**(1), pp. 1117–1126.

100. Grzelczak, M., Vermant, J., Furst, E. M. and Liz-Marzán, L. M. (2010). Directed self-assembly of nanoparticles, *ACS Nano*, **4**(7), pp. 3591–3605.

101. Wetterskog, E., Agthe, M., Mayence, A., Grins, J., Wang, D., Rana, S., Ahniyaz, A., Salazar-Alvarez, G. and Bergström, L. (2014). Precise control over shape and size of iron oxide nanocrystals suitable for assembly into ordered particle arrays, *Sci. Technol. Adv. Mater.*, **15**(5), p. 055010.

102. Pinchuk, A. O. (2012). Size-dependent hamaker constant for silver nanoparticles, *J. Phys. Chem. C*, **116**(37), pp. 20099–20102.

103. Zhang, L. and Wan, M. (2003). Polyaniline/TiO_2 composite nanotubes, *J. Phys. Chem. B*, **107**(28), pp. 6748–6753.

104. Cui, S., Wang, J. and Wang, X. (2015). Fabrication and design of a toxic gas sensor based on polyaniline/titanium dioxide nanocomposite film by layer-by-layer self-assembly, *RSC Adv.*, **5**(72), pp. 58211–58219.

105. Li, J., Zhu, L., Wu, Y., Harima, Y., Zhang, A. and Tang, H. (2006). Hybrid composites of conductive polyaniline and nanocrystalline titanium oxide prepared via self-assembling and graft polymerization, *Polymer*, **47**(21), pp. 7361–7367.

106. Xie, S., Gan, M., Ma, L., Li, Z., Yan, J., Yin, H., Shen, X., Xu, F., Zheng, J. and Zhang, J. (2014). Synthesis of polyaniline-titania nanotube arrays hybrid composite via self-assembling and graft polymerization for supercapacitor application, *Electrochim. Acta*, **120**, pp. 408–415.

107. Chen, X., Zhu, H. Y., Zhao, J. C., Zheng, Z. F. and Gao, X. P. (2008). Visible-light-driven oxidation of organic contaminants in air with gold nanoparticle catalysts on oxide supports, *Angew. Chem.*, **120**(29), pp. 5433–5436.

108. Heinrich, G. (2015). Thermoelasticity of rubbers, *Encycl. Polym. Nanomater.*, pp. 2505–2510.

109. Klumperman, B. (2014). The rationale behind sequence-controlled maleimide copolymers, in *Sequence-Controlled Polymers: Synthesis, Self-Assembly, and Properties* (ACS Publications), pp. 213–221.

110. Akashi, M. and Ajiro, H. (2015). Template polymerization (molecular templating), *Encycl. Polym. Nanomater.*, pp. 2498–2502.

111. Xiong, S., Wang, Q. and Xia, H. (2004). Template synthesis of polyaniline/TiO 2 bilayer microtubes, *Synth. Met.*, **146**(1), pp. 37–42.

112. Li, Q., Zhang, C. and Li, J. (2010). Photocatalysis and wave-absorbing properties of polyaniline/TiO_2 microbelts composite by in situ polymerization method, *Appl. Surf. Sci.*, **257**(3), pp. 944–948.

113. Su, L. and Gan, Y. X. (2012). Experimental study on synthesizing TiO_2 nanotube/polyaniline (PANI) nanocomposites and their thermoelectric

and photosensitive property characterization, *Composites Part B*, **43**, pp. 170–182.

114. Cheng, Y.-T. and Ouyang, Z.-B. (2009). Synthesis of PANI/TiO_2 nanorods composites by reverse micelle template and their properties, *Spectrosc. Spectral Anal.*, **29**(9), pp. 2509–2513.

115. El-Arnaouty, M., Eid, M., Salah, M., Soliman, E.-S. and Hegazy, E.-S. A. (2017). Synthesis of poly (aniline/glycidyl methacrylate)-TiO_2 nanocomposites via gamma irradiation and their electro-responsive characteristic, *J. Inorg. Organomet. Polym Mater.*, **27**(5), pp. 1482–1490.

116. Karim, M. R., Yeum, J. H., Lee, M. S. and Lim, K. T. (2008). Preparation of conducting polyaniline/TiO_2 composite submicron-rods by the γ-radiolysis oxidative polymerization method, *React. Funct. Polym.*, **68**(9), pp. 1371–1376.

117. Afify, T., Ghazy, O., Saleh, H. and Ali, Z. (2018). Efficient in situ synthetic routes of polyaniline/poly (vinyl alcohol)/TiO_2 nanocomposites using gamma irradiation, *J. Mol. Struct.*, **1153**, pp. 128–134.

118. Kim, B.-J., Oh, S.-G., Han, M.-G. and Im, S.-S. (2001). Synthesis and characterization of polyaniline nanoparticles in SDS micellar solutions, *Synth. Met.*, **122**(2), pp. 297–304.

119. Malik, M. A., Wani, M. Y. and Hashim, M. A. (2012). Microemulsion method: a novel route to synthesize organic and inorganic nanomaterials: 1st nano update, *Arabian J. Chem.*, **5**(4), pp. 397–417.

120. Li, Y., Yu, Y., Wu, L. and Zhi, J. (2013). Processable polyaniline/titania nanocomposites with good photocatalytic and conductivity properties prepared via peroxo-titanium complex catalyzed emulsion polymerization approach, *Appl. Surf. Sci.*, **273**, pp. 135–143.

121. Feng, W., Sun, E., Fujii, A., Wu, H., Niihara, K. and Yoshino, K. (2000). Synthesis and characterization of photoconducting polyaniline-TiO_2 nanocomposite, *Bull. Chem. Soc. Jpn.*, **73**(11), pp. 2627–2633.

122. Guo, Y., He, D., Xia, S., Xie, X., Gao, X. and Zhang, Q. (2012). Preparation of a novel nanocomposite of polyaniline core decorated with anatase-TiO_2 nanoparticles in ionic liquid/water microemulsion, *J. Nanomater.*, **2012**(1), pp. 1–7.

123. Karim, M. R., Lee, H. W., Cheong, I. W., Park, S. M., Oh, W. and Yeum, J. H. (2010). Conducting polyaniline-titanium dioxide nanocomposites prepared by inverted emulsion polymerization, *Polym. Compos.*, **31**(1), pp. 83–88.

124. Bhatkhande, V., Samel, M. and Bhatavadekar, S. A. (2017). polyaniline-TiO_2 nanocomposite formation and characterization, *Int. Res. J. Eng. Technol.*, **4**(6), pp. 3271–3273.

125. Wei, C., Zhu, Y., Yang, X. and Li, C. (2007). One-pot synthesis of polyaniline-doped in mesoporous TiO_2 and its electrorheological behaviour, *Mater. Sci. Eng., B*, **137**(1), pp. 213–216.

126. Han, Y.-G., Kusunose, T. and Sekino, T. (2008). One-pot preparation of core–shell structure titania/polyaniline hybrid materials: the effect of sodium dodecyl sulfate surfactant, *Chem. Lett.*, **37**(8), pp. 858–859.

127. He, Z., Liu, J., Miao, J., Liu, B. and Tan, T. T. Y. (2014). A one-pot solvothermal synthesis of hierarchical microspheres with radially assembled single-crystalline TiO_2-nanorods for high performance dye-sensitized solar cells, *J. Mater. Chem. C*, **2**(8), pp. 1381–1385.

128. Su, S. and Kuramoto, N. (2000). Processable polyaniline–titanium dioxide nanocomposites: effect of titanium dioxide on the conductivity, *Synth. Met.*, **114**(2), pp. 147–153.

129. Han, Y.-G., Kusunose, T. and Sekino, T. (2008). The preparation and characterization of organic solvent dispersible polyaniline coated titania hybrid nanocomposites, *Mater. Sci. Forum*, **569**, pp. 161–164.

130. Han, Y.-G., Kusunose, T. and Sekino, T. (2009). Facile one-pot synthesis and characterization of novel nanostructured organic dispersible polyaniline, *J. Polym. Sci., Part B: Polym. Phys.*, **47**(10), pp. 1024–1029.

131. Güleryüz, H., Filiâtre, C., Euvrard, M., Buron, C. and Lakard, B. (2013). Novel strategy to prepare polyaniline—Modified SiO_2/TiO_2 composite particles, *Synth. Met.*, **181**(1), pp. 104–109.

132. Yang, C., Dong, W., Cui, G., Zhao, Y., Shi, X., Xia, X., Tang, B. and Wang, W. (2017). Highly-efficient photocatalytic degradation of methylene blue by PoPD-modified TiO_2 nanocomposites due to photosensitization-synergetic effect of TiO_2 with PoPD, *Sci. Rep.*, **7**(1), p. 3973.

133. Zhang, H., Zong, R., Zhao, J. and Zhu, Y. (2008). Dramatic visible photocatalytic degradation performances due to synergetic effect of TiO_2 with PANI, *Environ. Sci. Technol.*, **42**(10), pp. 3803–3807.

134. Hao, Y., Yang, M., Yu, C., Cai, S., Liu, M., Fan, L. and Li, Y. (1998). Photoelectrochemical studies on acid-doped polyaniline as sensitizer for TiO_2 nanoporous film, *Sol. Energy Mater. Sol. Cells*, **56**(1), pp. 75–84.

135. Sarmah, S. and Kumar, A. (2011). Photocatalytic activity of polyaniline-TiO_2 nanocomposites, *Indian J. Phys.*, **85**(5), p. 713.

136. Gilja, V., Novakovíc, K., Travas-Sejdic, J., Hrnjak-Murgíc, Z., Rokovíc, M. K. and Žic, M. (2017). Stability and synergistic effect of polyaniline/TiO_2 photocatalysts in degradation of azo dye in wastewater, *Nanomaterials*, **7**, p. 412.

137. Manjunath, S., Koppalkar, R., Kumar, A., Siddappa, R. and Prasad, M. (2008). Frequency-dependent conductivity and dielectric permittivity of polyaniline/TiO_2 composites, *Ferroelectr. Lett.*, **35**(3), pp. 36–46.

138. Mo, C., Wang, H. W., Chen, S. Y. and Yeh, Y. C. (2008). Synthesis and dielectric properties of polyaniline/titanium dioxide nanocomposites, *Ceram. Int.*, **34**, pp. 1767–1771.

139. Li, X., Wang, D., Cheng, G., Luo, Q., An, J. and Wang, Y. (2008). Preparation of polyaniline-modified TiO_2 nanoparticles and their photocatalytic activity under visible light illumination, *Appl. Catal., B*, **81**(3), pp. 267–273.

140. Jinzhang, G., Shengying, L., Wu, Y., Guohu, Z., Lili, B. and Li, S. (2007). Preparation and photocatalytic activity of PANI/TiO_2 composite film, *Rare Met.*, **26**(1), pp. 1–7.

141. Wei, J., Zhang, Q., Liu, Y., Xiong, R., Pan, C. and Shi, J. (2011). Synthesis and photocatalytic activity of polyaniline–TiO_2 composites with bionic nanopapilla structure, *J. Nanopart. Res.*, **13**(8), pp. 3157–3165.

142. Li, S., Du, C., Zhao, D., Zheng, J., Liu, H. and Wang, Y. (2015). Preparation and application of a new-type of ordered macroporous PANI/TiO_2 photocatalyst, *Chem. Lett.*, **44**(4), pp. 568–570.

143. Radoičić, M., Ćirić-Marjanović, G., Spasojević, V., Ahrenkiel, P., Mitrić, M., Novaković, T. and Šaponjić, Z. (2017). Superior photocatalytic properties of carbonized PANI/TiO_2nanocomposites, *Appl. Catal., B*, **213**, pp. 155–166.

144. Khanmohammadi, M., Mizani, F., Khaleghi, M. B. and Garmarudi, A. B. (2013). Optimized synthesis of polyaniline-TiO_2 composites for corrosion protection of carbon steel using design of experiment (DOE), *Prot. Met. Phys. Chem. Surf.*, **49**(6), pp. 662–668.

145. Radhakrishnan, S., Siju, C., Mahanta, D., Patil, S. and Madras, G. (2009). Conducting polyaniline–nano-TiO_2 composites for smart corrosion resistant coatings, *Electrochim. Acta*, **54**(4), pp. 1249–1254.

146. Pagotto, J., Recio, F., Motheo, A. and Herrasti, P. (2016). Multilayers of PAni/n-TiO_2 and PAni on carbon steel and welded carbon steel for corrosion protection, *Surf. Coat. Technol.*, **289**, pp. 23–28.

147. Ates, M. and Uludag, N. (2016). Synthesis and application of conducting polymers and their nanocomposites as a corrosion protection performances, *Bulg. Chem. Commun.*, **48**, pp. 27–32.

148. Bairi, V. G., Bourdo, S. E., Sacre, N., Nair, D., Berry, B. C., Biris, A. S. and Viswanathan, T. (2015). Ammonia gas sensing behavior of tanninsulfonic acid doped polyaniline-TiO_2 composite, *Sensors*, **15**(10), pp. 26415–26429.

149. Bhadra, S., Khastgir, D., Singha, N. K. and Lee, J. H. (2009). Progress in preparation, processing and applications of polyaniline, *Prog. Polym. Sci.*, **34**(8), pp. 783–810.

150. Barsan, N. and Weimar, U. (2001). Conduction model of metal oxide gas sensors, *J. Electroceram.*, **7**(3), pp. 143–167.

151. Huyen, D. N., Tung, N. T., Thien, N. D. and Thanh, L. H. (2011). Effect of TiO_2 on the gas sensing features of TiO_2/PANI nanocomposites, *Sensors*, **11**(2), pp. 1924–1931.

152. Srivastava, S., Kumar, S., Singh, V., Singh, M. and Vijay, Y. (2011). Synthesis and characterization of TiO_2 doped polyaniline composites for hydrogen gas sensing, *Int. J. Hydrogen Energy*, **36**(10), pp. 6343–6355.

153. Ansari, M. O. and Mohammad, F. (2011). Thermal stability, electrical conductivity and ammonia sensing studies on p-toluenesulfonic acid doped polyaniline: titanium dioxide (pTSA/Pani: TiO_2) nanocomposites, *Sens. Actuators, B*, **157**(1), pp. 122–129.

154. Pawar, S., Chougule, M., Patil, S., Raut, B., Godse, P., Sen, S. and Patil, V. (2011). Room temperature ammonia gas sensor based on polyaniline-TiO_2 nanocomposite, *IEEE Sens. J.*, **11**(12), pp. 3417–3423.

155. Harb, M. E., Ebrahim, S., Soliman, M. and Shabana, M. (2017). Fabrication of organic transistors using nanomaterials for sensing applications, *J. Electron. Mater.*, pp. 1–6.

156. Winkler, K. and Grądzka, E. (2015). Electron transfer and charge storage in thin films of nanoparticles, in *Handbook of Nanoelectrochemistry*, Aliofkhazraei, M. and Makhlouf, A., eds. (Springer, Cham), pp. 1–62.

157. Reddy, A. L. M. and Ramaprabhu, S. (2007). Nanocrystalline metal oxides dispersed multiwalled carbon nanotubes as supercapacitor electrodes, *J. Phys. Chem. C*, **111**(21), pp. 7727–7734.

158. Shao, Z., Li, H., Li, M., Li, C., Qu, C. and Yang, B. (2015). Fabrication of polyaniline nanowire/TiO_2 nanotube array electrode for supercapacitors, *Energy*, **87**, pp. 578–585.

159. Bian, C., Yu, A. and Wu, H. (2009). Fibriformpolyaniline/nano-TiO_2 composite as an electrode material for aqueous redox supercapacitors, *Electrochem. Commun.*, **11**(2), pp. 266–269.

160. Zhang, Y., Zhuang, X., Su, Y., Zhang, F. and Feng, X. (2014). Polyaniline nanosheet derived B/N co-doped carbon nanosheets as efficient metal-free catalysts for oxygen reduction reaction, *J. Mater. Chem. A*, **2**(21), pp. 7742–7746.

161. Lin, Y., Cui, X., Yen, C. H. and Wai, C. M. (2005). PtRu/carbon nanotube nanocomposite synthesized in supercritical fluid: a novel electrocatalyst for direct methanol fuel cells, *Langmuir*, **21**(24), pp. 11474–11479.

162. Qiao, Y., Bao, S.-J., Li, C. M., Cui, X.-Q., Lu, Z.-S. and Guo, J. (2007). Nanostructured polyaniline/titanium dioxide composite anode for microbial fuel cells, *ACS Nano*, **2**(1), pp. 113–119.

163. Ganesan, R. and Gedanken, A. (2008). Organic–inorganic hybrid materials based on polyaniline/TiO_2 nanocomposites for ascorbic acid fuel cell systems, *Nanotechnology*, **19**(43), p. 435709.

164. Grätzel, M. (2009). Recent advances in sensitized mesoscopic solar cells, *Acc. Chem. Res.*, **42**(11), pp. 1788–1798.

165. Li, Q., Wu, J., Tang, Q., Lan, Z., Li, P., Lin, J. and Fan, L. (2008). Application of microporous polyaniline counter electrode for dye-sensitized solar cells, *Electrochem. Commun.*, **10**(9), pp. 1299–1302.

166. Zhang, X., Yan, G., Ding, H. and Shan, Y. (2007). Fabrication and photovoltaic properties of self-assembled sulfonated polyaniline/TiO_2 nanocomposite ultrathin films, *Mater. Chem. Phys.*, **102**(2), pp. 249–254.

167. Kawata, K., Gan, S. N., Ang, D. T. C., Sambasevam, K. P., Phang, S. W. and Kuramoto, N. (2013). Preparation of polyaniline/TiO_2 nanocomposite film with good adhesion behavior for dye-sensitized solar cell application, *Polym. Compos.*, **34**(11), pp. 1884–1891.

168. Al-Daghman, N. J. A., Ibrahim, K., Ahmed, N. M. and Zaidan, K. M. (2015). Effect of TiO_2 thin film morphology on polyaniline/TiO_2 solar cell efficiency, *World J. Nano Sci. Eng.*, **5**(2), pp. 41–48.

Chapter 3

Metal Oxide Nanocomposites: Cytotoxicity and Targeted Drug Delivery Applications

Jaison Jeevanandam,[a] Yen S. Chan,[a] Sharadwata Pan,[b] and Michael K. Danquah[c]
[a]*Department of Chemical Engineering, Curtin University, CDT250 Miri, Sarawak 98009, Malaysia*
[b]*School of Life Sciences Weihenstephan, Technical University of Munich, 85354 Freising, Germany*
[c]*Chemical Engineering Department, University of Tennessee, Chattanooga, TN 37403, USA*
michael-danquah@utc.edu

Nanocomposites have gained prominence in pharmaceutical formulation and delivery applications due to their tunable biophysical properties, including particulate size, surface characteristics, and morphology. Metal oxide nanocomposites with small size, uniformly dispersed particulates, edge surface sites, and lattice symmetry are mostly advantageous for biomedical applications. Additionally, these metal oxide nanocomposites are engineered at the molecular level to downgrade their toxicities using eclectic synthesis routes based on necessary biomedical

Hybrid Nanocomposites: Fundamentals, Synthesis, and Applications
Edited by Kaushik Pal

ISBN 978-981-4800-34-1 (Hardcover), 978-0-429-00096-6 (eBook)
www.panstanford.com

requirements. Metal oxides are classified into two broad categories: magnetic, which mostly includes iron oxides, and nonmagnetic. Iron oxide nanocomposites with superparamagnetic properties have demonstrated efficacy for enhanced magnetofection and targeted drug delivery. Amongst nonmagnetic nanocomposites, zinc, titanium, and copper oxide have garnered attention in drug formulation due to their low toxicity in mammalian systems. This chapter explores an overview idea of the significance of magnetic and nonmagnetic metal oxide nanocomposites, their cytotoxicity profiles, and recent advancements in targeted drug delivery applications in the treatment of cancer, diabetes, and renal ailments. A few potential applications of these nanocomposites in the development of next-generation pharmaceuticals for the treatment of rare and neurodegenerative diseases such as Alzheimer's disease, Parkinson's disease, Lafora, and progeria were also investigated. The chapter helps the reader to the information of different recently available magnetic and nonmagnetic nanocomposites, their cytotoxicity toward various cell lines, and their targeted drug delivery applications toward cancer, diabetes, and renal ailments and consists of a novel idea of using these metal oxide nanocomposites for the treatment of rare and neurodegenerative diseases such as Lafora, progeria, etc.

3.1 Introduction

Specific matrices with a single or multiple fillers act as the principal constituent of composites. Additionally, in these composites, a handful of constituents containing fibers, sheets, or particles make up the required combination involving multiple phases [1, 2]. Nanocomposites are defined as special composites that possess at least one phase with specifications ranging in the nanometer regime. These materials demonstrate extraordinary performance abilities in displaying exclusive opportunities for specific designs, including exhibiting unique characteristics [3]. These novel set of composites has emerged as a potential alternative to counter and overcome the challenges faced, while using microcomposites in varied applications. They often pose exceptional advantages in potentially scientific and technology-related applications involving structural diversity. The assorted structures of nanocomposites show an exclusive diversity of tunable properties that find credible applicability in electrical, mechanical, optical, catalytic, thermal, and electrochemical fields [3]. From biomedical solicitations to packaging, they present distinct advantages with a foreseeable yearly growth rate of about 25% in engineering plastics and elastomers. Properties of building components, morphologies, interfacial characteristics, and synthesis methods are the prominent factors that determine the properties of a nanocomposite for a desired application [2]. A distinguished property of a nanocomposite is that it may acquire a new characteristic property, while assembling two components, which may not be present in each component at the pure state. Recently, nanocomposites have signaled a novel research domain that offers business opportunities for all the industrial sectors involved. In addition, several nanocomposites are environmentally amenable, touting them as the materials of the 21st century [4]. Furthermore, the preparation, characterization, and application of nanocomposites have turned out to be a fascinating interdisciplinary research area, with imminent biomedical and pharmaceutical applications [5]. This chapter aims to catalog various groups of metal oxide nanocomposites into magnetic and nonmagnetic counterparts and expose their cytotoxicity and biomedical applications, especially in targeted drug delivery. Additionally, novel nanocomposites have been proposed for unique drug delivery applications for management of sporadic diseases such as Lafora, progeria, and Parkinson's disease. The next section investigates metal oxide nanocomposites and their types. Cytotoxicity of metal oxide nanocomposites is discussed in Section 3.3. The targeted drug delivery applications are recapitulated in Section 3.4. Finally, the future perspectives and the salient inferences of the current chapter are summarized in Sections 3.5 and 3.6, respectively.

3.2 Metal Oxide Nanocomposites and Their Types

Attributing to their structural robustness, metal oxide nanocomposites are custom-made for a plethora of biomedical applications. These nanocomposites are broadly categorized into either magnetic or nonmagnetic nanocomposites, depending on their magnetic properties, as shown in Fig. 3.1.

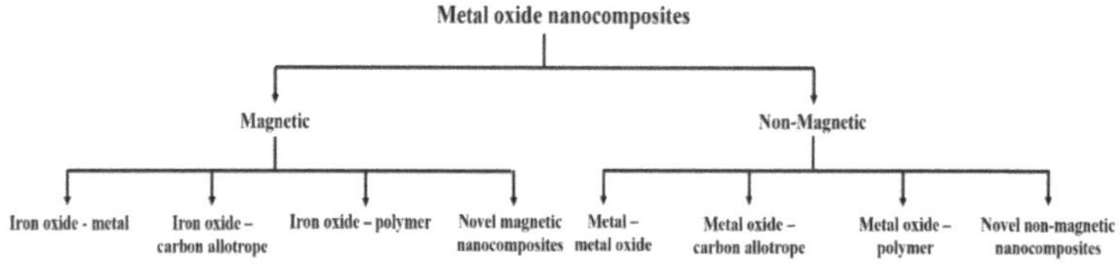

Figure 3.1 Classification of metal oxide nanocomposites on the basis of their magnetic properties.

3.2.1 Magnetic Nanocomposites

Among magnetic nanoparticles, iron oxide nanoparticles are especially crucial for the synthesis of magnetic nanocomposites. In its nanoconformation, iron oxide exhibits diverse forms of magnetisms such as dia-, para-, ferro-, ferri-, and antiferromagnetism, as shown in Fig. 3.2. Diamagnetic nanoparticles contain atoms with preoccupied orbital shells. Additionally, these do not possess a net magnetic moment due to the absence of any unpaired electron. In fact, the unpaired electrons in partially filled orbitals generate a net magnetic moment in the paramagnetic particles [6]. Ferromagnetic nanoparticles display a parallel magnetic moment alignment, causing a hefty net magnetization effect, irrespective of the magnetic field [7]. Ferrimagnetism occurs especially in oxides where the unique crystal structure leads to a complex magnetic ordering form [8]. In polycrystalline nanoparticles, the net magnetic moment is zero due to equal and opposite sublattice magnetic moments. Consequently, these are known as antiferromagnetic materials [9, 10]. For instance, Kim et al. have prepared 13–30 Å monodispersed iron oxide nanoparticles exhibiting diamagnetic property [11], as well as 9 nm ellipsoidal iron oxides coated with poly(vinyl alcohol) (PVA), exhibiting paramagnetism [12]. Similarly, iron oxide nanoparticles prepared by heating the coprecipitants of Co^{2+}, Fe^{2+}, and Fe^{3+} ions demonstrate ferromagnetism [13], 30 nm chitosan oligosaccharide–stabilized iron oxide nanoparticles prepared via standard Schlenk techniques possess ferrimagnetic properties [14], and ferrihydrite particles at low temperature exhibit antiferromagnetic properties [15, 16]. The magnetic properties of iron oxides vary when composites are prepared with metals, carbon allotropes, or polymers. Thus, they may be subclassified into iron oxide–metal composites, iron oxide–carbon allotrope composites, and iron oxide–polymer composites.

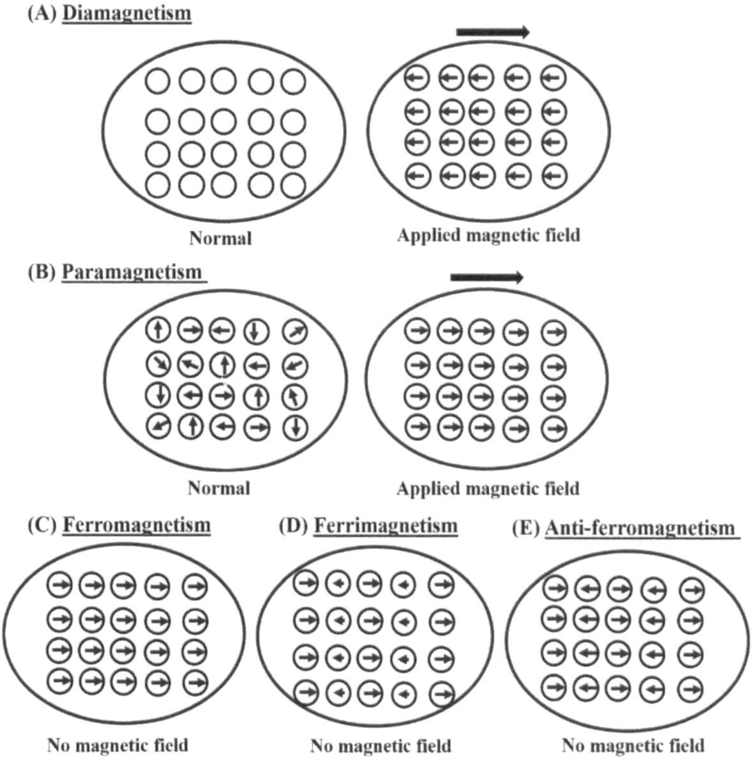

Figure 3.2 Schematics representing (A) dia-, (B) para-, (C) ferro-, (D) ferri-, and (E) antiferromagnetic materials

3.2.1.1 Iron oxide–metal nanocomposites

Catering to different applications, iron oxide has been prepared as composites along with metals and metal oxides to enhance their properties. Gold and silver are the most commonly used metal components to prepare iron oxide–metal composites. Sood et al. prepared 16 nm iron oxide–gold nanocomposites using ascorbic acid as the reducing agent, which are highly monodispersed and possess a superior superparamagnetic behavior at room temperature [17]. Similarly, gold–iron oxide magnetic nanocomposites were produced via a simple aqueous process, which are proposed to aid magnetic photothermal therapy and as recyclable nanocatalysts [18]. Silver, due to its superior antimicrobial activities, is used as a composite material along with iron oxides. Past studies have reported different methods to synthesize a wide array of silver–iron oxide nanocomposites with enhanced antimicrobial and physicochemical characters [19–21]. Other than silver and gold, silica was also used to fabricate composites along with iron oxides, as it is approved for clinical trials [22, 23]. Omar et al. synthesized 20–60 nm biodegradable magnetic silica–iron oxide nanocomposites using a modified thermal annealing procedure [24]. Recently, a novel synthesis method was reported via galvanic displacement and electroless deposition to incorporate palladium nanoparticle into a porous silica–iron oxide nanoparticle for photocatalytic activity [25]. In addition to silica, palladium–iron oxide nanocomposites were synthesized using a novel pepper extract–mediated green synthesis method for eliminating colored pollutants from water [26]. However, metal oxides were mostly used to composite along with iron oxide nanoparticles as they provide high stability in comparison to individual oxide particles.

Among metal oxides, combinations of titanium oxide, alumina (aluminum oxide), and zinc oxide are frequently employed used to fabricate iron oxide–metal oxide nanocomposites. Gold nanoparticle–functionalized iron–titanium oxide nanocomposites, with interwoven α-Fe_2O_3 dendritic structures, high porosity, and active area, as shown in Fig. 3.3A, were recently developed and designed by a plasma-assisted strategy for solar hydrogen production [27]. Similar work by Mirzoeva et al. yielded core–shell structures of 3,4-dihydroxyphenylacetic acid (DOPAC)-conjugated iron oxide–titanium dioxide nanocomposites for in vitro sensitization of cells from human neuroblastoma toward radiotherapy [28]. Iron–titanium oxide thin-film nanocomposites were synthesized via simple sol–gel route and are proposed to be useful as liquefied petroleum gas sensors and opto-electronic humidity [29]. Yusoff et al. fabricated novel titanium oxide–iron oxide nanocomposites, with incorporated silica in a reduced graphene oxide (rGO) nanohybrid, by the hydrothermal method for methanol electro-oxidation in an alkaline medium [30]. Other than titanium dioxide, alumina–iron oxide nanocomposites are significant to be the next-generation composite materials that are widely synthesized for various purposes, especially in wastewater treatment. Initially, iron–alumina nanocomposites were prepared by the ball milling method [31] and the modified wet impregnation procedure [32]. Later, Mahapatra et al. utilized the electrospinning process to synthesize iron oxide–alumina nanocomposite fibers in the 200–500 nm diameter range after sintering at 1000°C, which proved to be efficient for scavenging of heavy metals [33]. Electrochemical synthesis was reported by Dresvyannikov et al. for the fabrication of complex alumina–iron oxide nanocomposite materials, dispersed via anodic dissolution, which is highly suitable in battery and wastewater treatment [34]. Another metal oxide that is composited with iron oxide, especially for biomedical and wastewater treatment applications, is zinc oxide. Iron oxide–ZnO nanocomposites were produced by spin-coating and sol–gel techniques by Tang et al. for selective NH_3 gas sensing at room temperature [35]. Similarly, bifunctional iron oxide–zinc oxide nanocomposites were prepared via a facile two-step strategy for photocatalysts application [36]. Later, Wu et al. fabricated spindle-like mesoporous iron oxide–ZnO core–shell heteronanostructures via an ecofriendly, low-cost, and surfactant-free seed-mediated approach with postannealing treatment [37]. Similarly, ferromagnetic iron oxide–zinc oxide nanocomposites were synthesized via the sol–gel method by Hasanpour et al. [38]. It was revealed that their magnetism has an inverse relationship with annealing temperature. Singh et al. fabricated magnetic semiconductor iron oxide–embedded ZnO nanocomposites via a facile soft-chemical approach. The authors reported that the nanoadsorbent efficiency is beneficial as these pose as reusable and highly separable materials for fast elimination of organic dyes, foreign pathogens such as bacteria, and remediation of toxic metal ions [39]. Recently, Pend et al. fabricated iron oxide mesoporous nanocomposites using a template-free method. These are feasible for drug delivery and heat therapy applications, as they possess magnetic waves and microwaves to heat responsive properties [40].

Other than the common metal oxides, rare-earth metal oxides have also been supplemented as composites with iron oxides for various applications. Ma et al. synthesized and biofunctionalized multifunctional magnetic $\alpha\text{-}Fe_2O_3@Y_2O_3{:}Eu^{3+}$ nanocomposites via a facile homogeneous precipitation method. The composite material showed superparamagnetic property of α-Fe_2O_3@Y_2O_3 and exclusive europium with high emission intensity properties. The biofunctionalization with biotin and p-aminobenzoic acid (PABA) leads to a precise targeting of the polystyrene beads coupled with avidin [41]. Xia et al. fabricated a complex luminescent and magnetic composite material with α-Fe_2O_3@Y_2O_3:Eu^{3+} as a bifunctional hollow microsphere by incorporating template-assisted, coprecipitation, and high-temperature calcination processes. The composite material exhibited a noteworthy drug-loading capacity (126 mg/g) and a sustained drug release profile and has been proposed to be an essential nanodrug carrier for malignant tumor therapy [42]. Very recently, photoluminescence of α-Fe_2O_3@Y_2O_3:Eu^{3+} bifunctional composites was enhanced via Gd^{3+} codoping and the material properties were found to be adequate for biomedical applications [43]. Recently, an identical work was published by Zhang et al. in which Li^+ was doped to enhance the luminescent and magnetic properties of ~1 μm α-Fe_2O_3@Y_2O_3:Eu^{3+} core–shell bifunctional nanocomposites [44]. Other than yttrium, zirconia was extensively utilized to prepare nanocomposites along with iron oxide. Saraji et al. prepared zirconia magnetic nanocomposites via the one-step coprecipitation method, which shows higher efficiency in plasmid DNA purification [45]. In another study, Noormohamadi et al. prepared core–shell Fe_2O_3@ZrO_2/PAN nanocomposites membrane to reduce biofouling. The authors revealed that the addition of zirconia enhanced the membrane's hydrophilicity by 51% and that the iron oxide improves membrane porosity by 47% [46]. Recently, Fang et al. prepared superparamagnetic ZrO_2@Fe_2O_3 nanocomposites for phosphate recovery from treated sewage effluents to prevent eutrophication. The authors revealed that the composites possess 1.5-fold higher phosphorus adsorption capacity, in comparison to the ZrO_2@SiO_2@Fe_2O_3 composites [47]. Besides these, novel rare-earth nanocomposites such as lanthanum oxide–iron oxide nanocomposites for phosphate removal from wastewater [48], multifunctional nanocomposites of superparamagnetic iron oxide composited with near infrared–responsive rare-earth dopants such as NaYF4:Yb, Er as up-conversion fluorescent nanoparticles for biolabeling and cancer cell imaging via the fluorescent method [49], and scandium oxide–iron oxide nanocomposites [50] are synthesized and currently under research for various applications.

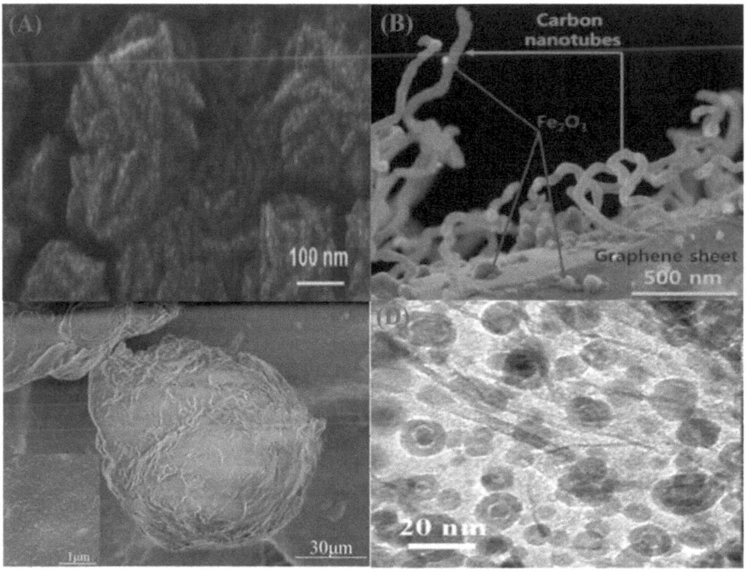

Figure 3.3 (A) Field emission scanning electron microscopy (FESEM) micrograph of iron oxide–titania–gold (iron oxide–metal) nanocomposite. Reproduced from Ref. [27] with permission from John Wiley and Sons. (B) Transmission electron microscopy (TEM) micrograph of graphene–CNT–iron oxide (carbon allotrope–iron oxide) 3D nanocomposite structure. Reprinted with permission from Ref. [95]. Copyright (2013) American Chemical Society. (C) Scanning electron microscopy (SEM) micrograph of iron oxide–LDPE (iron oxide–polymer) nanocomposite particles [96]. (D) TEM micrograph of graphene–iron oxide@iron core–shell nanoparticle–ZnO nanoparticle (novel magnetic) nanocomposite. Reprinted with permission from Ref. [89]. Copyright (2012) American Chemical Society.

3.2.1.2 Iron oxide–carbon allotrope nanocomposites

The pure form of carbon exists in different physical forms that are known as allotropes. Diamond and graphite are the most common and significant allotropes of carbon. The advancements in nanotechnology have enabled the addition of a few novel carbon allotropes such as graphene, carbon nanotubes (CNTs), and fullerenes [51]. Although CNTs are considered to be the strongest and, at the same time, the lightest material known to us, other allotropes are also considerably stronger than several polymeric and metallic particles [52]. Thus, incorporation of iron oxide with carbon allotropes yields a composite material with higher strength and increased magnetic properties. Magnetic iron oxide–graphene nanocomposites, prepared by Deng et al., showed that these nanocomposite materials possess enhanced inhibition property toward *E. coli* in an aqueous medium [53]. Similarly, rGO–iron carbide nanocomposites were prepared through iron-based intercalation of graphite oxide and have been proposed for magnetic and supercapacitor applications [54]. Recently, iron oxide–graphene oxide (GO) nanocomposites were synthesized via iron oxide coprecipitation on GO sheets, which are fabricated by the modified Hummer's method. The material exhibits enhanced adsorbent properties for arsenic removal [55]. Likewise, the levels of 2,4,6-trinitrotuluene in various water samples could be determined by utilizing GO–iron oxide nanocomposites as magnetic sorbents for solid-phase isolation in tandem with liquid chromatography [56]. Garg et al. listed the multifunctional applications of GO–iron oxide nanocomposites, for instance, magnetically photothermal therapy, directed drug transfer, and magnetic resonance imaging (MRI) [57].

CNTs are also composited with iron oxide nanoparticles to form high-strength nanocomposites. Magnetic graphene–CNT–iron composites, as shown in Fig. 3.3B, are highly suitable as smart adsorbents, which can inactivate viruses and bacteria and eliminate pollutants proficiently in water. Furthermore, with the help of an exterior magnet, the absorbents may be effortlessly removed from the treated water. Thus, these materials are of great assistance in bioremediation and wastewater-cleaning applications [58]. A facile controlled in situ fabrication process was introduced by Liu et al. for synthesizing monodispersed magnetic CNT nanocomposites using a mixture of water–ethylene glycol solvents [59]. Later, magnetic iron oxide–CNT nanocomposites fabricated via chemical vapor and alkali-activated methods were reported to possess higher adsorption properties toward toluene, ethyl benzene, and xylene and can be used to eliminate them from aqueous solutions [60]. Recently, trivial and adaptable nanocomposite aerogels, containing mesoporous iron oxide and strung by CNTs, were prepared through the in situ hydrothermal method and the material was reported to display enhanced microwave adsorption properties [61]. Another recent study also reported that magnetic iron oxide nanoparticle–multiwalled CNT composites, fabricated via the facile one-pot solvothermal method, possess highly effective elimination capability of aqueous Cr(VI), which is convenient and potentially applicable for targeted pollutant removal and wastewater treatment [62]. In addition to wastewater treatment applications, these composite materials are also used in biomedical imaging [63], targeted magnetic drug delivery, and cancer cell-imaging purposes [64].

Fullerenes are another set of hollow carbon allotropes that are supplemented with iron oxides as composites for biomedical applications, especially as targeted nanodrug carriers. Shi et al. synthesized and reported on the MRI, photodynamic treatment, and enhanced directed drug transfer capabilities of poly(ethylene PEI (PEG)ylated fullerene–iron oxide nanocomposites [65]. Cano et al. recently studied the manufacturing of superparamagnetic, multifunctional iron oxide nanoparticles–fullerene (C_{60}) nanocomposites, amassed by the fullerene–amine click chemistry process. These pose proposed applications for radical hunting applications, photodynamic remedy, and photocatalytic oxidation of organic contaminants [66]. Similarly, Gogoi et al. reported the preparation of novel and highly stable β-cyclodextrin-supported magnetic iron oxide–fullerene nanocomposites, which are proposed to be useful for the Fenton oxidation response as heterogeneous catalysts for destroying aqueous alizarin [67]. Other than these allotropes, novel carbon composites were also recently synthesized. A novel renewable source carbon–iron oxide nanocomposite, based on tannin, has been established recently, with a strong objective to eliminate arsenic from polluted water [68]. Likewise, magnetic iron oxide carboxylated nanodiamonds was prepared by Yilmaz et al., as solid-phase isolation adsorbents, for identifying ziram (zinc dimethyldithiocarbamate) in several water trials, food samples, and synthetic mixtures [69]. However, the cytotoxicity of magnetic iron oxide and carbon allotrope nanocomposites is a major drawback while considering them for biomedical applications.

3.2.1.3 Iron oxide–polymer nanocomposites

Iron oxides are added to polymers to pose them as magnetic composites with regular polymer characteristics. Novakova et al. listed the magnetic features of polymer nanocomposites comprising iron oxide nanoparticles, which has huge potential as a high-aptitude magnetic package and essential nanoscale circuits [70]. Oh et al. listed a wide array of superparamagnetic, iron oxide–based polymeric nanomaterials, incorporating their strategy, formulation, and biomedical claims. These magnetic polymer nanocomposites have proposed applicability in MRI disparity augmentation, directed drug distribution, hyperthermia, biological isolation, protein immobilization, and biosensors [71]. Recently, nanocomposites of highly homogeneous and condensed superparamagnetic iron oxide nanocrystals, equivalently disseminated in a poly(ethylene oxide) melt, were fabricated by Feld et al. These nanocomposite materials are proposed to be used as adaptive materials for magnetorheological applications [72]. Bonilla et al. reported a list of polymer nanocomposites, based on magnetic nanoparticles and with conductive properties, for fabricating a composite material with enhanced magnetic and electrical properties that would be suitable in absorbing microwaves and screen electromagnetic radiations [73]. Likewise, Chi et al. deposited iron oxide nanoparticles on low-density polyethylene (LDPE) particle surfaces via the solvothermal procedure, as shown in Fig. 3.3C, to enhance their thermal conductivity and dielectric properties [74]. Iron oxide–polymer nanocomposites also possess imaging properties encompassing multiple modes and magnetically enhanced and directed drug transport prospect [75]. It is clear from past available studies that polymer nanocomposites possess applications pertaining to a wide range, from the biomedical to the electronic field.

It is not advisable to engage synthetic polymers for biomedical applications as they may initiate toxic reactions in cells. Thus, biopolymers were incorporated with iron oxide nanoparticles to form less toxic magnetic polymer nanocomposites. Li et al. prepared iron oxide–chitosan nanocomposites via covalent binding hydrothermally synthesized magnetic iron oxide nanoparticles with chitosan at their surfaces, using H_2O_2 as an oxidizer, and revealed their superparamagnetic properties. This composite material may be useful in magnetic drug delivery systems and cell–enzyme immobilization [76]. Liu et al. manufactured magnetic chitosan nanocomposites that show higher adsorption of heavy metals and can be a useful recyclable tool for eliminating heavy metal ions from drinking water [77]. Kaushik et al. also synthesized iron oxide nanoparticle–chitosan composites, with a shelf life of around eight weeks in chilled environments, and utilized them as a glucose biosensor [78]. In another work, Singh et al. dispersed hydrothermally prepared magnetic α-iron oxide nanoparticles in chitosan solution to construct a hybrid, monodispersed, nanocomposite film for commercial and biomedical solicitations [79]. Carbofuran is a toxic carbamate product that finds usage as an agricultural insecticide. Jeyapragasam et al. used acetylcholinesterase arrested onto chitosan–iron oxide nanocomposites as an electrochemical biosensor to detect carbofuran [80]. Another work by Kaushik et al. showed enhanced urea-sensing ability of iron oxide–chitosan nanobiocomposites [81]. Recent studies have also revealed the anticarcinogenic properties of iron oxide–chitosan nanocomposites [82] and their peroxidase purification capabilities [83].

3.2.1.4 Novel magnetic nanocomposites

Many novel magnetic nanocomposites have been synthesized for a range of applications. Superparamagnetic core–shell iron oxide–*N*-chloramine nanocomposites were fabricated by Haham et al. for water purification applications [84]. Similarly, Taufik et al. fabricated novel iron (II, III) oxide–zinc oxide–copper (II) oxide nanocomposites and revealed their photosonocatalytic characteristic for eliminating organic dyes [85]. Likewise, hexavalent chromium was removed from wastewater using magnetic nanocomposites supplemented with manganese dioxide–iron oxide–acid-oxidized multiwalled CNTs [86] and using polypyrrole–iron oxide magnetic nanocomposites [87]. These polypyrrole–iron oxide magnetic nanocomposites are also used as gas and humidity sensors [88]. Also, quaternary nanocomposites made up of graphene, iron oxide@iron core@shell, and zinc oxide nanoparticles, as shown in Fig. 3.3D, with outstanding electromagnetic engrossment characteristics [89], differentially structured yttrium oxide–cerium oxide nanocomposites as UV light–stimulated photocatalytic deprivation and as agents advancing catalytic reduction [90], magnetic iron–zirconium binary oxide nanocomposites for phosphate elimination from aqueous solutions [91], and iron oxide–bone char nanocomposites captured in a chitosan biopolymer for decontamination of an arsenic (V)-contained liquid phase [92], are recently synthesized and employed in various applications. Recently, an innovative yttria-stabilized tetragonal zirconia–nickel nanocomposite was prepared via the altered interior reduction technique [93] and iron (III) sulfide–ferritin bioinorganic nanocomposites were synthesized via preorganized biomolecular architectures [94], which may be of significance for various future prospects.

3.2.2 Nonmagnetic Nanocomposites

Nonmagnetic nanocomposites are usually manufactured to enhance the catalytic or mechanical properties of a material. These types of nanocomposites are subclassified into metal–metal oxide nanocomposites, metal oxide–carbon-based nanocomposites, metal oxide–polymer nanocomposites, and novel combinations.

3.2.2.1 Metal–metal oxide nanocomposites

Generally, metal and metal oxides are not incorporated together due to the lack of structural stabilities. However, a few past studies are available reporting the synthesis of metal–metal oxide nanocomposites for unique applications. Takacs fabricated a metal–metal oxide system for generation of nanocomposites using a high-energy ball milling approach. The experimental results showed that chromium oxide reduction by aluminum or zinc possesses the capability to form nanocomposites [97]. Richter et al. fabricated stabilizer-free metal–metal oxide nanocomposites, without any stabilizer and with prolonged steadiness, by accumulation of physical vapor into ionic liquids [98]. Later, Wang et al. used the one-pot synthesis process, facilitated by microwaves, to fabricate metal (PtRu)–metal oxide (SnO_2) nanocomposites over graphene, which has higher supercapacitance compared to unadulterated graphene. Additionally, they also showed electrocatalytic actions for oxidation of methanol in contrast to the commercially available E-TEK PtRu/C electrocatalysts [99]. In 2013, Kochuveedu et al. reported a list of noble metal–metal oxide semiconductor nanocomposites, studied their interactions with light, and studied their mechanisms for various photophysical applications. It was evident that TiO_2–noble metal and ZnO–noble metal combination of nanocomposites show enhanced photocatalysts, photoluminescence, photochromism, and photovoltaic abilities in comparison to other metal–metal oxide combinations [100]. TiO_2–gold semiconductor nanocomposites for photocatalysis applications [101], nickel–zirconium oxide (Fig. 3.4A) for natural gas utilization applications [102], and platinum–cerium oxide composites as photocatalysts [103] are some of the unique metal–metal oxide combinations of nanocomposites that find use in specific applications.

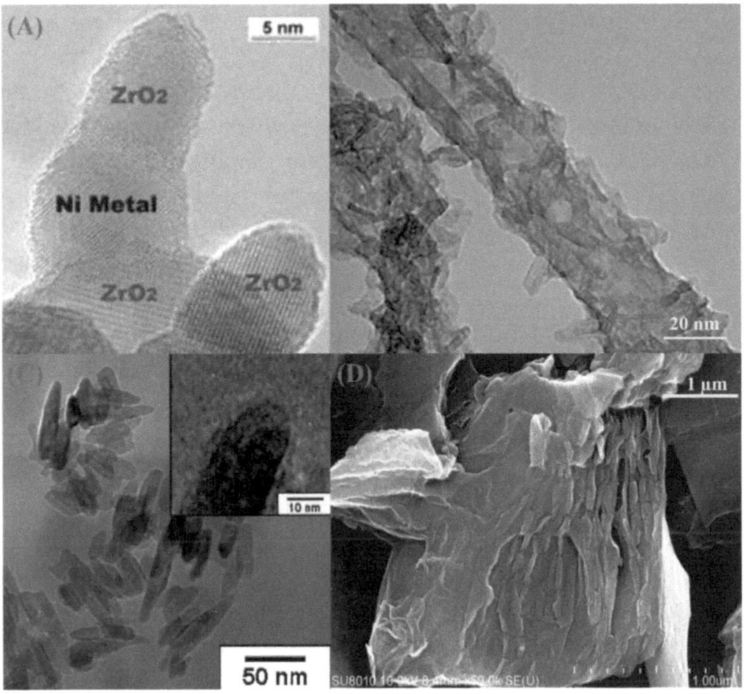

Figure 3.4 (A) High-resolution transmission electron microscopy (HRTEM) micrograph of nickel–zirconium oxide (metal–metal oxide) nanocomposites. Reprinted with permission from Ref. [102]. Copyright (2003) American Chemical Society. (B) TEM micrograph of manganese oxide–functionalized CNT (carbon allotrope–metal oxide) nanocomposites. Reprinted from Ref. [123], Copyright (2015), with permission from Elsevier. (C) TEM micrograph of titania–polymer (metal oxide–polymer) core–shell nanocomposites. Reprinted with permission from Ref. [132]. Copyright (2010) American Chemical Society. (D) SEM micrograph of GO–C_{60} buckyball (novel nonmagnetic metal oxide) nanocomposites. Reprinted from Ref. [148], Copyright (2016), with permission from Elsevier.

3.2.2.2 Metal oxide–carbon allotrope nanocomposites

Mostly, composite materials made up of carbon allotropes are magnetic in nature since the nanoform of carbon allotropes tends to orient their electrons in a spinning condition that leads to magnetism. However, a few past studies show the preparation of metal oxide–carbon allotrope combinations of nanocomposites for various purposes. In 2002, Wang et al. utilized tin oxide–graphite nanocomposites as a cathode material in lithium-ion batteries using the microemulsion method [104]. Similarly, in 2005, Wang et al. synthesized tin oxide–graphite nanocomposites through a microwave-assisted method for the same application with enhanced electrode properties [105]. In 2007, Chang et al. synthesized nanocomposites of tin oxide/tin elements over a graphite exterior as the positive electrode using the argon atmosphere pyrolysis technique for lithium-ion batteries [106]. The requirement for enhancing the electrode properties leads to the fabrication of graphene along with metal oxides as nanocomposites for lithium-ion battery applications. In 2010, Wang et al. synthesized ordered metal oxide–graphene nanocomposites via a ternary self-assembly process for electrochemical energy storage purposes [107]. Similarly, in 2011, Baek et al. manufactured tin oxide–graphene nanocomposites through the nonaqueous, one-pot sol–gel tactic, facilitated by microwaves, for lithium-ion battery applications [108]. In 2012, Su et al. synthesized 2D graphene–metal oxide hybrid nanocomposites, coated with carbon, for improved lithium accommodation [109]. In 2010, cobalt oxide–graphene nanocomposites were synthesized via in situ assembly and chemical reduction by Yang et al. for high-performance anode materials [110]. Titanium oxide–graphene hybrid nanocomposites were synthesized and utilized for lithium-ion application [111] and as photocatalyst materials [112, 113]. Also, graphene–tungsten oxide nanocomposites were synthesized using the peroxopolytungstic acid (PTA) Kudo method and find applications in visible light–induced antiviral properties [114]. Furthermore, metal oxide and graphene-incorporated nanocomposites are employed in water treatment applications as sterilizers, photocatalysts, and adsorbents [115].

These advances have clearly demonstrated the efficacy of the incorporation of synthesized CNT with metal oxides as composites for lithium-ion battery applications. In 2005, nickel oxide–CNT nanocomposites were prepared via a simple chemical precipitation method by Lee et al. for electrochemical capacitance applications [116]. In the same year, Ye et al. prepared aligned CNT–ruthenium oxide nanocomposites via the magnetic sputtering method for using them as supercapacitors [117]. Tin oxide–CNT nanocomposites were also fabricated in supercritical fluids for usage as chemical sensors and as positive electrodes for lithium-ion batteries [118]. Later, manganese oxide was used to prepare a composite material along with CNTs for energy storage applications. In 2007, Ma et al. used the simple immersion method to coat manganese dioxide onto CNTs into a $KMnO_4$ aqueous solution [119]. Similarly, in 2008, Ma et al. used a direct redox reaction for depositing birnessite type of manganese dioxide on CNTs. These nanocomposite combinations exhibited improved structural and electrochemical reversibilities after heat treatment for enhanced energy storage [120]. Recently, many metal oxide–CNT composites were prepared for lithium-ion storage applications [121, 122]. Also, novel manganese oxide–functionalized CNT nanocomposites, as shown in Fig. 3.4B, have found usage in bioelectrochemical systems, such as biological fuel cells, as oxygen-reducing enhancers [123]. In addition, aqueous nanoparticles of ruthenium oxide were attached to CNTs to form hybrid nanocomposites for supercapacitor applications [124]. The buckyball form of carbon allotropes, such as fullerenes, was also composited with metal oxides for solar cell and energy storage applications. In 2008, Liu et al. prepared innovative cuprous oxide–fullerene core–shell nanocomposites using the copper (I)-facilitated fullerene polymerization process for photocatalytic applications [125]. Similarly, in 2010, Motoyoshi et al. used the spin-coating procedure to fabricate cuprous oxide–fullerene-based nanocomposite solar cells and showed that the nanocomposite material has efficient solar cell properties compared to the cuprous nanoparticles [126]. These constitute a few of the nanocomposite combinations with metal oxide and carbon allotropes that are under extensive research for foreseeable applications.

3.2.2.3 Metal oxide–polymer nanocomposites

Metal oxides are added as composite materials along with polymers to enhance their properties, in addition to incorporating the property of the concerned metal oxide. This eventually results in the manifestation of a novel characteristic feature of the composite material. In 2006, Wu et al. fabricated polymer electrolytes of poly(vinylidene-fluoride-co-hexafluoropropylene) (PVdF-HFP) with metal oxides such as titanium oxide, magnesium oxide, and zinc oxide as nanocomposites, which are highly porous with salient conductive properties. The prepared nanocomposites were proposed to be useful as low-leakage porous polymer electrolytes [127]. It can be noted that nanocomposites with an oxide center that is nonconductive and coated with poly(methyl methacrylate) (PMMA) polymer exhibit stout light emission. Also, when the insulating oxide core is replaced with semiconducting oxides, such as zinc oxide, it enhances the luminescence effect additional to the quantum confinement phenomena [128]. Many such polymer–metal oxide nanocomposites for solicitations in lithium-ion batteries were compiled earlier by Croce et al. [129] and recently by Sarkar et al. [130]. In 2007, Boucle et al. listed a set of hybrid polymer–metal oxide thin nanocomposite films that have high potential in photovoltaic applications [131]. In 2010, Kong et al. prepared novel titanium oxide–biocidal polymer nanocomposites via the surface-commenced photopolymerization process, as shown in Fig. 3.4C, and are proven to possess photocatalytic antibacterial abilities [132]. Similarly, novel transparent gadolinium oxide–polymer nanocomposites were fabricated by Cai et al. and are proposed to be useful in γ-ray spectroscopy [133]. Recently, innovative metal oxide–polymer nanocomposite films from disposable scarp tire powder and poly(Îµ-caprolactone) were reported for advanced electrical capacitor applications [134].

Besides synthetic polymers, biopolymers are also added to metal oxides for several applications. In 2006, Allouche et al. manufactured core–shell and biomimetic nanocomposites with silica and gelatin nanoparticles, which paved the way for the development of new biopolymer-based nanocomposites for drug delivery systems [135]. Similarly, the solvent-casting procedure was adopted to synthesize gelatin–zinc oxide nanocomposite films, which exhibit strong UV-screening effects and counteractions against both gram-negative and gram-positive bacteria [136]. Later, in 2007, Kim et al. prepared phosphonic acid–modified barium titanate polymer nanocomposites, which showed enhanced permittivity and dielectric strength for energy storage applications [137]. Subsequently, chitosan was utilized to prepare nanocomposites as it possesses superior biocompatibility in body fluids. In 2012, Cai et al. fabricated cellulose–silica nanocomposite aerogels based on organic silicates using an innovative sol–gel technique that included supercritical CO_2-mediated drying. The nanocomposite aerogels showed enhanced mechanical potency, semitransparency, elasticity, a hefty surface area, and diminished heat conductivity [138]. Likewise, a chemical vapor technique enhanced by plasma was employed to manufacture chitosan–titania nanocomposites using planar, compressed, magnetron equipment. The nanocomposites showed a better dispersion rate of titania, exhibiting higher antimicrobial abilities [139]. As biopolymers are highly biocompatible and are highly bioactive in nature, nanocomposites fabricated using biopolymers may be utilized for biomedical applications.

3.2.2.4 Novel metal oxide nanocomposites

A few nonmagnetic novel metal oxide nanocomposites have been fabricated especially for photocatalytic applications, which can further be enhanced and utilized in biomedical applications. In 2009, Su et al. synthesized copper oxide–titanium oxide core–shell nanocomposites via the solution synthesis method, which was proposed to have potential as a photocatalyst and photoelectric transition material [140]. Novel zinc–biochar nanocomposites were synthesized by using sugarcane bagasse and proved to possess significant prospects in Cr (VI) elimination from wastewater [141]. Recently, Thirumalraj et al. (2016) fabricated extraordinarily steady fullerene over GO nanocomposites, as shown in Fig. 3.4D, for subtle, electrochemical dopamine recognition in therapeutic tests on rat brains [142]. In 2017, novel nanocomposites using hydroxyethyl cellulose and graphene [143], poly(ethylene-co-vinyl acetate) with hydroxyapatite, multiwalled CNTs and ammonium polyphosphate [144], polypyrrole doped with fullerene [145], biodegradable nanocomposites with Natureplast poly(butylene succinate) (PBE), poly(butylene adipate-co-terephthalate) (PBAT), various toxic-free expanded organoclays (EOCs) [146], and bacterial cellulose–zinc oxide nanocomposites, utilizable for innovative dressing purposes in combating burn injuries [147], were synthesized for unique and distinct applications. This shows that many novel nanocomposites are under research, which may potentially replace nanoparticles in many applications, due to the combined properties of a wide range of particles in a single composite material.

3.3 Cytotoxicity of Metal Oxide Nanocomposites

Cytotoxicity of nanocomposites plays a major role in their utilization as potential agents for biomedical applications. The less toxic nanocomposites, along with their less toxic dosage, are identified via cytotoxicity analysis, which helps to predict their applications in biological systems. Attributing to the same, cytotoxicity analysis has been performed for various nanocomposites before proposing them for biomedical and pharmaceutical applications.

3.3.1 Cytotoxicity of Magnetic Metal Oxide Nanocomposites

In 2006, Yi et al. fabricated silica–iron oxide magnetic nanocomposites with ~25-nm-thick shells. They measured their cytotoxicity in the human hepatocarcinoma (HepG2) cell line and NIH3T3 mouse fibroblast cells using the 3-(4,5-dimethylthiazol-2-yl)-2,5-diphenyltetrazolium bromide (MTT) assay over a nanocomposite dosage of 10–100 μg/mL for three days. The outcomes revealed that the inhibitory activity of the nanocomposite depends on the dosage and the exposure period. It was observed that the particle dosage of 100 μg/mL on day 3 shows the strongest inhibitory activity [149]. In 2012, a complex multifunctional magnetic nanocomposite, namely carboxymethyl chitosan-capped magnetic nanoparticle-intercalated montmorillonite nanocomposites, were developed as a novel drug delivery system for doxorubicin (DOX) by Anirudhan et al. This study showed that while the complex magnetic nanocomposite system is toxic to cancer cells (MCF-7), it is less toxic to H9c2 cardiac muscle cells, even after adding 75 mg/g for 96 h (~85% viable cells), compared to the free DOX drug [150]. In 2013, another twofold surface-functionalized and multiplex Janus nanocomposites were made up of polystyrene–iron oxide–silica. The target was to achieve precise focusing of the tumor cells and stimulus-promoted secretion of drugs such as folic acid and DOX simultaneously. This complex nanocomposite showed fivefold less cytotoxicity toward MDA-MB-231 breast cancer cell line. This is a widely acknowledged in vitro prototype for breast cancer cells, without being hormone supplemented and excessively manifesting the folate receptors compared to free DOX. This reveals that the nanocomposite is less toxic to cells than the drug and isare applicable for controlled drug delivery due to its enhanced drug release abilities [151]. In the same year, Zhu et al. fabricated superparamagnetic iron oxide nanocomposites, which are pH responsive and possessing a wide array of functions, for directed drug transfer and MRI applications. The cytotoxicity of the nanocomposites was investigated in NIH/3T3 cells using the MTT assay for 24 and 48 h, and the results revealed that these nanocomposites possess low in vitro cytotoxicity. Even for a higher dosage of the nanocomposite (1 mg/mL), the relative cell viability was 90% ± 8% after 24 h and 73% ± 3% after 48 h of incubation [152].

In addition to iron oxide nanocomposites, GO composites in nanoform were also subjected to cytotoxicity analysis. In 2011, Chen et al. synthesized nanocomposites of aminodextran-laminated iron oxide nanoparticles and GO for cellular MRI applications. The 8(2-(2-methoxy-4-nitrophenyl)-3-(4-nitrophenyl)-5-(2,4-disulfophenyl)-2H-tetrazolium (WST) assay was used to examine the cytotoxicity of the composite material toward HeLa cells, which revealed that the nanocomposite showed availability of ~90%, even at a concentration of 80 μg/mL [153]. Later, in 2012, Wen et al. fabricated redox-responsive PEGylated nano-GO composites for drug penetration within the cells. It was reported via WST assay that the PEGylated nanocomposites do not significantly affect cell proliferation up to 1 mg/mL of dosage and result in diminished HeLa cell viability, depending upon the dosage [154]. Similarly, a relatively easy short emulsification procedure was employed to synthesize manganese ferrite–GO nanocomposites along with evaporation of the solvent. The cell viability of this nanocomposite was determined after 12 h of incubation in MCF-7 cancer cells. No noteworthy cytotoxicity could be observed up to 117 μg/mL, indicating that the material is unsuitable for biomedical applications [155]. In 2014, bifurcated and low-molecular-weight rGO–hydrophilic polyethylene glycol (PEG) nanocomposites, reduced by polyethylenimine (BPEI), were prepared by Kim et al. for photothermally controlled gene delivery applications. The MTT cytotoxicity assay revealed that the composite material has less cytotoxicity (90% viable cells) toward the prostate cancer (PC-3) cell line as compared to the individual materials [156]. In 2015, Chang et al. showed that magnetic GO–titania nanocomposites exhibit novel solar light-irradiated antibacterial activities. The composites were synthesized by simple dispersion and ultrasonication, and the results showed that the composite material inhibits *E. coli* growth and that its 4log removal was obtained within a time frame of 41 min. [157]. It is clear from the above literature that magnetic metal oxide nanoparticles are less toxic to human cells and are more noxious to cancer cells and disease-causing bacteria. Novel nanocomposites, such as poly(*N*-isopropylacrylamide) (pNIPAAm)–iron oxide-based magnetic hydrogels and manganese-doped superparamagnetic iron oxides, were subjected to cytotoxic assays. pNIPAAm–iron oxide–based magnetic hydrogel composites were synthesized via UV polymerization, and their cytotoxicity was tested in NIH 3T3 murine fibroblasts [158]. On the other hand, the cytotoxicity of manganese-incorporated superparamagnetic iron oxide nanocomposites were tested in both HepG2 and mouse macrophage cell lines (RAW 264.7) [159]. Both these studies showed that the nanocomposites show less toxicity toward the cells as compared to individual particles.

3.3.2 Cytotoxicity of Nonmagnetic Metal Oxide Nanocomposites

Recently, several nonmagnetic nanocomposites were subjected to cytotoxicity analysis, as they are widely proposed for unique biomedical and pharmaceutical applications. In 2016, zinc oxide–silver core shell nanocomposites were fabricated using wild ginger essential oil and their in vitro cytotoxicity was analyzed using VERO cells via the MTT assay by Azizi et al. No remarkable cellular cytotoxicity, depending on the dosage, could be observed up to a concentration of 100 μg/mL of the nanocomposite material [160]. In the same year, Chaturvedi et al. prepared PVA-incorporated cryogel–zinc oxide nanocomposites via in situ condensation of zinc oxide nanoparticles in tandem with a reiterated freeze-thaw procedure with a cryogel network. The in vitro cytotoxicity of the polymer–ZnO composite materials was studied using L-929 mouse fibroblast cells. The results showed that the composites are nonreactive and less toxic to the cells up to 24 h of incubation. Additionally, the composite material was found to be biocompatible via in vitro biocompatible tests, such as hemolysis and protein adsorption [161]. Silver–titania–bentonite nanocomposites were synthesized through facile thermal decomposition and their cytotoxicity analysis was carried out via the MTT assay in the human embryonic kidney (HEK 293) cell line. The composite material demonstrated concentration-dependent cytotoxicity and was shown to possess high cell viability as compared to the individual materials [162]. Similarly, magnesium–titania nanocomposites were subjected to the MTT cytotoxicity assay, which showed that magnesium–titania (2.5 vol%) has low cytotoxicity toward the murine-derived preosteoblast cell line (MC3T3-E1), even after five days of incubation [163]. Also, carboxylated nanodiamond–cobalt oxide nanocomposites were synthesized via in situ chemical reduction and their cytogenotoxicity was studied using *Allium Cepa* L. (onion). The results clearly showed that the composite material possesses less genotoxicity, which is concentration dependent, compared to cobalt oxide and carboxylated nanodiamonds [164]. Very recently, for the first time, Bhowmick et al. showed promise by incorporating zirconium oxide nanoparticles in chitosan-included mixed-breed organic-inorganic composites, nanohydroxyapatite (CS-PEG-HA), and PEG to advance nanocomposites with orthopedic applications. Human osteoblastic MG-63 cells were involved in the MTT cytotoxicity assay of the composite material. The results revealed that they are less toxic to osteoblastic cells, which are utilized for bone tissue engineering applications [165].

Besides metal–metal oxide nanocomposites, GO–based nanocomposites were also extensively subjected to cytotoxicity analysis. Very recently, Rasoulzadeh et al. fabricated carboxymethyl cellulose–GO bionanocomposite hydrogel beads via the physical crosslinking process. The cytotoxicity of the composite material was studied using human colon cancer cells (SW480) via the MTT assay. The results showed that the nanocomposite does not possess obvious toxicity toward cancer cells. However, the nanocomposite with DOX showed increased cytotoxicity toward cancer cells, indicating suitability of the composite material as an anticarcinogenic drug carrier agent [166]. Similarly, silver–GO nanocomposites were prepared using the modified Hummer's method, under microwave radiation, for cancer drug delivery applications. The cytotoxicity analysis was performed via the MTT assay in human glioblastoma cancer cells (U87MG). The outcomes demonstrated a dosage-based reduction in cancer cell viability enforced by the nanocomposite, highlighting its anticancer abilities [167]. Recently, rGO–titanium oxide nanotube composites were prepared via hydrothermal treatment and their cellular toxicity was identified using Raw264.7 mouse monocyte-macrophage cells via the MTT assay. In sync with other studies, the nanocomposite showed less cytotoxicity at a lower dosage in relation to the individual particles [168]. It can be noted that compared to the individual materials, while the nanocomposites are less toxic toward normal cells, they show higher toxicity toward the cancer cells. Hence, it is evident that nanocomposites, especially with metal oxides, will be a better choice for biomedical applications. Table 3.1 summarizes the specific physicochemical characters and cytotoxicity of magnetic and nonmagnetic metal oxide nanocomposites toward various cell lines.

3.4 Metal Oxide Nanocomposites for Targeted Drug Delivery Applications

To date, several simple and complex metal oxide nanocomposites have been synthesized and employed toward directed drug penetration and transfer applications against a wide variety of diseases. The components for the nanocomposites are selected on the basis of the synthesis routes and their potential cytotoxicity toward cells. On the basis of the advances in nanocomposite preparation, a wide range of novel nanocomposite materials has been introduced for drug delivery applications. This ultimately aids in countering the menace caused by a plethora of serious lifestyle-related ailments, such as diabetes, cancer, renal problems, and neurodegenerative diseases.

3.4.1 Targeted Drug Delivery for Cancer Treatment

Metal oxide nanocomposites, both magnetic and nonmagnetic, have been proposed and studied for targeted drug delivery applications to enhance cancer diagnosis and prognosis. In 2008, Lv et al. fabricated novel nano–iron oxide–polylactide nanofiber composites for drug delivery applications in the K562 leukemia cancer cell line. The results showed that the composite carrier served as a beneficial tool to deliver daunorubicin and to effectively facilitate the interactions

Table 3.1 Summary of cytotoxicity of magnetic and nonmagnetic nanocomposites toward various cell lines

Nanocomposite	Characteristics	Cell lines	Cytotoxic outcomes	Reference(s)
		Magnetic Metal Oxide Nanocomposites		
Silica–iron oxide	~25-nm-thick shells	HepG2 human liver cancer cell line, NiH3T3 mouse fibroblast	100 μg/mL particle dosage on day 3 with strongest inhibitory activity	[149]
Carboxymethyl chitosan-capped magnetic nanoparticle-intercalated montmorillonite nanocomposites	Complex multifunctional, 132–164 nm	MCF 7, H9c2 cardiac muscle cells	Less toxic (85% cells are viable) to H9c2 cardiac muscle cells than free DOX drug, even after adding 75 mg/g for 96 h	[150]
Polystyrene–iron oxide–silica	313 nm, superparamagnetic Janus nanocomposites	MDA-MB-231 breast cancer cell lines	Fivefold less cytotoxicity toward breast cancer cell lines than DOX	[151]
Iron oxide/cystamine tert-acylhydrazine/DOX/PEG	Superparamagnetic, pH-responsive, 145–185 nm	NIH/3T3 cells	Even at higher concentration of 1 mg/mL, cell viability 90% ± 8% after 24 h and 73% ± 3% after 48 h	[152]

(*Continued*)

Table 3.1 (*Continued*)

Nanocomposite	Characteristics	Cell lines	Cytotoxic outcomes	Reference(s)
Aminodextran-laminated iron oxide nanoparticles and GO	5 nm uniform-sized spherical aminodextran-coated iron oxide nanoparticles assembled onto the GO surface	HeLa cells	90% of cells viable at 80 μg/mL concentration	[153]
Nano-GO composites	Redox-responsive, PEGylated, < 200 nm size GO sheets	HeLa cells	No significant cytotoxicity up to 1 mg/mL and toxicity depending on dosage	[154]
Manganese ferrite–GO	~55 nm nanocomposite loaded with 6, 11, and 14 nm manganese ferrite nanoparticles	MCF-7 cancer cells	No significant toxicity up to 117 μg/mL after 12 h	[155]
rGO–hydrophilic PEG	Bifurcated, low molecular weight, reduced by polyethylenimine, round shaped, <100 nm with thickness of 10 nm	PC-3 (prostate cancer) cell lines	90% of cells viable compared to individual components of nanocomposites	[156]
GO–titania	Nanocomposite sheets with wrinkles with 13.69 nm	*Escherichia coli* bacterial cell	4log removal obtained within 41 min.	[157]

Nanocomposite	Characteristics	Cell lines	Cytotoxic outcomes	Reference(s)
pNIPAAm–iron oxide hydrogels	Iron oxide nanoparticles with 20–30 nm diameter incorporated into hydrogels	NIH 3T3 murine fibroblasts	~95% of cells viable after 90 M of nanocomposites added 90 M representing the mole percentage of NIPAAm present in the system, where M refers to system with magnetic particles at 3 wt%	[158]
Manganese ferrite–incorporated superparamagnetic iron oxide	Hydrophobic nanocomposite of mean diameter 80 nm self-assembled into small clusters inside micelles	Human hepatocarcinoma cells (HepG2) and mouse macrophage cell lines (RAW 264.7)	No significant toxicity found after 4 h, when nanocomposites added up to 140 μg/mL concentration	[169]
Nonmagnetic Metal Oxide Nanocomposites				
Zinc oxide–silver	Core–shell structure, synthesized using wild ginger essential oil	VERO kidney epithelial cells	No significant cytotoxicity up to 100 μg/mL concentration	[170]
Cryogel–zinc oxide	PVA-incorporated, nearly spherical morphology with diameter <100 nm with well-defined pores	L-929 mouse fibroblast cells	No significant cytotoxicity after incubating nanocomposite for 24 h	[171]

(*Continued*)

Table 3.1 (*Continued*)

Nanocomposite	Characteristics	Cell lines	Cytotoxic outcomes	Reference(s)
Silver–titania–bentonite	Ag/TiO_2 nanoparticles clusters of size ~5–50 nm uniformly distributed on the surface of bentonite clay	Human embryonic kidney (HEK 293) cell lines	High cell viability of 98.3% when 2.5 μg/mL concentration of nanocomposite added	[162]
Magnesium-titania	Layered sandwich structure synthesized using disintegrated melt deposition technique	Murine-derived preosteoblast cell line (MC3T3-E1)	No significant cytotoxicity observed when 0.97 and 1.98 vol% of TiO_2 added to magnesium to form nanocomposites after 5 days of incubation	[172]
Nanodiamond–cobalt oxide	Nanocomposite 16.95 nm	*Allium cepa* L. (onion) cells	Chromosomal aberration insignificant for nanocomposite at 10 μg/mL concentration compared to 20 μg/mL	[164]
Zirconium oxide nanoparticle in chitosan–nanohydroxyapatite–PEG	Mixed-breed organic-inorganic composites, highly porous structure with 1–10 μm pore size along with 13–21 nm zirconium oxide nanoparticles	Human osteoblastic MG-63 cells	Less toxic to cells Increase in nanohydroxyapatite–zirconium oxide from 5 to 15 wt% increasing cell proliferation	[173]

Nanocomposite	Characteristics	Cell lines	Cytotoxic outcomes	Reference(s)
Carboxymethyl cellulose– GO bionanocomposite hydrogel beads	Surface morphology via SEM showing a rough structure and severe wrinkles GO well exfoliated within hydrogel nanocomposite network without agglomeration	Human colon cancer cells (SW480)	No significant toxicity observed even at high concentration (64 μM) of nanocomposite	[174]
Silver–GO	Synthesized by Hummer's method using microwave irradiation Flake-like shape of GO with slight wrinkles with densely covered, subnanometer-sized dots of silver over their surface	Human glioblastoma cancer cells (U87MG)	Dose-dependent (10–300 μg/mL) decrease in cancer cell viability observed	[175]
rGO–titanium oxide nanotube composites	GO nanosheets smooth with less defects Titanium dioxide nanotube in the shape of siphonate with 100–110 nm length Nanotubes closely coated with thin unilamellar rGO layer	Raw264.7 mouse monocyte-macrophage cells	Cell viability 85% even after incubated with 100 μg/mL for 48 h	[176]

of the drug in cancer cells to induce cancer cell apoptosis [177]. Likewise, carboxymethyl chitosan-coated and magnetic iron oxide nanoparticle–inserted montmorillonite nanocomposites have been employed to induce sustained excretion of the drug DOX to cancer cells, thereby advancing drug targeting applications. The study demonstrated that the multifunctional nanocomposite is a decent cancer drug transfer scheme, and revealed their promising strategy in protecting cardiac cells against oxidative damage perceived in cancer drug–provoked toxicity against the same [150]. Moreover, in 2013, Zhu et al. fabricated superparamagnetic iron oxide nanocomposites, which are pH responsive and possess several functionalities, for specific transport of a cancer drug. The results clearly shows that the cancer drug, formulated in magnetic nanocomposites, exhibits better inhibition of cancer cell growth under a magnetic field compared to the nonformulated cancer drug [152]. Later, in the same year, Wang et al. synthesized iron oxide–polypyrrole (PPy) core–shell nanocomposites functionalized with PEG for imaging-directed, tenuously organized cancer permutation rehabilitation. This nanocomposite scheme contains an iron oxide center, which facilitates drug transfer via magnetic intervention. Additionally, these are beneficial as MRI contrast, where the PPy shell has drug-loading abilities and strong photothermal activities, which are useful for chemotherapy enhancement, with exceptional in vivo synergistic cooperative antitumor actions under in vivo conditions [178].

Besides iron oxide nanocomposites, GO–silver-reduced nanoparticle nanocomposites, synthesized by incorporating extracts from the plant *Tiliaamurensis*, were engaged in anticancer nanotherapy. The results showed the sensitive restraining influence on cell viability compared to the individual GO, rGO, and silver nanoparticles, and are proposed to be an effective and less toxic cancer therapeutic agent [179]. In 2016, a complex iron oxide–copper oxide–titanium oxide nanocomposite has been prepared using the sol–gel method and was subjected to test its anticancer abilities using the MCF-7 human breast cancer cell line. The MTT cytotoxicity assay result showed enhanced anticancer activities of the composite material compared to tamoxifen (a common cancer drug) and undoped titanium oxide and is proposed to be a potential, alternative agent for human breast cancer therapy [180]. Recently, several novel nanocomposites have been tested and proved to possess significant anticancer characteristics. Very recently, Wu et al. fabricated iron oxide–layered double hydroxide with a methotrexate cancer drug using the hydrothermal method. Additionally, its abilities to combat cancer cells were tested in different cancer cell lines such as human umbilical vein endothelial cell (HUVEC), MCF-7, and HepG2 via the WST-1 assay. The results revealed that the composite material acts as a perfect cancer drug delivery system against liver cancer (HepG2) and breast cancer (MCF-7) cell lines and is less toxic to the HUVEC line [181]. Also, nickel oxide–manganese dioxide nanocomposites were synthesized by the anchoring method as carriers of anticancer drugs such as (3-chloropropyl)triethoxysilane and guanidine nitrate. The results clearly showed that the nanocomposites possess higher potential as a carrier for drug delivery applications [182]. Likewise, in 2017, Wei et al. synthesized monodispersed, mesoporous, walnut kernel-like silicon dioxide–iron oxide nanocomposites with an intense magnetic property for sustained drug release pattern. The outcomes of the MTT cytotoxicity assay of the novel nanocomposite material, performed in rat liver (BRL-3A) cells, showed that it is low on toxicity. The drug release mechanism was studied using vancomycin overloaded inside the composite substance. Outcomes demonstrated that the composite material exhibits a drug-enhanced preliminary spray discharge, trailed by a sluggish continuous discharge procedure. Consequently, it is proposed to be promising targeted drug delivery carriers with synergistic properties [183].

3.4.2 Targeted Insulin Delivery for Diabetes Treatment

The natural prognosis for treating both type I and type II diabetes is insulin therapy. However, recent alarming reports have indicated that high insulin concentrations during insulin therapy lead to serious side effects such as hypoglycemia (decreased glucose level) and weight gain in diabetic patients [184, 185]. Contextually, nanoparticles are often employed as carriers for controlled insulin transport in order to reduce these toxic side effects. Recently, nanocomposites have been utilized for controlled insulin therapy applications due to their low toxicity and enhanced drug delivery properties. Initially, a bioinorganic nanohydrid glucose-responsive membrane was fabricated by crosslinking bovine serum albumin–

manganese dioxide nanocomposites along with glucose oxidase, catalase, and poly(*N*-isopropylacrylamide-co-methacrylic acid) nanoparticles. The study showed that the composite membrane increases insulin permeation by almost four times by enhancing the glucose concentration from a normal to a hyperglycemic level. Consequently, it is proposed to be a self-controlled insulin transport corresponding to a fit human pancreas [186]. Later, in 2011, Gordijo et al. synthesized a novel, bioinorganic nanocomposite membrane, with wide functionality, for self-regulating the insulin secretion adjustable to the plasma glucose quantities. An identical manganese dioxide composite material was used in addition to the pH-responsive hydrogel nanoparticles. The results revealed that the insulin secretion is regulated catering to the in vitro glucose concentration. Furthermore, the insulin transfer instrument, fabricated with a fastened loop scheme and incorporated with nanocomposites, demonstrated potential in regulating standard plasma glucose concentrations in diabetes-infected mice [187].

Several novel nanocomposites were also subjected to diabetic studies, on the controlled delivery of insulin, for maintaining a normal glucose level in diabetic patients. In 2012, Shen et al. fabricated chitosan-constructed, radiant, hybrid magnetic nanogels, facilitated by uninterrupted gelation of cadmium telluride, chitosan, and superparamagnetic iron oxide, for antidiabetic and insulin transport solicitations. The results demonstrated that the nanogel composite has better insulin responsiveness in L02 cell lines, along with increasing the specifying incorporation of insulin [188]. Similarly, iron oxide nanocomposites were also employed in insulin delivery applications. In 2012, iron oxide–chitosan nanocomposites were synthesized using the laser ablation technique by Kebede et al. for insulin transport via the oral route. The nanocomposites loaded with insulin showed a 51% reduction in the plasma glucose concentration, compared to the control group, in the test subjects (mice), depending on the degree and severity of diabetic manifestation [189]. Past studies have also suggested that a mucoadhesive, silica-dependent, nanoporous, oral insulin composite preparation, with capped insulin and cell-diffusing peptides, are used to overwhelm the obstacles presented by the enzyme and mucosal insulin transfer via oral administration [190]. Also, magnetic GO–iron nanoparticle composites, dumped with insulin, were proposed for the in vivo insulin delivery applications. The study gave a clear insight that the nanocomposite protects insulin, ensure its stability even at low pH, and are highly useful for controlled delivery of insulin [191]. Recently, montmorillonite–insulin–titanium oxide hybrid nanocomposites [192] and chitosan–zinc oxide nanocomposite hydrogels [193] were also fabricated for oral insulin delivery applications, which showed promising advancements in controlled insulin delivery. Extensive research are still underway to select a superior composite combination that can sense and target diabetic cells, deliver insulin, and control delivery for maintaining a normal blood glucose level.

3.4.3 Nanocomposites for Diagnosis and Prognosis of Renal Ailments

In addition to drug delivery applications for cancer and diabetes, nanocomposite materials are also used to detect compounds that can manifest kidney malfunction or failure. In 2009, a novel silylated chitosan–silicon dioxide–multiwalled CNT nanocomposite matrix was fabricated and its ability to immobilize urease was investigated. The majority of commercially available urea biosensors utilize the enzyme urease to sense and detect urea in blood serum, which directly refers to kidney function and is a potent indicator of renal ailments [194]. An increase in urea concentration leads to urinary tract obstruction in conjunction with gastrointestinal bleeding. On the other hand, hepatic failure, nephritic syndrome, and cachexia are caused due to a decrease in serum urea concentration [195]. This nanocomposite has been proposed as a potential bioelectrode for biosensors that have the ability to identify urea in serum or urine samples, which helps in preventing renal failure [196]. Similarly, glucose oxidase–immobilized cellulose–tin oxide hybrid nanocomposites have been studied and tested as glucose biosensors. The nanocomposite was fabricated using the liquid-phase deposition technique, and glucose oxidase enzyme was immobilized via the physical absorption method. The results indicated that the fusion nanocomposite can be used as a low-cost, adaptable, and nonrenewable glucose biosensor [197]. Moreover, titanium oxide–cellulose hybrid nanocomposites were studied and proved to possess abilities to be utilized as conductometric glucose sensors. These glucose sensors will help to control diabetes by keeping the blood glucose level in check, thereby preventing renal failure [198]. Apart from urea and glucose, uric acid is a significant biomolecule found in urine and blood plasma, which acts as a marker for renal failure as well as toxicity. A magnetic core–shell iron oxide–silicon dioxide–multiwalled CNT nanocomposite-amended carbon paste electrode was studied for augmented electrochemical detection of uric acid in order to avoid renal failure [199]. Mercury (Hg^{2+}) has accumulative characteristics and environmentally toxic properties, which makes it a highly toxic heavy metal ion. Additionally, its long-term exposure may lead to kidney failure and other chronic diseases. Thus, in 2016, Zhang et al. fabricated an electrochemical biosensor, based on DNA-amended, 3D-reduced chitosan–GO nanocomposite, for the extraordinarily delicate identification of mercury ions [200]. A graphene–tin oxide nanocomposite for saline water desalination using capacitive deionization [201]; a manganese-doped iron oxide–N-doped CNT composite as a novel electrochemical sensor for caffeine, acetaminophen, and ascorbic acid determination [202]; and a GO–poly(ether imide) nanocomposite membrane for blood purification application [203] are some of the latest composite materials that can be helpful to detect and cure renal malfunctions and diseases. Recently, a novel fluorescent imaging assay was proposed to analyze kidneys using quantum dot–iron oxide nanocomposites, which may find potential applications as a lab diagnostic tool for renal ailments in the foreseeable future [204]. Figure 3.5 summarizes different drug delivery applications of metal oxide nanocomposites.

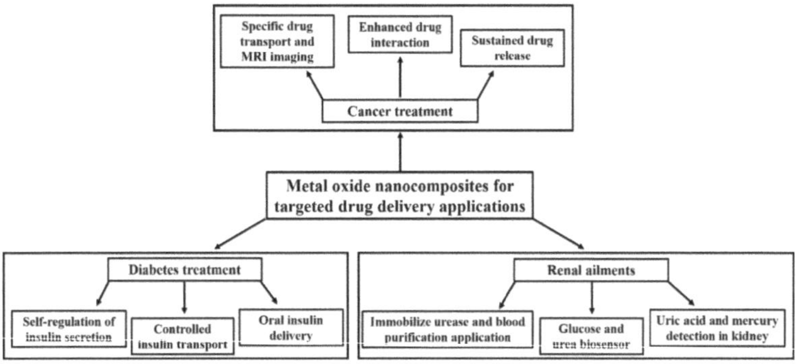

Figure 3.5 Targeted drug delivery applications of metal oxide nanocomposites.

3.5 Future Perspectives

Of late, nanocomposites have gained considerable attention as prominent vehicles for targeted drug delivery, especially as part of treatments against ailments such as diabetes and cancer. As a part of cancer treatment, several novel nanocomposites have been fabricated and employed for controlled, directed delivery of cancer drugs. A new innovative method called nanocryosurgery was introduced to treat and kill cancer cells in concerned patients. In the context of the latter, biodegradable MgO nanoparticles are extensively applied for better cancer prognosis [205]. In addition, ferromagnetic iron–magnesium oxide nanocomposites have been used as potential hyperthermia agents for cancer management [206–208]. Such novel nanocomposite materials may also be used for nanocryosurgery applications in the future, which may help to overcome excessive dependency on traditional forms of cancer therapy, that is, by posing as viable alternatives to hyperthermia and chemotherapy. In terms of diabetes, nanocomposites are used as sensors for glucose quantification and for controlled insulin delivery to regulate blood glucose levels. Recently, MgO nanoparticles have been proposed to possess enhanced insulin resistance reversal abilities [209], which helps to treat type II diabetes. Nanocomposites of MgO nanoparticles, either with CNTs or biopolymers, may further enhance their insulin resistance reversal ability. This may present itself as an integral part of future diabetes prognosis as an alternative to the insulin administration. On the other hand, magnetic nanocomposites may also be fabricated along with MgO nanoparticles for magnetic, targeted delivery of MgO nanoparticles in diabetic cells.

Besides cancer and diabetes management, several hydrogel-based nanocomposites have been fabricated and recommended to be valuable to treat neurodegenerative diseases [210, 211]. Iron oxide–based hydrogel nanocomposites are prospective in magnetically targeting and reaching the nerve cells, followed by drug delivery, which will reduce the amount of drug used and control the toxicity exhibited by those drugs toward normal cells. Thus, nanocomposites will strengthen their position as a potent, drug delivery entity in the future for timely prognosis of neurodegenerative diseases, such as Alzheimer's disease, Parkinson's disease, and Lafora. Lafora is

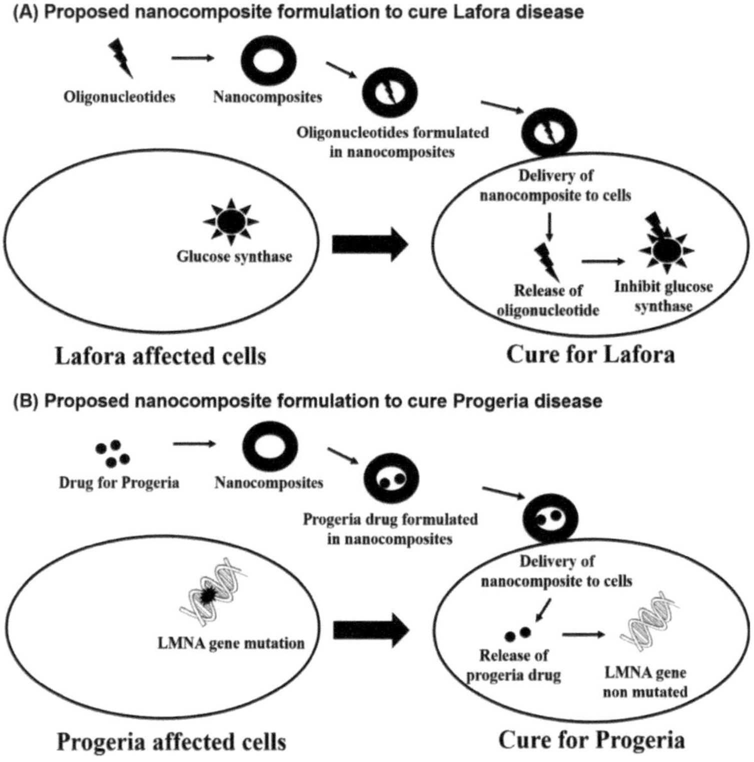

Figure 3.6 Proposed nanocomposite formulation and its possible mechanism to cure (A) Lafora and (B) progeria.

a rare, autosomal recessive, neurodegenerative genetic syndrome that is illustrated by the occurrence of inclusion bodies within the cellular cytoplasm in the lung, liver, muscle, and/or skin [212, 213]. Unfortunately, with the manifestation of this disease, the majority of affected patients pass away before attaining the age of 25 [214]. It has been found out that the antisense oligonucleotides and molecules that inhibit glycogen synthase happen to be the key factor for recovery from this ailment [215–217]. Thus, nanocomposites may be employed for the targeted delivery of the glycogen synthase–inhibiting molecule, which will act as a potential nanomedicine for the treatment of Lafora in the future, as shown in Fig. 3.6A. Similarly, a rare and early maturing syndrome, progeria, is manifested by a point mutation in the *LMNA* gene, which encodes for a protein called prelamin A [218, 219]. This syndrome is often discovered in children and leads to aging-related problems at a very early age [220]. Recently, the drug lonafarnib [221] and a novel combination of mevinolin to induce defarnesylation of progerin protein and chromatin-modifying agents, such as 5-azadeoxycytidine, anacardic acid, and trichostatin, were used for progeria treatment [222]. In future, nanocomposites may be employed for specific and regulated drug transport, as shown in Fig. 3.6B. It is also possible to use nanocomposites for *LMNA* gene delivery, which may possibly alleviate the severity of the disorder.

3.6 Conclusions and Outlook

Vast opportunities are available for nanocomposites to be efficiently explored and utilized from the perspective of drug targeting and transfer solicitations, especially from the outlook of prognosis of a wide variety of ailments. Recent advances in the nanocomposite domain lead to innovative, novel, and hybrid synthesis methods that can yield composites with unique properties. Amongst many other unique characteristics, nanocomposites possess a special feature in their ability to mimic the properties of individual materials, in addition to enhanced properties that are unavailable in individual materials. This unique nature of nanocomposites makes them a perfect candidate in biomedical applications, especially in the area of drug delivery. However, the cytotoxicity of nanocomposites is a major concern in their full realization as effective entities in biomedical applications. In addition, while the majority of nanocomposite cytotoxic studies are carried out in vitro, the in vivo findings will further improve our knowledge to efficiently design them as a potentially less toxic, biomedical formulation agent. Moreover, scarce studies have been carried out targeting the large-scale synthesis of nanocomposites, and only a handful of composite materials have been sanctioned by the US Food and Drug Administration (FDA) for therapeutic transport applications. This shortcoming can be solved by carefully comprehending the cytotoxicity and biomedical mechanisms of these composite materials. This may be achieved via the realization of nanocomposite research in a broader range by involving large-scale animal studies and clinical trials and by adopting unique strategies for necessary drug delivery applications with minimum toxicity.

Acknowledgments

The authors would like to thank Curtin University, Malaysia, for the support provided to complete this book chapter.

References

1. Roy, R., Roy, R. A. and Roy, D. M. (1986). Alternative perspectives on "quasi-crystallinity": non-uniformity and nanocomposites, *Mater. Lett.*, **4**(8–9), pp. 323–328.
2. Ebrahimi, F. (2012). *Nanocomposites: New Trends and Developments* (InTech, Crotia).
3. Guifu Zou, H. L., Sun, B., Jain, M. and Peng, H. (2011). Nanocomposites, *J. Nanotechnol.*, **2011**, p. 2.
4. Benítez, A. and Walther, A. (2017). Cellulose nanofibril nanopapers and bioinspired nanocomposites: a review to understand the mechanical property space, *J. Mater. Chem. A*, **5**, pp. 16003–16024.
5. Wong, W. and Gupta, M. (2016). High performance lightweight magnesium nanocomposites for engineering and biomedical applications, *NanoWorld J.*, **2**(4), pp. 78–83.
6. Raoux, A., et al. (20140. From dia-to paramagnetic orbital susceptibility of massless fermions, *Phys. Rev. Lett.*, **112**(2), p. 026402.
7. Bozorth, R. M. (1993). *Ferromagnetism* (Wiley-VCH), p. 992, ISBN 0-7803-1032-2.
8. Smart, J. S. (1955). The Néel theory of ferrimagnetism, *Am. J. Phys.*, **23**(6), pp. 356–370.
9. Anderson, P. (1950). Antiferromagnetism. Theory of superexchange interaction, *Phys. Rev.*, **79**(2), pp. 350–356.
10. Doniach, S. (1977). The Kondo lattice and weak antiferromagnetism, *Physica B+C*, **91**, pp. 231–234.
11. Kim, D. K., et al. (2001). Synthesis and characterization of surfactant-coated superparamagnetic monodispersed iron oxide nanoparticles, *J. Magn. Magn. Mater.*, **225**(1), pp. 30–36.
12. Petri-Fink, A., et al. (2005). Development of functionalized superparamagnetic iron oxide nanoparticles for interaction with human cancer cells, *Biomaterials*, **26**(15), pp. 2685–2694.

13. Kita, E., et al. (2010). Heating characteristics of ferromagnetic iron oxide nanoparticles for magnetic hyperthermia, *J. Appl. Phys.*, **107**(9), p. 09B321.
14. Bae, K. H., et al. (2012). Chitosan oligosaccharide-stabilized ferrimagnetic iron oxide nanocubes for magnetically modulated cancer hyperthermia, *ACS Nano*, **6**(6), pp. 5266–5273.
15. Levy, M., et al. (2011). Long term in vivo biotransformation of iron oxide nanoparticles, *Biomaterials*, **32**(16), pp. 3988–3999.
16. Brem, F., Stamm, G. and Hirt, A. M. (2006). Modeling the magnetic behavior of horse spleen ferritin with a two-phase core structure, *J. Appl. Phys.*, **99**(12), p. 123906.
17. Sood, A., et al. (2016). Ascorbic acid-mediated synthesis and characterisation of iron oxide/gold core–shell nanoparticles, *J. Exp. Nanosci.*, **11**(5), pp. 370–382.
18. Li, L., et al. (2014). Synthesis and morphology control of gold/iron oxide magnetic nanocomposites via a simple aqueous method, *IEEE Trans. Magn.*, **50**(1), pp. 1–5.
19. Trang, V. T., et al. (2017). Functional iron oxide–silver hetero-nanocomposites: controlled synthesis and antibacterial activity, *J. Electron. Mater.*, **46**(6), pp. 3381–3389.
20. Chiou, J.-R., et al. (2013). One-pot green synthesis of silver/iron oxide composite nanoparticles for 4-nitrophenol reduction, *J. Hazard. Mater.*, **248**, pp. 394–400.
21. Bonyani, M., et al. (2015). Electrochemical properties of Ag@ iron oxide nanocomposite for application as nitrate sensor, *Electroanalysis*, **27**(11), pp. 2654–2662.
22. Allen, T. M. and Cullis, P. R. (2013). Liposomal drug delivery systems: from concept to clinical applications, *Adv. Drug Delivery Rev.*, **65**(1), pp. 36–48.
23. Phillips, E., et al. (2014). Clinical translation of an ultrasmall inorganic optical-PET imaging nanoparticle probe, *Sci. Transl. Med.*, **6**(260), pp. 260ra149–260ra149.
24. Omar, H., et al. (2017). Biodegradable magnetic silica@iron oxide nanovectors with ultra-large mesopores for high protein loading, magnetothermal release, and delivery, *J. Controlled Release*, **259**, pp. 187–194.
25. Kim, T., et al. (2017). Size-controlled Pd nanoparticle catalysts prepared by galvanic displacement into a porous Si-iron oxide nanoparticle host, *ACS Nano*, **11**(3), pp. 2773–2784.

26. Khaghani, S., Ghanbari, D. and Khaghani, S. (2017). Green synthesis of iron oxide-palladium nanocomposites by pepper extract and its application in removing of colored pollutants from water, *J. Nanostruct.*, **7**(3), pp. 175–182.

27. Barreca, D., et al. (2016). Iron–titanium oxide nanocomposites functionalized with gold particles: from design to solar hydrogen production, *Adv. Mater. Interfaces*, **3**(16), p. 1600348.

28. Mirzoeva, S., et al. (2014). In vitro radiosensitization of human neuroblastoma cells with iron oxide-titanium dioxide core-shell nanocomposites conjugated with 3, 4-Dihydroxyphenylacetic acid (DOPAC), *Proceedings AACR Annual Meeting 2014*, San Diego, CA.

29. Verma, N., et al. (2014). Fabrication of iron titanium oxide thin film and its application as opto-electronic humidity and liquefied petroleum gas sensors, *Opt. Laser Technol.*, **57**(Suppl C), pp. 181–188.

30. Yusoff, N., et al. (2016). A facile preparation of titanium dioxide-iron oxide@silicon dioxide incorporated reduced graphene oxide nanohybrid for electrooxidation of methanol in alkaline medium, *Electrochim. Acta*, **192**(Suppl C), pp. 167–176.

31. Pardavi-Horvath, M. and Takacs, L. (1992). Iron-alumina nanocomposites prepared by ball milling, *IEEE Trans. Magn.*, **28**(5), pp. 3186–3188.

32. Karakassides, M. A., et al. (2003). Magnetic Fe_2O_3–Al_2O_3 composites prepared by a modified wet impregnation method, *J. Mater. Chem.*, **13**(4), pp. 871–876.

33. Mahapatra, A., Mishra, B. G. and Hota, G. (2013). Electrospun Fe_2O_3–Al_2O_3 nanocomposite fibers as efficient adsorbent for removal of heavy metal ions from aqueous solution, *J. Hazard. Mater.*, **258–259**(Suppl C), pp. 116–123.

34. Dresvyannikov, A., et al. (2016). Electrochemical synthesis of nanosized iron oxide-alumina system, *J. Adv. Ceram.*, **5**(1), p. 70.

35. Tang, H., et al. (2006). A selective NH3 gas sensor based on Fe_2O_3–ZnO nanocomposites at room temperature, *Sens. Actuators, B*, **114**(2), pp. 910–915.

36. Xia, J., et al. (2011). Preparation and characterization of bifunctional, Fe_3O_4/ZnO nanocomposites and their use as photocatalysts, *Appl. Surf. Sci.*, **257**(23), pp. 9724–9732.

37. Wu, W., et al. (2012). Controllable synthesis, magnetic properties, and enhanced photocatalytic activity of spindlelike mesoporous α-Fe_2O_3/

ZnO core–shell heterostructures, *ACS Appl. Mater. Interfaces*, **4**(7), pp. 3602–3609.

38. Hasanpour, A., et al. (2013). Synthesis and characterization of Fe_3O_4 and ZnO nanocomposites by the sol–gelmethod, *J. Magn. Magn. Mater.*, **334**(Suppl C), pp. 41–44.
39. Singh, S., Barick, K. and Bahadur, D. (2013). Fe_3O_4 embedded ZnO nanocomposites for the removal of toxic metal ions, organic dyes and bacterial pathogens, *J. Mater. Chem. A*, **1**(10), pp. 3325–3333.
40. Peng, H., et al. (2016). Fe_3O_4@mZnO nanoparticles as magnetic and microwave responsive drug carriers, *Microporous Mesoporous Mater.*, **226**(Suppl C), pp. 140–145.
41. Ma, Z. Y., et al. (2009). Synthesis and bio-functionalization of multifunctional magnetic Fe_3O_4@Y_2O_3: Eu nanocomposites, *J. Mater. Chem.*, **19**(27), pp. 4695–4700.
42. Xia, L., et al. (2017). Luminescent and magnetic α-Fe_2O_3@Y_2O_3: Eu^{3+} bifunctional hollow microspheres for drug delivery, *J. Phys. Chem. C*, **121**(37), pp. 20279–20286.
43. Wu, T., et al. (2016). Enhanced photoluminescence of Fe_3O_4@Y_2O_3: Eu^{3+} bifunctional nanoparticles by the Gd3+ co-doping, *J. Alloys Compd.*, **666**(Suppl C), pp. 507–512.
44. Zhang, H., et al. (2017). Enhancing the magnetic–luminescent properties of Fe_3O_4@ Y_2O_3: Eu^{3+} core–shell bifunctional particles using Li+ doping, *Aust. J. Chem.*, **70**(8), pp. 896–901.
45. Saraji, M., Yousefi, S. and Talebi, M. (2017). Plasmid DNA purification by zirconia magnetic nanocomposite, *Anal. Biochem.*, **539**(Suppl C), pp. 33–38.
46. Noormohamadi, A., et al. (2017). Synergistic effect of concurrent presence of zirconium oxide and iron oxide in the form of core-shell nanoparticles on the performance of Fe_3O_4@ZrO_2/PAN nanocomposite membrane, *Ceram. Int.*, **43**(18), pp. 17174–17185.
47. Fang, L., Wu, B. and Lo, I. M. C. (2017). Fabrication of silica-free superparamagnetic ZrO_2@Fe_3O_4 with enhanced phosphate recovery from sewage: performance and adsorption mechanism, *Chem. Eng. J.*, **319**(Suppl C), pp. 258–267.
48. Wu, B., et al. (2017). Highly efficient and selective phosphate removal from wastewater by magnetically recoverable $La(OH)_3$/ Fe_3O_4 nanocomposites, *Water Res.*, **126**(Suppl C), pp. 179–188.
49. Mi, C., et al. (2010). Multifunctional nanocomposites of superparamagnetic (Fe_3O_4) and NIR-responsive rare earth-doped

up-conversion fluorescent (NaYF4:Yb, Er) nanoparticles and their applications in biolabeling and fluorescent imaging of cancer cells, *Nanoscale*, **2**(7), pp. 1141–1148.

50. Allwes, M., Mekaoui, M. and Sorescu, M. (2017). Mutual solubility of scandium oxide-hematite magnetic nanocomposites: Mössbauer spectroscopy investigation, *Solid State Commun.*, **261**(Suppl C), pp. 46–49.
51. Hirsch, A. (2010). The era of carbon allotropes, *Nat. Mater.*, **9**(11), pp. 868–871.
52. Shim, B. S., et al. (2008). Smart electronic yarns and wearable fabrics for human biomonitoring made by carbon nanotube coating with polyelectrolytes, *Nano Lett.*, **8**(12), pp. 4151–4157.
53. Deng, C.-H., et al. (2014). Inactivation performance and mechanism of Escherichia coli in aqueous system exposed to iron oxide loaded graphene nanocomposites, *J. Hazard. Mater.*, **276**(Suppl C), pp. 66–76.
54. Vermisoglou, E., et al. (2014). Reduced graphene oxide/iron carbide nanocomposites for magnetic and supercapacitor applications, *J. Alloys Compd.*, **590**, pp. 102–109.
55. Su, H., Ye, Z. and Hmidi, N. (2017). High-performance iron oxide–graphene oxide nanocomposite adsorbents for arsenic removal, *Colloids Surf., A*, **522**(Suppl C), pp. 161–172.
56. dos Reis, L. C., Vidal, L. and Canals, A. (2017). Graphene oxide/ Fe_3O_4 as sorbent for magnetic solid-phase extraction coupled with liquid chromatography to determine 2, 4, 6-trinitrotoluene in water samples, *Anal. Bioanal. Chem.*, **409**(10), pp. 2665–2674.
57. Garg, B., Bisht, T. and Thomas, K. J. (2017). Magnetic graphene nanocomposites for multifunctional applications, in *Complex Magnetic Nanostructures* (Springer), pp. 317–357.
58. Sharma, V. K., et al. (2015). Magnetic graphene–carbon nanotube iron nanocomposites as adsorbents and antibacterial agents for water purification, *Adv. Colloid Interface Sci.*, **225**(Suppl C), pp. 229–240.
59. Liu, R., et al. (2016). A facile controlled in-situ synthesis of monodisperse magnetic carbon nanotubes nanocomposites using water-ethylene glycol mixed solvents, *J. Alloys Compd.*, **657**(Suppl C), pp. 138–143.
60. Yu, F., et al. (2016). Magnetic iron oxide nanoparticles functionalized multi-walled carbon nanotubes for toluene, ethylbenzene and xylene removal from aqueous solution, *Chemosphere*, **146**(Suppl C), pp. 162–172.

61. Jia, X., et al. (2017). Synthesis of lightweight and flexible composite aerogel of mesoporous iron oxide threaded by carbon nanotubes for microwave absorption, *J. Alloys Compd.*, **697**(Suppl C), pp. 138–146.

62. Lu, W., et al. (2017). One-pot synthesis of magnetic iron oxide nanoparticle-multiwalled carbon nanotube composites for enhanced removal of Cr(VI) from aqueous solution, *J. Colloid Interface Sci.*, **505**(Suppl C), pp. 1134–1146.

63. Choi, J. H., et al. (2007). Multimodal biomedical imaging with asymmetric single-walled carbon nanotube/iron oxide nanoparticle complexes, *Nano Lett.*, **7**(4), pp. 861–867.

64. Chen, M.-L., et al. (2012). Quantum dots conjugated with Fe_3O_4-filled carbon nanotubes for cancer-targeted imaging and magnetically guided drug delivery, *Langmuir*, **28**(47), pp. 16469–16476.

65. Shi, J., et al. (2013). PEGylated fullerene/iron oxide nanocomposites for photodynamic therapy, targeted drug delivery and MR imaging, *Biomaterials*, **34**(37), pp. 9666–9677.

66. Cano, M., et al. (2016). Synthesis and characterization of multifunctional superparamagnetic iron oxide nanoparticles (SPION)/C60 nanocomposites assembled by fullerene–amine click chemistry, *RSC Adv.*, **6**(74), pp. 70374–70382.

67. Gogoi, A., et al. (2017). Novel highly stable β-cyclodextrin fullerene mixed valent Fe-metal framework for quick Fenton degradation of alizarin, *RSC Adv.*, **7**(64), pp. 40371–40382.

68. Viswanathan, T., et al. (2011). Evaluation of a renewable resource-based carbon-iron oxide nanocomposite for removal of arsenic from contaminated water, *J. Macromol. Sci., Part A*, **48**(5), pp. 348–354.

69. Yılmaz, E. and Soylak, M. (2016). Preparation and characterization of magnetic carboxylated nanodiamonds for vortex-assisted magnetic solid-phase extraction of ziram in food and water samples, *Talanta*, **158**(Suppl C), pp. 152–158.

70. Novakova, A. A., et al. (2003). Magnetic properties of polymer nanocomposites containing iron oxide nanoparticles, *J. Magn. Magn. Mater.*, **258–259**(Suppl C), pp. 354–357.

71. Oh, J. K. and Park, J. M. (2011). Iron oxide-based superparamagnetic polymeric nanomaterials: design, preparation, and biomedical application, *Prog. Polym. Sci.*, **36**(1), pp. 168–189.

72. Feld, A., et al. (2017). Nanocomposites of highly monodisperse encapsulated superparamagnetic iron oxide nanocrystals

homogeneously dispersed in a poly (ethylene oxide) melt, *ACS Nano*, **11**(4), pp. 3767–3775.

73. Muñoz-Bonilla, A., Sánchez-Marcos, J. and Herrasti, P. (2017). Magnetic nanoparticles-based conducting polymer nanocomposites, in *Conducting Polymer Hybrids* (Springer), pp. 45–80.

74. Chi, Q., et al. (2017). Enhanced thermal conductivity and dielectric properties of iron oxide/polyethylene nanocomposites induced by a magnetic field, *Sci. Rep.*, **7**, p. 3072.

75. Xu, H., et al. (2011). Polymer encapsulated upconversion nanoparticle/iron oxide nanocomposites for multimodal imaging and magnetic targeted drug delivery, *Biomaterials*, **32**(35), pp. 9364–9373.

76. Li, G.-y., et al. (2008). Preparation and properties of magnetic Fe_3O_4–chitosan nanoparticles, *J. Alloys Compd.*, **466**(1), pp. 451–456.

77. Liu, X., et al. (2009). Magnetic chitosan nanocomposites: a useful recyclable tool for heavy metal ion removal, *Langmuir*, **25**(1), pp. 3–8.

78. Kaushik, A., et al. (2008). Iron oxide nanoparticles–chitosan composite based glucose biosensor, *Biosens. Bioelectron.*, **24**(4), pp. 676–683.

79. Singh, J., et al. (2011). Preparation and properties of hybrid monodispersed magnetic α-Fe_2O_3 based chitosan nanocomposite film for industrial and biomedical applications, *Int. J. Biol. Macromol.*, **48**(1), pp. 170–176.

80. Jeyapragasam, T. and Saraswathi, R. (2014). Electrochemical biosensing of carbofuran based on acetylcholinesterase immobilized onto iron oxide–chitosan nanocomposite, *Sens. Actuators, B*, **191**, pp. 681–687.

81. Kaushik, A., et al. (2009). Iron oxide-chitosan nanobiocomposite for urea sensor, *Sens. Actuators, B*, **138**(2), pp. 572–580.

82. Badry, M.D., et al. (2017). Synthesis, characterization, and in vitro anticancer evaluation of iron oxide/chitosan nanocomposites, *Inorg. Nano-Met. Chem.*, **47**(3), pp. 405–411.

83. Kurt, B. Z., Uckaya, F. and Durmus, Z. (2017). Chitosan and carboxymethyl cellulose based magnetic nanocomposites for application of peroxidase purification, *Int. J. Biol. Macromol.*, **96**, pp. 149–160.

84. Haham, H., et al. (2016). Engineering of superparamagnetic core–shell iron oxide/N-chloramine nanoparticles for water purification, *ACS Appl. Mater. Interfaces*, **8**(28), pp. 18488–18495.

85. Taufik, A. and Saleh, R. (2017). Synthesis of iron (II, III) oxide/zinc oxide/copper (II) oxide (Fe_3O_4/ZnO/CuO) nanocomposites and

their photosonocatalytic property for organic dye removal, *J. Colloid Interface Sci.*, **491**, pp. 27–36.

86. Luo, C., et al. (2013). Manganese dioxide/iron oxide/acid oxidized multi-walled carbon nanotube magnetic nanocomposite for enhanced hexavalent chromium removal, *Chem. Eng. J.*, **234**, pp. 256–265.
87. Bhaumik, M., et al. (2011). Enhanced removal of Cr (VI) from aqueous solution using polypyrrole/Fe_3O_4 magnetic nanocomposite, *J. Hazard. Mater.*, **190**(1), pp. 381–390.
88. Suri, K., et al. (2002). Gas and humidity sensors based on iron oxide–polypyrrole nanocomposites, *Sens. Actuators, B*, **81**(2), pp. 277–282.
89. Ren, Y.-L., et al. (2012). Quaternary nanocomposites consisting of graphene, Fe_3O_4@ Fe core@ shell, and ZnO nanoparticles: synthesis and excellent electromagnetic absorption properties, *ACS Appl. Mater. Interfaces*, **4**(12), pp. 6436–6442.
90. Magdalane, C. M., et al. (2017). Facile synthesis of heterostructured cerium oxide/yttrium oxide nanocomposite in UV light induced photocatalytic degradation and catalytic reduction: synergistic effect of antimicrobial studies, *J. Photochem. Photobiol., B*, **173**, pp. 23–34.
91. Long, F., et al. (2011). Removal of phosphate from aqueous solution by magnetic Fe–Zr binary oxide, *Chem. Eng. J.*, **171**(2), pp. 448–455.
92. Soltani, R. D. C., et al. (2017). Decontamination of arsenic (V)-contained liquid phase utilizing Fe_3O_4/bone char nanocomposite encapsulated in chitosan biopolymer, *Environ. Sci. Pollut. Res.*, pp. 1–10.
93. Kondo, H., et al. (2005). Mechanical and magnetic properties of novel yttria-stabilized tetragonal zirconia/Ni nanocomposite prepared by the modified internal reduction method, *J. Am. Ceram. Soc.*, **88**(6), pp. 1468–1473.
94. Douglas, T., et al. (1995). Synthesis and structure of an iron (III) sulfide-ferritin bioinorganic nanocomposite, *Science*, **269**(5220), p. 54.
95. Vadahanambi, S., et al. (2013). Arsenic removal from contaminated water using three-dimensional graphene-carbon nanotube-iron oxide nanostructures, *Environ. Sci. Technol.*, **47**(18), pp. 10510–10517.
96. Chi, Q., et al. (2017). Enhanced thermal conductivity and dielectric properties of iron oxide/polyethylene nanocomposites induced by a magnetic field, *Sci. Rep.*, **7**, p. 3072.
97. Takacs, L. (1993). Metal-metal oxide systems for nanocomposite formation by reaction milling, *Nanostruct. Mater.*, **2**(3), pp. 241–249.
98. Richter, K., Birkner, A. and Mudring, A. V. (2010). Stabilizer-free metal nanoparticles and metal–metal oxide nanocomposites with long-term

stability prepared by physical vapor deposition into ionic liquids, *Angew. Chem. Int. Ed.*, **49**(13), pp. 2431–2435.

99. Wang, S., Jiang, S. P. and Wang, X. (2011). Microwave-assisted one-pot synthesis of metal/metal oxide nanoparticles on graphene and their electrochemical applications, *Electrochim. Acta*, **56**(9), pp. 3338–3344.
100. Kochuveedu, S. T., Jang, Y. H. and Kim, D. H. (2013). A study on the mechanism for the interaction of light with noble metal-metal oxide semiconductor nanostructures for various photophysical applications, *Chem. Soc. Rev.*, **42**(21), pp. 8467–8493.
101. Dawson, A. and Kamat, P. V. (2001). Semiconductor–metal nanocomposites. Photoinduced fusion and photocatalysis of gold-capped TiO_2 (TiO_2/gold) nanoparticles, *J. Phys. Chem. B*, **105**(5), pp. 960–966.
102. Xu, B.-Q., et al. (2003). Size limit of support particles in an oxide-supported metal catalyst: nanocomposite Ni/ZrO2 for utilization of natural gas, *J. Phys. Chem. B*, **107**(22), pp. 5203–5207.
103. Zhang, N., Liu, S. and Xu, Y.-J. (2012). Recent progress on metal core@ semiconductor shell nanocomposites as a promising type of photocatalyst, *Nanoscale*, **4**(7), pp. 2227–2238.
104. Wang, Y., Lee, J. Y. and Chen, B.-H. (2003). Microemulsion synthesis of tin oxide-graphite nanocomposites as negative electrode materials for lithium-ion batteries, *Electrochem. Solid-State Lett.*, **6**(1), pp. A19–A22.
105. Wang, Y. and Lee, J. Y. (2005). Microwave-assisted synthesis of SnO2–graphite nanocomposites for Li-ion battery applications, *J. Power Sources*, **144**(1), pp. 220–225.
106. Chang, C.-C., et al. (2007). Nano-tin oxide/tin particles on a graphite surface as an anode material for lithium-ion batteries, *J. Phys. Chem. C*, **111**(44), pp. 16423–16427.
107. Wang, D., et al. (2010). Ternary self-assembly of ordered metal oxide–graphene nanocomposites for electrochemical energy storage, *ACS Nano*, **4**(3), pp. 1587–1595.
108. Baek, S., et al. (2011). A one-pot microwave-assisted non-aqueous sol–gel approach to metal oxide/graphene nanocomposites for Li-ion batteries, *RSC Adv.*, **1**(9), pp. 1687–1690.
109. Su, Y., et al. (2012). Two-dimensional carbon-coated graphene/metal oxide hybrids for enhanced lithium storage, *ACS Nano*, **6**(9), pp. 8349–8356.

110. Yang, S., et al. (2010). Fabrication of cobalt and cobalt oxide/graphene composites: towards high-performance anode materials for lithium ion batteries, *ChemSusChem*, **3**(2), pp. 236–239.

111. Wang, D., et al. (2009). Self-assembled TiO_2–graphene hybrid nanostructures for enhanced Li-ion insertion, *ACS Nano*, **3**(4), pp. 907–914.

112. Liang, Y., et al. (2010). *TiO_2 nanocrystals grown on graphene as advanced photocatalytic hybrid materials, Nano Res.*, **3**(10), pp. 701–705.

113. Sher Shah, M. S. A., et al. (2012). Green synthesis of biphasic TiO_2–reduced graphene oxide nanocomposites with highly enhanced photocatalytic activity, *ACS Appl. Mater. Interfaces*, **4**(8), pp. 3893–3901.

114. Akhavan, O., Choobtashani, M. and Ghaderi, E. (2012). Protein degradation and RNA efflux of viruses photocatalyzed by graphene–tungsten oxide composite under visible light irradiation, *J. Phys. Chem. C*, **116**(17), pp. 9653–9659.

115. Upadhyay, R. K., Soin, N. and Roy, S. S. (2014). Role of graphene/metal oxide composites as photocatalysts, adsorbents and disinfectants in water treatment: a review, *RSC Adv.*, **4**(8), pp. 3823–3851.

116. Lee, J. Y., et al. (2005). Nickel oxide/carbon nanotubes nanocomposite for electrochemical capacitance, *Synth. Met.*, **150**(2), pp. 153–157.

117. Ye, J. S., et al. (2005). Preparation and characterization of aligned carbon nanotube–ruthenium oxide nanocomposites for supercapacitors, *Small*, **1**(5), pp. 560–565.

118. An, G., et al. (2007). SnO2/carbon nanotube nanocomposites synthesized in supercritical fluids: highly efficient materials for use as a chemical sensor and as the anode of a lithium-ion battery, *Nanotechnology*, **18**(43), p. 435707.

119. Ma, S.-B., et al. (2007). Synthesis and characterization of manganese dioxide spontaneously coated on carbon nanotubes, *Carbon*, **45**(2), pp. 375–382.

120. Ma, S.-B., et al. (2008). Electrochemical properties of manganese oxide coated onto carbon nanotubes for energy-storage applications, *J. Power Sources*, **178**(1), pp. 483–489.

121. Choi, S. H., Lee, J.-H. and Kang, Y. C. (2015). Perforated metal oxide–carbon nanotube composite microspheres with enhanced lithium-ion storage properties, *ACS Nano*, **9**(10), pp. 10173–10185.

122. Mallakpour, S. and Khadem, E. (2016). Carbon nanotube–metal oxide nanocomposites: fabrication, properties and applications, *Chem. Eng. J.*, **302**, pp. 344–367.

123. Liew, K. B., et al. (2015). Manganese oxide/functionalised carbon nanotubes nanocomposite as catalyst for oxygen reduction reaction in microbial fuel cell, *Int. J. Hydrogen Energy*, **40**(35), pp. 11625–11632.

124. Wang, W., et al. (2014). Hydrous ruthenium oxide nanoparticles anchored to graphene and carbon nanotube hybrid foam for supercapacitors, *Sci. Rep.*, **4**, p. 4452.

125. Liu, D., Yang, S. and Lee, S.-T. (2008). Preparation of novel cuprous oxide–fullerene [60] core–shell nanowires and nanoparticles via a copper (I)-assisted fullerene-polymerization reaction, *J. Phys. Chem. C*, **112**(18), pp. 7110–7118.

126. Motoyoshi, R., et al. (2010). Fabrication and characterization of cuprous oxide: fullerene solar cells, *Synth. Met.*, **160**(11), pp. 1219–1222.

127. Wu, C.-G., et al. (2006). PVdF-HFP/metal oxide nanocomposites: the matrices for high-conducting, low-leakage porous polymer electrolytes, *J. Power Sources*, **159**(1), pp. 295–300.

128. Vollath, D., Szabo, D. and Schlabach, S. (2004). Oxide/polymer nanocomposites as new luminescent materials, *J. Nanopart. Res.*, **6**(2), pp. 181–191.

129. Croce, F., et al. (1998). Nanocomposite polymer electrolytes for lithium batteries, *Nature*, **394**(6692), pp. 456–458.

130. Sarkar, S., et al. (2012). Polymer-supported metals and metal oxide nanoparticles: synthesis, characterization, and applications, *J. Nanopart. Res.*, **14**(2), p. 715.

131. Bouclé, J., Ravirajan, P. and Nelson, J. (2007). Hybrid polymer–metal oxide thin films for photovoltaic applications, *J. Mater. Chem.*, **17**(30), pp. 3141–3153.

132. Kong, H., Song, J. and Jang, J. (2010). Photocatalytic antibacterial capabilities of TiO_2– biocidal polymer nanocomposites synthesized by a surface-initiated photopolymerization, *Environ. Sci. Technol.*, **44**(14), pp. 5672–5676.

133. Cai, W., et al. (2013). Synthesis of bulk-size transparent gadolinium oxide–polymer nanocomposites for gamma ray spectroscopy, *J. Mater. Chem. C*, **1**(10), pp. 1970–1976.

134. Bhagavathula, S. (2017). Metal-oxide polymer nanocomposite films from disposable scrap tire powder/poly-Îµ-caprolactone for advanced

electrical energy (capacitor) applications, *J. Cleaner Prod.*, **161**, pp. 888–895.

135. Allouche, J., et al. (2006). Biomimetic core–shell gelatine/silica nanoparticles: a new example of biopolymer-based nanocomposites, *J. Mater. Chem.*, **16**(30), pp. 3120–3125.

136. Shankar, S., et al. (2015). Preparation, characterization, and antimicrobial activity of gelatin/ZnO nanocomposite films, *Food Hydrocolloids*, **45**(Suppl C), pp. 264–271.

137. Kim, P., et al. (2007). Phosphonic acid-modified barium titanate polymer nanocomposites with high permittivity and dielectric strength, *Adv. Mater.*, **19**(7), pp. 1001–1005.

138. Cai, J., et al. (2012). Cellulose–silica nanocomposite aerogels by in situ formation of silica in cellulose gel, *Angew. Chem. Int. Ed.*, **124**(9), pp. 2118–2121.

139. Blantocas, G.Q., Alaboodi, A. S. and Abdel-baset, H. M. (2017). Synthesis of chitosan–TiO_2 antimicrobial composites via a 2-step process of electrospinning and plasma sputtering, *Arabian J. Sci. Eng.*, pp. 1–10.

140. Su, X., et al. (2009). Solution synthesis of Cu_2O/ TiO_2 core-shell nanocomposites, *Colloids Surf., A*, **349**(1), pp. 151–155.

141. Gan, C., et al. (2015). Effect of porous zinc–biochar nanocomposites on Cr (vi) adsorption from aqueous solution, *RSC Adv.*, **5**(44), pp. 35107–35115.

142. Thirumalraj, B., et al. (2016). Preparation of highly stable fullerene C60 decorated graphene oxide nanocomposite and its sensitive electrochemical detection of dopamine in rat brain and pharmaceutical samples, *J. Colloid Interface Sci.*, **462**(Suppl C), pp. 375–381.

143. Zhang, Z. and Yang, R. (2017). Novel nanocomposites based on hydroxyethyl cellulose and graphene oxide, *Fibers Polym.*, **18**(2), pp. 334–341.

144. Vahabi, H., et al. (2017). Novel nanocomposites based on poly(ethylene-co-vinyl acetate) for coating applications: the complementary actions of hydroxyapatite, MWCNTs and ammonium polyphosphate on flame retardancy, *Prog. Org. Coat.*, **113**(Suppl C), pp. 207–217.

145. Ozkazanc, H. and Ozkazanc, E. (2017). Novel nanocomposites of polypyrrole doped with fullerene C60, *J. Macromol. Sci., Part B*, **56**(2), pp. 83–96.

146. Moustafa, H., et al. (2017). Facile modification of organoclay and its effect on the compatibility and properties of novel biodegradable PBE/PBAT nanocomposites, *Eur. Polym. J.*, **87**(Suppl C), pp. 188–199.

147. Khalid, A., et al. (2017). Bacterial cellulose-zinc oxide nanocomposites as a novel dressing system for burn wounds, *Carbohydr. Polym.*, **164**, pp. 214–221.

148. Thirumalraj, B., et al. (2016). Preparation of highly stable fullerene C60 decorated graphene oxide nanocomposite and its sensitive electrochemical detection of dopamine in rat brain and pharmaceutical samples, *J. Colloid Interface Sci.*, **462**, pp. 375–381.

149. Yi, D. K., et al. (2006). Nanoparticle architectures templated by SiO_2/ Fe_2O_3 nanocomposites, *Chem. Mater.*, **18**(3), pp. 614–619.

150. Anirudhan, T. and Sandeep, S. (2012). Synthesis, characterization, cellular uptake and cytotoxicity of a multi-functional magnetic nanocomposite for the targeted delivery and controlled release of doxorubicin to cancer cells, *J. Mater. Chem.*, **22**(25), pp. 12888–12899.

151. Wang, F., et al. (2013). Dual surface-functionalized Janus nanocomposites of polystyrene/ Fe_3O_4@ SiO_2 for simultaneous tumor cell targeting and stimulus-induced drug release, *Adv. Mater.*, **25**(25), pp. 3485–3489.

152. Zhu, L., et al. (2013). Multifunctional pH-sensitive superparamagnetic iron-oxide nanocomposites for targeted drug delivery and MR imaging, *J. Controlled Release*, **169**(3), pp. 228–238.

153. Chen, W., et al. (2011). Composites of aminodextran-coated Fe_3O_4 nanoparticles and graphene oxide for cellular magnetic resonance imaging, *ACS Appl. Mater. Interfaces*, **3**(10), pp. 4085–4091.

154. Wen, H., et al. (2012). Engineered redox-responsive PEG detachment mechanism in PEGylated nano-graphene oxide for intracellular drug delivery, *Small*, **8**(5), pp. 760–769.

155. Peng, E., et al. (2012). Synthesis of manganese ferrite/graphene oxide nanocomposites for biomedical applications, *Small*, **8**(23), pp. 3620–3630.

156. Kim, H. and Kim, W. J. (2014). Photothermally controlled gene delivery by reduced graphene oxide–polyethylenimine nanocomposite, *Small*, **10**(1), pp. 117–126.

157. Chang, Y.-N., et al. (2015). Synthesis of magnetic graphene oxide–TiO 2 and their antibacterial properties under solar irradiation, *Appl. Surf. Sci.*, **343**, pp. 1–10.

158. Meenach, S. A., et al. (2009). Biocompatibility analysis of magnetic hydrogel nanocomposites based on poly (N-isopropylacrylamide) and iron oxide, *J. Biomed. Mater. Res. Part A*, **91**(3), pp. 903–909.

159. Lu, J., et al. (2009). Manganese ferrite nanoparticle micellar nanocomposites as MRI contrast agent for liver imaging, *Biomaterials*, **30**(15), pp. 2919–2928.

160. Azizi, S., et al. (2016). ZnO-Ag core shell nanocomposite formed by green method using essential oil of wild ginger and their bactericidal and cytotoxic effects, *Appl. Surf. Sci.*, **384**(Suppl C), pp. 517–524.

161. Chaturvedi, A., et al. (2016). Evaluation of poly (vinyl alcohol) based cryogel–zinc oxide nanocomposites for possible applications as wound dressing materials, *Mater. Sci. Eng., C*, **65**(Suppl C), pp. 408–418.

162. Krishnan, B. and Mahalingam, S. (2017). Ag/ TiO_2/bentonite nanocomposite for biological applications: synthesis, characterization, antibacterial and cytotoxic investigations, *Adv. Powder Technol.*, **28**(9), pp. 2265–2280.

163. Ong, T. H. D., et al. (2017). Insight into cytotoxicity of Mg nanocomposites using MTT assay technique, *Mater. Sci. Eng., C*, **78**(Suppl C), pp. 647–652.

164. Syam Sundar, L., et al. (2017). Biocompatibility and biotoxicity of in-situ synthesized carboxylated nanodiamond-cobalt oxide nanocomposite, *J. Mater. Sci. Technol.*, **33**(8), pp. 879–888.

165. Bhowmick, A., et al. (2017). Development of bone-like zirconium oxide nanoceramic modified chitosan based porous nanocomposites for biomedical application, *Int. J. Biol. Macromol.*, **95**(Suppl C), pp. 348–356.

166. Rasoulzadeh, M. and Namazi, H. (2017). Carboxymethyl cellulose/ graphene oxide bio-nanocomposite hydrogel beads as anticancer drug carrier agent, *Carbohydr. Polym.*, **168**(Suppl C), pp. 320–326.

167. Moghayedi, M., et al. (2017). Kinetics and mechanism of antibacterial activity and cytotoxicity of Ag-RGO nanocomposite, *Colloids Surf., B*, **159**(Suppl C), pp. 366–374.

168. Su, X., et al. (2018). Interactions between rGO/TNT nanocomposites and cells: regulation of cell morphology, uptake, cytotoxicity, adhesion and migration, *J. Mech. Behav. Biomed. Mater.*, **77**(Suppl C), pp. 510–518.

169. Lu, J., et al. (2009). Manganese ferrite nanoparticle micellar nanocomposites as MRI contrast agent for liver imaging, *Biomaterials*, **30**(15), pp. 2919–2928.

170. Azizi, S., et al. (2016). ZnO-Ag core shell nanocomposite formed by green method using essential oil of wild ginger and their bactericidal and cytotoxic effects, *Appl. Surf. Sci.*, **384**, pp. 517–524.

171. Chaturvedi, A., et al. (2016). Evaluation of poly (vinyl alcohol) based cryogel–zinc oxide nanocomposites for possible applications as wound dressing materials, *Mater. Sci. Eng., C*, **65**, pp. 408–418.

172. Ong, T. H. D., et al. (2017). Insight into cytotoxicity of Mg nanocomposites using MTT assay technique, *Mater. Sci. Eng., C*, **78**, pp. 647–652.

173. Bhowmick, A., et al. (2017). Development of bone-like zirconium oxide nanoceramic modified chitosan based porous nanocomposites for biomedical application, *Int. J. Biol. Macromol.*, **95**, pp. 348–356.

174. Rasoulzadeh, M. and Namazi, H. (2017). Carboxymethyl cellulose/graphene oxide bio-nanocomposite hydrogel beads as anticancer drug carrier agent, *Carbohydr. Polym.*, **168**, pp. 320–326.

175. Moghayedi, M., et al. (2017). Kinetics and mechanism of antibacterial activity and cytotoxicity of Ag-RGO nanocomposite, *Colloids Surf., B*, **159**, pp. 366–374.

176. Su, X., et al. (2018). Interactions between rGO/TNT nanocomposites and cells: regulation of cell morphology, uptake, cytotoxicity, adhesion and migration, *J. Mech. Behav. Biomed. Mater.*, **77**, pp. 510–518.

177. Lv, G., et al. (2008). Novel nanocomposite of nano Fe_3O_4 and polylactide nanofibers for application in drug uptake and induction of cell death of leukemia cancer cells, *Langmuir*, **24**(5), pp. 2151–2156.

178. Wang, C., et al. (2013). Iron oxide@ polypyrrole nanoparticles as a multifunctional drug carrier for remotely controlled cancer therapy with synergistic antitumor effect, *ACS Nano*, **7**(8), pp. 6782–6795.

179. Gurunathan, S., et al. (2015). Reduced graphene oxide–silver nanoparticle nanocomposite: a potential anticancer nanotherapy, *Int. J. Nanomed.*, **10**, pp. 6257–6276.

180. Babji, P. and Sri, D. R. (2016). Synthesis, characterization and antitumor activity of Fe_2O_3-Cu2O-TiO_2 nanocomposite on MCF-7 human breast cancer cells, *J. Chem. Chem. Sci.*, **6**(7), pp. 616–623.

181. Wu, J., et al. (2017). Synthesis and in vitro evaluation of pH-sensitive magnetic nanocomposites as methotrexate delivery system for targeted cancer therapy, *Mater. Sci. Eng., C*, **71**(Suppl C), pp. 132–140.

182. Gupta, V. K., et al. (2017). Synthesis and characterization of MnO2/NiO nanocomposites for photocatalysis of tetracycline antibiotic and modification with guanidine for carriers of caffeic acid phenethyl ester-an anticancer drug, *J. Photochem. Photobiol., B*, **174**(Suppl C), pp. 235–242.

183. Wei, W., et al. (2017). Monodisperse and mesoporous walnut kernel-like SiO_2/γ-Fe_2O_3 nanocomposite: synthesis, magnetic properties, and

application in drug delivery, *J. Alloys Compd.*, **728**(Suppl C), pp. 585–591.

184. Mudaliar, S. and Edelman, S. V. (2001). Insulin therapy in type 2 diabetes, *Endocrinol. Metab. Clinics*, **30**(4), pp. 935–982.

185. Swinnen, S. G., Hoekstra, J. B. and DeVries, J. H. (2009). Insulin therapy for type 2 diabetes, *Diabetes Care*, **32**(Suppl 2), pp. S253–S259.

186. Gordijo, C. R., Shuhendler, A. J. and Wu, X. Y. (2010). Glucose-responsive bioinorganic nanohybrid membrane for self-regulated insulin release, *Adv. Funct. Mater.*, **20**(9), pp. 1404–1412.

187. Gordijo, C. R., et al. (2011). Nanotechnology-enabled closed loop insulin delivery device: in vitro and in vivo evaluation of glucose-regulated insulin release for diabetes control, *Adv. Funct. Mater.*, **21**(1), pp. 73–82.

188. Shen, J.-M., et al. (2012). Chitosan-based luminescent/magnetic hybrid nanogels for insulin delivery, cell imaging, and antidiabetic research of dietary supplements, *Int. J. Pharm.*, **427**(2), pp. 400–409.

189. Kebede, A., et al. (2013). Controlled synthesis, characterization, and application of iron oxide nanoparticles for oral delivery of insulin, *Lasers Med. Sci.*, **28**(2), pp. 579–587.

190. He, H., et al. (2013). Overcoming oral insulin delivery barriers: application of cell penetrating peptide and silica-based nanoporous composites, *Front. Chem. Sci. Eng.*, **7**(1), pp. 9–19.

191. Turcheniuk, K., et al. (2014). Insulin loaded iron magnetic nanoparticle–graphene oxide composites: synthesis, characterization and application for in vivo delivery of insulin, *RSC Adv.*, **4**(2), pp. 865–875.

192. Kamari, Y., Ghiaci, P. and Ghiaci, M. (2017). Study on montmorillonite/insulin/TiO_2 hybrid nanocomposite as a new oral drug-delivery system, *Mater. Sci. Eng., C*, **75**(Suppl C), pp. 822–828.

193. El-Mekawy, R. E. and Jassas, R. S. (2017). Recent trends in smart and flexible three-dimensional cross-linked polymers: synthesis of chitosan–ZnO nanocomposite hydrogels for insulin drug delivery, *MedChemComm*, **8**(5), pp. 897–906.

194. Aronson, D., Mittleman, M. A. and Burger, A. J. (2004). Elevated blood urea nitrogen level as a predictor of mortality in patients admitted for decompensated heart failure, *Am. J. Med.*, **116**(7), pp. 466–473.

195. Ronco, C., et al. (2000). Effects of different doses in continuous veno-venous haemofiltration on outcomes of acute renal failure: a prospective randomised trial, *Lancet*, **356**(9223), pp. 26–30.

196. Tiwari, A. (2009). A novel nanocomposite matrix based on silylated chitosan and multiwall carban nanotubes for the immobilization of urease, *J. Inorg. Organomet. Polym Mater.*, **19**(3), pp. 361–366.

197. Mahadeva, S. K. and Kim, J. (2011). Conductometric glucose biosensor made with cellulose and tin oxide hybrid nanocomposite, *Sens. Actuators, B*, **157**(1), pp. 177–182.

198. Maniruzzaman, M., Jang, S.-D. and Kim, J. (2012). Titanium dioxide–cellulose hybrid nanocomposite and its glucose biosensor application, *Mater. Sci. Eng., B*, **177**(11), pp. 844–848.

199. Arvand, M. and Hassannezhad, M. (2014). Magnetic core–shell Fe_3O_4@ SiO_2/MWCNT nanocomposite modified carbon paste electrode for amplified electrochemical sensing of uric acid, *Mater. Sci. Eng., C*, **36**(Suppl C), pp. 160–167.

200. Zhang, Z., et al. (2016). One-step fabrication of electrochemical biosensor based on DNA-modified three-dimensional reduced graphene oxide and chitosan nanocomposite for highly sensitive detection of Hg(II), *Sens. Actuators, B*, **225**(Suppl C), pp. 453–462.

201. El-Deen, A. G., et al. (2014). Graphene/SnO2 nanocomposite as an effective electrode material for saline water desalination using capacitive deionization, *Ceram. Int.*, **40**(9, Part B), pp. 14627–14634.

202. Fernandes, D. M., et al. (205). MnFe2O4@CNT-N as novel electrochemical nanosensor for determination of caffeine, acetaminophen and ascorbic acid, *Sens. Actuators, B*, **218**(Suppl C), pp. 128–136.

203. Kaleekkal, N. J., et al. (2015). Graphene oxide nanocomposite incorporated poly (ether imide) mixed matrix membranes for in vitro evaluation of its efficacy in blood purification applications, *Ind. Eng. Chem. Res.*, **54**(32), pp. 7899–7913.

204. Jiang, D., et al. (2016). A fluorescent imaging assay of cast in renal disease based on graphene quantum dots and Fe_3O_4 nanoparticles, *Clin. Chim. Acta*, **454**(Suppl C), pp. 94–101.

205. Di, D.-R., et al. (2012). A new nano-cryosurgical modality for tumor treatment using biodegradable MgO nanoparticles, *Nanomed. Nanotechnol. Biol. Med.*, **8**(8), pp. 1233–1241.

206. Martinez-Boubeta, C., et al. (2012). Adjustable hyperthermia response of self-assembled ferromagnetic Fe-MgO core–shell nanoparticles by tuning dipole–dipole interactions, *Adv. Funct. Mater.*, **22**(17), pp. 3737–3744.

207. Chalkidou, A., et al. (2011). In vitro application of Fe/MgO nanoparticles as magnetically mediated hyperthermia agents for cancer treatment, *J. Magn. Magn. Mater.*, **323**(6), pp. 775–780.

208. Martinez-Boubeta, C., et al. (2010). Self-assembled multifunctional Fe/MgO nanospheres for magnetic resonance imaging and hyperthermia, *Nanomed. Nanotechnol. Biol. Med.*, **6**(2), pp. 362–370.

209. Jeevanandam, J., et al. (2015). Opportunities for nano-formulations in type 2 diabetes mellitus treatments, *Current Pharm. Biotechnol.*, **16**(10), pp. 853–870.

210. Giordano, C., et al. (2011). Nanocomposites for neurodegenerative diseases: hydrogel-nanoparticle combinations for a challenging drug delivery, *Int. J. Artif. Organs*, **34**(12), p. 1115.

211. Albani, D., et al. (2013). Hydrogel-based nanocomposites and mesenchymal stem cells: a promising synergistic strategy for neurodegenerative disorders therapy, *Sci. World J.*, **2013**, p. 270260.

212. Ianzano, L., et al. (2005). Lafora progressive myoclonus epilepsy mutation database-EPM2A and NHLRC1 (EMP2B) genes, *Hum. Mutat.*, **26**(4), pp. 397–397.

213. Wrone, D. A. and Kwon, N. J. (2001). Andrews' diseases of the skin: clinical dermatology, *Arch. Dermatol.*, **137**(4), pp. 518–518.

214. Minassian, B. A. (2001). Lafora's disease: towards a clinical, pathologic, and molecular synthesis, *Pediatr. Neurol.*, **25**(1), pp. 21–29.

215. Pederson, B. A., et al. (2013). Inhibiting glycogen synthesis prevents Lafora disease in a mouse model, *Ann. Neurol.*, **74**(2), pp. 297–300.

216. Tagliabracci, V. S., et al. (2007). Laforin is a glycogen phosphatase, deficiency of which leads to elevated phosphorylation of glycogen in vivo, *Proc. Natl. Acad. Sci. U.S.A.*, **104**(49), pp. 19262–19266.

217. Nguyen, V.-H. V., et al. (2017). Clinical center for the lafora epilepsy cure initiative (LECI): clinical trial readiness for a global curative drug trial (P4. 245), *Neurology*, **88**(16 Suppl), P4. 245.

218. Wydner, K. L., et al. (1996). Chromosomal assignment of human nuclear envelope protein genes LMNA, LMNB1, and LBR by fluorescence in situ hybridization, *Genomics*, **32**(3), pp. 474–478.

219. De Sandre-Giovannoli, A., et al. (2003). Lamin a truncation in Hutchinson-Gilford progeria, *Science*, **300**(5628), pp. 2055–2055.

220. Sinha, J. K., Ghosh, S. and Raghunath, M. (2014). Progeria: a rare genetic premature ageing disorder, *Indian J. Med. Res.*, **139**(5), p. 667.

221. Ullrich, N. J., et al. (2013). Neurologic features of Hutchinson-Gilford progeria syndrome after lonafarnib treatment, *Neurology*, **81**(5), pp. 427–430.

222. Columbaro, M., et al. (2005). Rescue of heterochromatin organization in Hutchinson-Gilford progeria by drug treatment, *Cell. Mol. Life Sci.*, **62**(22), p. 2669.

Chapter 4

Polymer Matrix Nanocomposites: Recent Advancements and Applications

Amit Rastogi[a] and Kaushik Pal[b]
[a]*Sam Higginbottom University of Agriculture, Technology and Sciences, Post Agricultural Institute, Rewa Rd, Naini, Allahabad 211007, Uttar Pradesh, India*
[b]*Bharath Institute of Higher Education and Research, Bharath University, Department of Nanotechnology, Research Park, 173 Agharam Road, Selaiyur, Chennai 600073, Tamil Nadu, India*
amitrastogi1986@gmail.com; kaushikpaul.nano@bharathuniv.ac.in

4.1 Introduction

Nanocomposite materials have gained much attention and interest of scientists in recent years because of their improved properties compared to single-metal nanoparticles. A nanocomposite is a matrix in which nanoparticles have been incorporated to improve a particular property of the material. The properties of nanocomposites have caused researchers and industries to consider using this material in several fields [1, 2].

According to Azonano (2009) nanocomposites are "materials with a nanoscale structure that improve the macroscopic properties

Hybrid Nanocomposites: Fundamentals, Synthesis, and Applications
Edited by Kaushik Pal

ISBN 978-981-4800-34-1 (Hardcover), 978-0-429-00096-6 (eBook)
www.panstanford.com

of products." Nanocomposites are composite materials that have one of the phases with dimension in the nanometer range. These are high-performance materials that exhibit unusual property combinations and unique design possibilities and are thought of as the materials of the 21st century. With an estimated annual growth rate of about 25% and a huge demand for engineering polymers, their potential is so promising that they are useful in several applications ranging from packaging to biomedicine. Nanocomposites are possible alternatives to microcomposites and monolithic due to their outstanding properties. However, the preparation technique of nanocomposites presents challenges due to the control of elemental composition and stoichiometry in the nanophase. Moreover, the use of carbon nanotubes (CNTs) in the fabrication of nanocomposites added a new interesting feature to this area [3].

Kamigaito (1994) defined a nanocomposite as a multiphase solid material where one of the phases has one, two, or three dimensions of less than 100 nm, or structures having nanoscale repeat distances between the different phases that make up the material. The mechanical, electrical, thermal, optical, electrochemical, and catalytic properties of the nanocomposite will differ markedly from that of the component materials [4].

According to Cammarata [5] nanocomposites can be described as "multiphase materials in which one or more of the phases have at least one dimension of order 100 nm or less." He observed that as with conventional composites, the properties of nanocomposites can display synergistic improvements over those of the component phases individually.

However, Okpala [6] explained that nanocomposites' "rapidly expanding field is generating many exciting new materials with novel properties." He pointed out that the latter can be derived by combining properties from the parent constituents into a single material and that there is also the possibility of new properties that are unknown in the parent constituent materials.

Nanocomposites are materials in which nanosized filler components are incorporated in order to improve the properties of the resulting materials. Nanocomposites are made up of two or more distinct constituents or phases having different physical and chemical properties and are separated by a distinct interface. Their unique properties are not depicted by any of the constituents. The constituent that is generally present in greater quantity is called the matrix. The constituent that is embedded into the matrix material in order to improve the mechanical properties of nanocomposites is called reinforcement (or nanomaterials). Reinforcement is generally in the form of nanosized filler materials. Generally, nanocomposites show anisotropy (properties are directionally dependent) because of the distinct properties of constituents and inhomogeneous distribution of the reinforcement [7, 8].

The term "nanocomposite" is extensively used to describe an extremely broad range of materials, where one of the dimensions is in the nanorange. Nanocomposites are a very good alternative to conventional composite materials because of their outstanding properties and due to this are finding a wide range of applications in various fields. Nanocomposite technology with CNTs has been a topic of recent research and development. Nanocomposites are also found in nature, that is, in the structure of the abalone shell and bone. Nanocomposites differ from conventional composites because of a high surface-area-to-volume ratio of the reinforcing nanoparticles and their exceptionally high aspect ratio. The reinforcing material can be made up of particles, sheets, or fibers. The area of the interface between matrix and nanoreinforcement is typically an order of magnitude higher than the conventional composites. Clays are a group of nanofiller materials that have been extensively used for the preparation of polymer matrix nanocomposites (PMNCs). Polymer/clay nanocomposites have been receiving tremendous attention recently in academia and in industries due to their improved properties compared to conventional composites. Polymer/clay nanocomposites can be fabricated in a number of ways, including in situ polymerization, melt blending, solution mixing, and latex methods [9].

4.2 Characteristics of Nanocomposites

The following characteristics of nanocomposites are observed by many researchers [10, 11]:

- Strong change of fracture mode from intergranular fracture of monolithic alumina to transgranular fracture of nanocomposites
- Moderate to considerable improvement in strength

- Enhancement of fracture toughness
- Improvement of several mechanical properties

4.3 Why Nanocomposites?

Nanocomposites are generally more advantageous than conventional composites in many aspects. The advantages of nanocomposites are as follows [12]:

- The improvement in properties of the matrix material in nanocomposites can be achieved by the addition of a small amount of nanofiller materials compared to conventional composites that require a high concentration of microparticles in order to improve properties.
- Due to the addition of a small percentage of nanofiller materials, nanocomposites are much lighter in weight compared to conventional composites.
- Nanomaterials having size-dependent properties enhance thermal, chemical, mechanical, optical, magnetic, and electrical properties to a much greater extent than conventional composites.

Nanocomposites are an extremely good alternative to conventional composite materials due to their outstanding properties and are finding a wide range of applications in various fields.

4.4 Types of Nanocomposites

Nanocomposite materials can be classified as per their matrix materials into three different categories given in the Table 4.1 [3].

Table 4.1 Types of nanocomposites

Class	Examples
Metal	Fe-Cr/Al_2O_3, Ni/Al_2O_3, Co/Cr, Fe/MgO, Al/CNT, Mg/CNT
Ceramic	Al_2O_3/SiO_2, SiO_2/Ni, Al_2O_3/TiO_2, Al_2O_3/SiC, Al_2O_3/CNT
Polymer	Thermoplastic/thermoset polymer/layered silicates, polyester/TiO_2, polymer/CNT, polymer/layered double hydroxides

4.5 Polymer Matrix Nanocomposites

Many methods have been described for the preparation of polymer nanocomposites, including layered materials and those containing CNTs. The most important ones are (i) intercalation of the polymer or prepolymer from solution, (ii) in situ intercalative polymerization, (iii) melt intercalation, (iv) direct mixture of polymer and particulates, (v) template synthesis, vi) in situ polymerization, and (vii) sol–gel process. Melt blending is a relatively new process of nanocomposite fabrication. This process involves melting a polymer to form a viscous liquid. The nanoparticles are dispersed into the polymer matrix by means of a high shear rate along with diffusion at high temperature. The nanocomposites are then fabricated by either compression molding or injection molding [13–17].

The preparation of CNT-reinforced polymer nanocomposites is generally performed by different methods, including direct mixing, solution mixing, melt mixing, and in situ polymerization. These are applicable to various systems. Similarly, different processing techniques, mostly chemical and electrochemical methods, have been employed for the preparation of conducting polymer nanocomposites [13, 18–19].

The in situ method can be used with mineral/vegetal fibers, with the possibility to attach the polymer to the grafted surface through linking groups (coupling agents), which optimize the interface bonding and, consequently, the mechanical properties [20].

Despite the successful use of these different methods for the preparation of polymer-based nanocomposites, information on various factors is still lacking, such as (i) the use of an appropriate method for a specific matrix–reinforcement combination and (ii) the maximum amount of reinforcements to give optimum property combinations and lower the cost of the processes, etc. Therefore, it is still necessary to look into these aspects, including the use of simulation and modeling techniques [9].

Polymer nanocomposites are materials that have a polymer as a matrix material and nanoadditives are used as reinforcement material. The additives can be 1D (nanotubes and fibers), 2D (layered materials like clay), or 3D (spherical particles). Polymer nanocomposites have been gaining considerably attention both in academia and in industries due to their outstanding mechanical properties such as high elastic stiffness and strength with a small concentration of nanoadditives. The other excellent properties of polymer nanocomposites are barrier resistance, flame retardancy, wear resistance, and magnetic, electrical, and optical properties [21].

A typical polymer composite is a combination of a polymer (matrix) and a filler (reinforcement). Polyamide is a thermoplastic polymer, and carbon and glass fibers are generally used as reinforcement materials. In the aerospace industry, carbon fibers are extensively used as reinforcement materials. The choice of the reinforcement material depends on the applications. The polymer matrix and fillers are generally bonded by weak intermolecular forces; however, in some cases chemical bonding is employed as well. If the filler material in the composite is dispersed on an atomic or molecular level (nanometer level) and chemical bonding with the matrix material is created, then remarkable improvements in the mechanical properties of the composite material and some new and unexpected or exotic properties can be achieved. To achieve high strength, clay minerals (montmorillonite [MMT], saponite, hectorite, etc.) are being used as filler materials [22].

Nanocomposite development led to significant progress in the chemical processing of nanoparticles. The recent interest in polymer nanocomposite began initially with fascinating observations involving exfoliated clay and CNTs, exfoliated graphite, and nanocrystalline metals. In polymer nanocomposites, the synergistic advantage of a "nanoeffect" relative to large-scale modification is an important factor affecting the properties of the nanocomposites. Experimental investigations have shown that nanoscale additives bring new phenomena, which changes the material properties. The inclusion of nanoparticles with high surface area, high surface energy, and often anisotropic geometry in the polymer matrix decreases the interparticle distance and increases polymer matrix interaction strength. Therefore, polymer nanocomposites with a completely new set of properties suitable for new applications also expand the performance space of traditionally filled polymers. Fabricating good PMNCs is a challenging issue. Researchers have tried many techniques to fabricate PMNCs, such as melt mixing, in situ polymerization, mixing filler materials during electrospinning, and many other approaches. Vollenberg and Heikens produced

Table 4.2 Processing methods for polymer-based nanocomposite systems

Process	System	Procedure	References
Intercalation/prepolymer from solution	Clay with PCL, PLA, HDPE, PEO, PVA, PVP, PVA, etc.	Employed for layered reinforcing material in which the polymer may intercalate Mostly for layered silicates, with intercalation of the polymer or prepolymer from solution Use of a solvent in which the polymer or prepolymer is soluble and the silicate layers are swellable	[17, 27–29]
In situ intercalative polymerization	MMT with N6/PCL/PMMA/PU/epoxy	Encasing of the layered silicate within the liquid monomer or a monomer solution → formation of polymer between the intercalated sheets Polymerization by heat or radiation, by diffusion of a suitable initiator, or by a catalyst fixed through cation exchange inside the interlayer, before the swelling step	[30, 31]
Melt intercalation	MMT with PS/PEO/PP/PVP, clay-PVPH	Annealing of a mixture of the polymer and the layered host above the softening point of the polymer, statically or under shear Diffusion of polymer chains from the bulk polymer melt into the galleries between the host layers during annealing (Fig. 4.1)	[32, 33]
Template synthesis	Hectorite with PVPR, HPMC, PAN, PDDA, and PANI	In situ formation of the layered structure of the inorganic material in an aqueous solution containing the polymer Water-soluble polymer as a template for the formation of layers Widely used for the synthesis of LDH nanocomposites, but less developed for layered silicates	[34, 35]

(*Continued*)

Table 4.2 (*Continued*)

Process	System	Procedure	References
(a) Mixing (b) In situ polymerization	PVA/Ag; PMMA/Pd polyester/TiO_2 PET/$CaCO_3$, epoxy vinyl ester/Fe_3O_4; epoxy vinyl ester/γ-Fe_2O_3; PAA/Ag, PAA/Ni, and PAA/Cu; $AgNO_3$, NiSO4, and $CuSO_4$	(a) Mixing of either polymer or monomer with reinforcing materials (b1) Dispersion of inorganic particles into a precursor of the polymeric matrix (monomer) (b2) Polymerization of the mixture by addition of an appropriate catalyst (b3) Processing of this material by conventional molding technologies Use of ultrasonics for dispersion in epoxy systems Exposition of AG systems to 60 Co γ-ray to promote simultaneous polymerization and metal nanoparticle formation	[36, 37]
Sol–gel process	Polyimide/SiO_2; 2-hydroxyethyl acrylate (HEA)/SiO_2, polyimide/silica PMMA/SiO_2, polyethylacrylate/SiO_2, PC/SiO_2, and poly(amide-imide)/TiO_2	Embedding of organic molecules and monomers on sol–gel matrices Introduction of organic groups by formation of chemical bonds → in situ formation of sol–gel matrix within the polymer and/or simultaneous generation of inorganic/organic networks	[38–40]

PLA, poly(lactic acid); PVA, poly(vinyl alcohol); N6, nylon-6; PEO, poly(ethylene oxide); PAA, poly(acrylic acid); PMMA, poly(methyl methacrylate); PCL, poly(ε-caprolactone); LDH, layered double hydroxide; PVPR, polyvinylpyrrolidone reactant; HPMC, hydroxypropyl methylcellulose; PAN, peroxyacytyl nitrate; PDDA, polydiallyldimethylammonium; PANI, polyaniline; PU, polyurethane; PLA, polylactide; HDPE, high-density polyethylene; PEO, poly(ethylene oxide); PVP, polyvinylpyrrolidone; MMT, montmorillonite.

nanocomposite samples by thoroughly mixing filler particles with a polymer matrix. They used polystyrene (PS), styrene–acrylonitrile-co-polymer (SAN), polycarbonate (PC), and polypropylene (PPy) with alumina beads having 35 nm and 400 nm dimensions as filler materials [9, 23–26]. Various processing methods for polymer-based nanocomposite systems are given in Table 4.2.

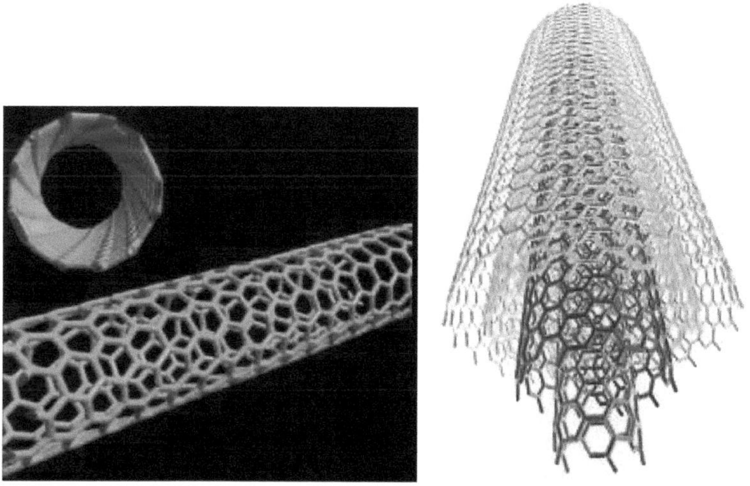

(A) (B)

Figure 4.1 (A) Single-walled carbon nanotubes and (B) multiwalled carbon nanotubes [22] (https://creativecommons.org/licenses/by/4.0/).

4.6 Types of Polymer Nanocomposites Based on Basic Material Used

4.6.1 Polymer Nanocomposites Based on Layered Silicates

Silicates could absorb catalysts and initiate monomers to make polymer silicate nanocomposites [41]. The polymer adsorption onto the silicate surface plays a significant role in affecting the diffusion rate of the polymer [42]. A silica coating provides an efficient barrier to quenching of any flurophore by the magnetic cores [43]. Layered silicates are prepared by incorporating finely dispersed layered silicate materials in a polymer matrix. Different approaches can integrate the ion-exchange layered silicates in polymer hosts by in situ polymerization, simple melt mixing, or solution intercalation [44, 45]. The use of layered silicates has immense potential for producing materials with enhanced flame retardancy along with better physical properties [46]. A stronger interface between polymer and silicate clay produces enhanced materials with improved mechanical properties [47]. Layered silicates used in the production of nanocomposites are natural or synthetic minerals consisting of extremely thin layers that are generally bound together with counterions [48]. Silica nanoparticles have been widely used in different engineering fields such as biomarkers, drug carriers, nanosensors, and nanosized fillers because of their biocompatibility and chemical stability [49]. The most commonly used layered silicates are MMT, saponite, and hectorite.

4.6.1.1 Properties of polymer nanocomposites based on layered silicates

Nanocomposites consisting of a polymer and a layered silicate remarkably enhance mechanical and materials properties. The main reason for enhancement of properties in nanocomposites is the greater interfacial interaction between the matrix and the layered silicate compared to conventional filler-reinforced systems.

4.6.1.1.1 *Tensile properties*

The tensile modulus of a polymeric material has been revealed to be extremely improved when nanocomposites are formed with layered silicates. Nylon-6 (N6) nanocomposites prepared through in situ intercalative ring-opening polymerization of 1-caprolactam, leading to the formation of exfoliated nanocomposites, show a drastic increase in the tensile properties at quite a low filler content. The major cause for the strong improvement in the tensile modulus in N6 nanocomposites is the strong interaction between the matrix and silicate layers via the formation of hydrogen bonds [50].

4.6.1.1.2 *Heat distortion temperature*

The heat distortion temperature (HDT) of a polymeric material is an indicator of heat resistance toward applied load [51].

4.6.1.1.3 *Thermal stability*

The thermal stability of polymeric materials is studied by thermogravimetric analysis (TGA). The weight loss because of the formation of volatile products after degradation at high temperature is examined as a function of temperature. When the heating occurs under an inert gas flow, non-oxidative degradation occurs, and when air or oxygen is used, oxidative degradation of the samples occurs. Commonly, the incorporation of clay into the polymer matrix was found to improve thermal stability by acting as a superior insulator and mass transport barrier to the volatile products generated during decomposition [52].

4.6.1.1.4 *Optical transparency*

Layered silicates are microns in size they are about 1 nm thick. Thus, when a single layer is diffused in a polymer matrix, the resulting nanocomposite is optically clear in visible light [28].

4.6.1.1.5 *Ionic conductivity*

Solvent-free electrolytes are of great interest because of their charge transport mechanism and their probable applications in electrochemical devices [53].

4.6.2 Polymer Nanocomposites Based on Nanotubes/ Nanofibers

CNTs are allotropes of carbon with a cylindrical nanostructure. Nanotubes have been constructed with a length-to-diameter ratio of up to 132,000,000:1, much larger than for any other material. These cylindrical carbon molecules have properties that are important for nanotechnology, optics, electronics, and other fields of materials science and technology [54]. In particular, due to their extraordinary thermal conductivity, electrical properties, and mechanical properties, CNTs may find applications as additives to a variety of structural materials. The development of nanotubes of carbon has involved a major deal of consideration in the past decade because of various attractive properties such as high anisotropy, small dimensions, and absorbing tube-like structures. CNTs and nanofibers (Fig. 4.2) developed by diffusion of carbon (by way of catalytic decomposition of carbon-containing gases vaporized as carbon from arc discharge or laser ablation) through catalysts and its consequent precipitation as a graphitic filament [55]. The synthesis of nanotubes other than carbon atoms may be possible, and they are relatively small, 15-nm-diameter tubes made of molybdenum disulfide and tungsten. Other layered materials synthesized as nanotubes, tube-like forms, have been reported, like boron nitride nanotubes with diameters of a few nanometers. There are three common types of CNTs that have high structural perfection: single-walled nanotubes (SWNTs), multiwalled nanotubes (MWNTs), and carbon nanofibers (CNFs). SWNTs consist of a single graphite sheet wrapped into a cylindrical tube. SWNTs show a range of remarkable mechanical and electronic properties because of their unique structure and dimensions [56]. MWNTs comprise a group of such nanotubes that are concentrically nested like rings of a tree trunk [57]. SWNTs reveal a range of remarkable physical, electronic, and mechanical properties because of their unique structure and dimensions. CNTs have commercial interest due to their conductivity at very low loading levels [42]. See Table 4.3 for a list of methods of preparing polymer–CNT nanocomposite systems.

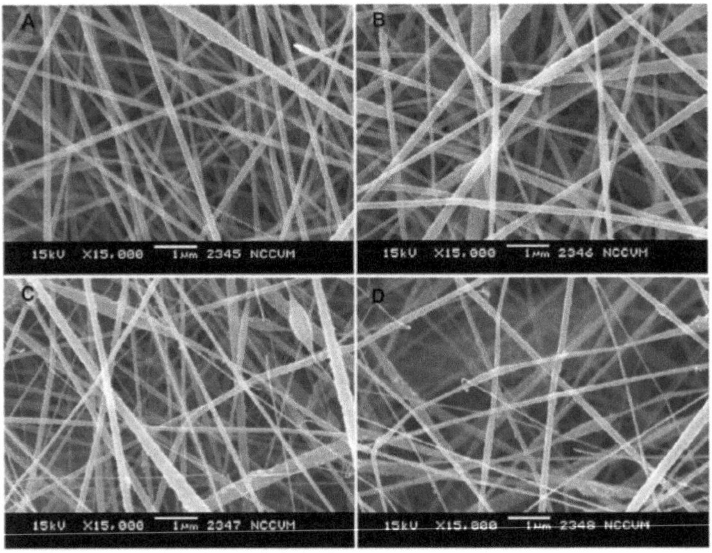

Figure 4.2 (A–D) Field emission scanning electron microscopy (FESEM) images of nanofibers at different magnifications. Reprinted from Ref. [55b], Copyright (2008), with permission from Elsevier.

Table 4.3 Processing methods for polymer–CNT nanocomposite systems

Process	System	Procedure	Remark	References
Direct mixing Solution mixing	Thermoset resins, thermoplastic resins (PS/epoxy)	Dispersion of CNTs; curing Dispersion of 0.2%–1% CNTs, (100 nm diameter, 10 μm length) Removal of solvent or precipitation of polymer; curing	Modification of polymer behavior, synergistic effect, shape memory nanocomposites	[18, 58, 59]
Melt mixing	Polymers, N6	Mechanical mixing of CNTs with prepolymer melt, followed by extrusion, injection, or compression molding	Use of 0.2%–2.0% MWNTs, twin-screw mixer	[60–62]
In situ polymerization	PANI/CNT, MMA/CNT, epoxy/CNT, poly(ether-ester)	Use of ultrasonics for dispersion in monomer/matrix; curing	Preparation of the polymer with CNTs, good chemical bonding	[19, 63]
Others	PP/CNT, PVK/SWNT, iPP/SWNT, PANI/SWNT	Solid-state mechanochemical pulverization; blending + sonication; melt blending; vapor deposition process (VDP)	0–10 wt% CNTs	[64, 65]

4.6.2.1 Properties of nanotubes/nanofibers

The properties of CNTs are important to understand for designing composite materials.

4.6.2.1.1 *Elastic and thermal properties*

Nanotubes are the stiffest-known fibers. The tensile strength ranges up to 50 GPa or more. When compared to carbon-reinforcing fibers, the strength-to-weight ratio of nanotubes in the axial direction is up to four times greater. The greatest strain of SWNTs is 10%, which is superior than most structural materials. These strong mechanical properties are due to the C–C covalent bonding and the seamless hexagonal network. Thermal conductivity is also very high in the direction of the nanotube axis, typically about 1750–5800 W/mK [66, 67].

4.6.2.1.2 *Electrical conductivity*

Electronically, CNTs can be metallic or semiconducting, depending on the chirality. CNTs also have been predicted to conduct current ballistically without dissipating heat. Metallic SWNTs behave as long ballistic quantum conductors with the charge carriers, and they have the strongest electrochemical properties perhaps because the lower resistance allows a greater double-layer charge buildup [68].

4.6.2.1.3 *Magnetoresistance*

CNTs also have spin-dependent transport properties or magnetoresistance. The direction of magnetization of the ferromagnetic electrodes used to contact the nanotubes defines the spin direction of the charge carriers into and out of the nanotubes and a change in the resistivity of the nanotubes. The nanotube–metallic junction appears to have a strong effect on the spin-dependent transport [69].

4.6.2.1.4 *Piezoresistance*

A pioneering experiment shows that the conductance of a metallic CNT might decrease by orders of magnitude when strained by an atomic force microscope tip. It appears that the band structure of a CNT is dramatically changed by mechanical strain and that the conductance of the CNT can increase or decrease depending on the chirality of the nanotube [70].

4.6.2.1.5 *Electrokinetics of nanotubes*

The electrical properties (conductivity and dielectric constant) of a nanotube are generally different than those of a fluid. Therefore, when a nanotube is in an electrolyte, it will attract ions of opposite electrical polarity, forming an electrical double layer [71].

4.6.3 Polymer Nanocomposites Based on Inorganic Materials

Polymeric composites consisting of inorganic nanoparticles and organic polymers represent a new class of materials, which enhance performance compared to their microparticle counterparts. Polymer–inorganic nanocomposite materials are inorganic nanofiller dispersed at a nanometer level in a polymer matrix. The inorganic particles provide mechanical and thermal stability and new functionalities that depend on the chemical nature, structure, size, and crystallinity of the inorganic nanoparticles (silica, transition metal oxides, nanoclays, metallic phosphates, metal chalcogenides, and nanometals). The inorganic particles can execute or get better mechanical, thermal, magnetic, and electronic properties; density and refractive index; etc. [28, 72, 73].

4.6.3.1 Methods for the preparation of polymer-based inorganic nanoparticle composites

Various methods are used for preparation of polymer-based inorganic nanocomposites. The important ones are as follows:

- Intercalation of nanoparticles with the polymer or prepolymer from solution [73]
- In situ intercalative polymerization [74]
- Melt intercalation [75]
- Direct mixture of polymer and particulates [76]
- Template synthesis [77]
- In situ polymerization [78]
- Sol–gel process [28]

The most important one is possibly the incorporation of inorganic nanoparticles in polymers. Polymer-based nanocomposite synthesis is affected by the size, content, molecular weight, and properties of inorganic particles. Maleic anhydride (MA)-grafted polyethylene (PE)/clay nanocomposites were prepared by simple melt compounding. The exfoliation and intercalation behaviors depended on the hydrophilicity of PE grafted with MA and the chain length of the organic modifier in the clay. When PE has a higher grafting level of MA than the critical grafting level of MA (0.1 wt%) and the number of methylene groups in alkylamine chain is more than 16, PE/clay nanocomposites are completely exfoliated [66].

4.7 Advantages and Limitations of Polymer-Based Nanocomposite Processing Methods

There are various advantages and limitations of processing methods for polymer-based nanocomposites. These are listed in Table 4.4. Also see Table 4.5 for examples of polymer matrix nanocomposites and their properties.

Table 4.4 Advantages and limitations of polymer-based nanocomposite processing methods

Process	Advantages	Limitations	References
Intercalation/prepolymer from solution	Synthesis of intercalated nanocomposites based on polymers with low or even no polarity Preparation of homogeneous dispersions of the filler	Industrial use of large amounts of solvents	[17, 27, 29, 79]
In situ intercalative polymerization	Easy procedure based on the dispersion of the filler in the polymer precursors	Difficult control of intragallery polymerization Limited applications	[30, 31]

Process	Advantages	Limitations	References
Melt intercalation	Environmentally benign, use of polymers not suited for other processes, compatible with industrial polymer processes	Limited applications to polyolefins, which represent the majority of used polymers	[32, 33]
Template synthesis	Large-scale production, easy procedure	Limited applications, based mainly in water-soluble polymers, contaminated by side products	[28, 34]
Sol–gel process	Simple, low processing temperature, versatile, high chemical homogeneity, rigorous stoichiometry control, high-purity products, formation of 3D polymers containing metal–oxygen bonds Single or multiple matrices Applicable specifically for the production of composite materials with liquids or with viscous fluids that are derived from alkoxides	Greater shrinkage and lower amount of voids compared to the mixing method	[35, 80]

Table 4.5 Examples of polymer matrix nanocomposites and their properties [38, 81–84]

Matrix/ reinforcement	Properties
PPy/MMT	Improved tensile strength, strain at break, stiffness, Young's modulus, and tensile stress
N6/layered silicates	Improved storage modulus, tensile modulus, HDT, tensile stress, and reduced flammability

(*Continued*)

Table 4.5 (*Continued*)

Matrix/ reinforcement	Properties
Polylactide/ layered silicates	Improved bending modulus, bending strength, distortion at break, storage modulus, gas barrier properties, and biodegradability
Epoxy/layered silicates	Improved tensile strength and modulus
Polyimide/MMT	Improved tensile strength, elongation at break, and gas barrier properties
PS/layered silicates	Improved tensile stress and reduced flammability
PEO/layered silicates	Improved ionic conductivity
PMMA/Pd	Improved thermal stability
Polyester/TiO_2	Improved fracture toughness and tensile strength
Epoxy/SiC	Improved microhardness, storage modulus, and elastic moduli

4.8 Properties of Polymer-Based Nanocomposites

Polymer-based nanoparticle nanocomposites were prepared via a range of processes and show the following mechanical, optical, electrical, and thermal properties.

4.8.1 Mechanical Properties

To get better mechanical properties such as tensile strength, modulus, or stiffness, inorganic materials can be added. Poor compatibility between the polymer matrix and inorganic particles in nanocomposites prepared by simple physical mixing will produce intrinsic defects, which, consequently, will result in harmful effects on the mechanical properties of the nanocomposites. The mechanical properties of nanocomposites, prepared from various polymers and inorganic particles, are not always enhanced. In some cases, the properties of nanocomposites decrease by the addition of inorganic particles because of aggregation in polymer matrices. To resolve this difficulty, the load amounts of inorganic particles were optimized with organic materials [85, 86].

4.8.2 Optical Properties

The optical properties of metal nanoparticles consisting of dielectrics have long been of concern. The expansion of materials for optical applications with functionality and clearness increases. Polymer-based inorganic nanocomposites show great promise as they can give the necessary stability and easy processability with interesting optical properties. The transparency of these composites depends upon the size and spatial distribution of inorganic particles in the polymer matrix [87–89].

4.8.3 Electrical Properties

Nanocomposites are strongly related to the invention of advanced devices for electronic and optoelectronic applications. The dimensional scale for electronic devices has entered the nanorange. The efficacy of polymer/inorganic particle nanocomposites in these areas is quite varied, involving many possible applications as well as types of nanocomposites [90].

4.8.4 Thermal Properties

For structural application at elevated temperatures, the dimensional constancy of low thermal expansion coefficient of these nanocomposites is also very important. The high thermal expansion coefficient of neat polymers causes dimensional changes during the molding procedure [91].

4.9 Polymer Nanocomposites for Biomedical Applications

4.9.1 Introduction and Challenges

Bionanocomposites form a fascinating interdisciplinary area that brings together biology, materials science, and nanotechnology.

New bionanocomposites are impacting diverse areas, in particular biomedical science. Generally, polymer nanocomposites are the result of the combination of polymers and inorganic/organic fillers at the nanometer scale. The extraordinary versatility of these new materials springs from the large selection of biopolymers and fillers available to researchers. Existing biopolymers include, but are not limited to, polysaccharides, aliphatic polyesters, polypeptides and proteins, and polynucleic acids, whereas fillers include clays, hydroxyapatite (HAp), and metal nanoparticles [92].

The interaction between filler components of nanocomposites at the nanometer scale enables them to act as molecular bridges in the polymer matrix. This is the basis for enhanced mechanical properties of the nanocomposite as compared to conventional microcomposites [93]. Bionanocomposites add a new dimension to these enhanced properties in that they are biocompatible and/or biodegradable materials. Biodegradable materials can be described as materials degraded and gradually absorbed and/or eliminated from the body, whether degradation is caused mainly by hydrolysis or mediated by metabolic processes [94] Therefore, these nanocomposites are of immense interest to biomedical technologies such as tissue engineering, bone replacement/repair, dental applications, and controlled drug delivery.

Current opportunities for polymer nanocomposites in the biomedical arena arise from the multitude of applications and the vastly different functional requirements for each of these applications. For example, the screws and rods that are used for internal bone fixation bring the bone surfaces in close proximity to promote healing. This stabilization must persist for weeks to months without loosening or breaking. The modulus of the implant must be close to that of the bone for efficient load transfer [95, 96]. The screws and rods must be noncorrosive, nontoxic, and easy to remove, if necessary. Thus, a polymer nanocomposite implant must meet certain design and functional criteria, including biocompatibility, biodegradability, mechanical properties, and, in some cases, aesthetic demands. The underlying solution to the use of polymer nanocomposites in vastly differing applications is the correct choice of matrix polymer chemistry, filler type, and matrix–filler interaction [97].

4.9.2 Hydroxyapatite–Polymer Nanocomposites

Producing bionanocomposites based on biomimetic approaches has been a recent focus of researchers. Among these materials, HAp–polymer nanocomposites have been used as a biocompatible and osteoconductive substitute for bone repair and implantation [98, 99]. As the main inorganic component of hard tissue, HAp [$Ca_{10}(PO_4)_6(OH)_2$] has long been used in orthopedic surgery. The incorporation of HAp in polymeric nanocomposites to overcome processing and dispersion challenges is of great interest to the biomedical community. Consequently, a desirable material for use in clinical orthopedics would be a biodegradable structure that induces and promotes new bone formation at the required site. To date, primarily polysaccharide and polypeptidic matrices have been used with HAp nanoparticles for composite formation. Yamaguchi and coworkers synthesized and studied flexible chitosan–HAp nanocomposites [100]. The matrix used for this study, chitosan (a cationic, biodegradable polysaccharide), is flexible and has high resistance upon heating because of intramolecular hydrogen bonds formed between the hydroxyl and amino groups [101, 102]. The resulting nanocomposite, prepared by the coprecipitation method, is mechanically flexible and can be formed into any desired shape. Nanocomposites formed from gelatine and HAp nanocrystals are conducive to the attachment, growth, and proliferation of human osteoblast cells (Fig. 4.3) [103]. Collagen-based polypeptidic gelatin has a high number of functional groups and is currently being used in wound dressings and pharmaceutical adhesives in clinics [104]. The flexibility and cost-effectiveness of gelatin can be combined with the bioactivity and osteoconductivity of HAp to generate potential engineering biomaterials. The traditional problem of HAp aggregation was overcome by precipitation of the apatite crystals within a polymer solution [105, 106]. Polysaccharides, such as alginate, provide a natural polymeric sponge structure that has been used in tissue engineering scaffold design. The weak, soft alginate scaffolds can be strengthened with HAp and have widespread applications [107]. Composite membranes from HAp nanoparticles and chitosan/collagen sols have also been synthesized to study connective tissue reactions [108, 109].

Specifically, amino acids like aspartic acid and glutamic acid have been used as the matrix protein. Both of these amino acids are known to play an important role in intercellular communication and osteoblast differentiation that increases extracellular mineralization. Related studies have also highlighted the significance of aspartic acid in the treatment of osteoporosis and other bone dysfunctions [110].

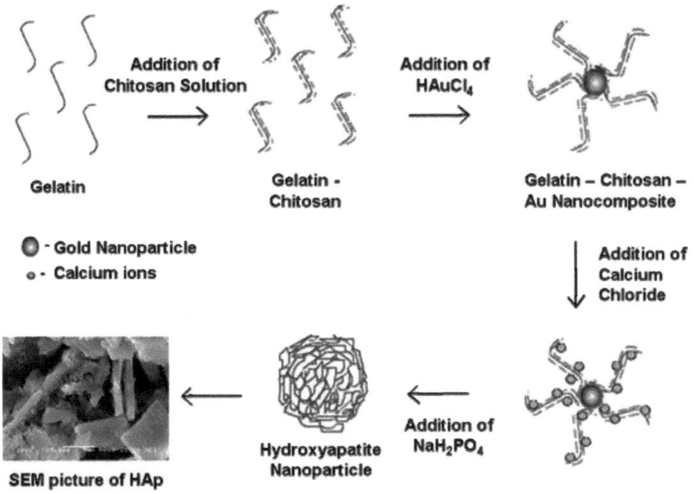

Figure 4.3 Low-magnification scanning electron micrograph of a hydroxyapatite (HAp) polymer/chitosan–gelatin scaffold prepared from a HAp/chitosan–gelatin/acetic acid mixture. Reprinted by permission from Springer Nature Customer Service Centre GmbH: Springer Nature, *Journal of Nanoparticle Research*, Ref. [103b], copyright (2008).

4.9.3 Aliphatic Polyester Nanocomposites

Aliphatic polyesters have been the predominant choice for materials in degradable drug delivery systems. Homopolymers and copolymers derived from glycolic acid or glycolide, lactic acid or lactide, and ε-caprolactone form the bulk of this research. Most of these polymers degrade by acid- or base-catalyzed hydrolytic cleavage of the backbone or by enzymatic activity [111].

Of the polyesters that show promise in biomedical fields, poly(L-lactic acid) (PLLA) is the most prevalent. Although reports of the use of PLLA can be found in the 1960s, a phenomenal amount of work has been performed recently. PLLA has widespread applications in sutures, drug delivery devices, prosthetics, scaffolds, vascular grafts, bone screws, pins, and plates for temporary internal fixation. Good mechanical properties and degradation into nontoxic products are the main reasons for such an array of applications [112, 113].

As a material, PLLA is relatively hard and crystalline with a melting point between 170°C–180°C and a glass transition temperature T_g of −60°C to 67°C. Another factor contributing to the use of PLLA in biomedical applications is its ability to be copolymerized and blended to obtain a nanocomposite-like product with desirable properties. The racemic version of the polymer, poly(D,L-lactide) (PDLLA), sometimes simply called poly(lactic acid) (PLA), is amorphous with a T_g of around 50°C–60°C [113].

Studies also have been done to evaluate the potential of PLLA as a bone reinforcement material. The mechanical properties of neat PLLA might be inadequate for high load-bearing applications. This necessitates the incorporation of reinforcements like oriented fibers, HAp, or clays to form nanocomposites. Such studies report an increase in the flexural modulus, strength, and moduli values commensurate with bone replacement implants [114, 115].

Poly(glycolic acid) (PGA) is another aliphatic polyester with applicability for biomedical use. However, unlike PLLA, PGA is readily soluble in water. Mechanical properties of self-reinforced PGA have been investigated and are found to worsen on exposure to distilled water [111]. Thus, water solubility and its high melting point limit the use of PGA in bionanocomposites. Another biodegradable and nontoxic aliphatic polyester discussed in the biomedical nanocomposite literature is poly(ε-caprolactone) (PCL). Synthesized by the ring-opening polymerization of ε-caprolactone, it has a melting temperature of 61°C and a T_g of −60°C. The rubbery state permits the diffusion of low-molecular-weight species at body temperature, thus making PCL a promising candidate for controlled release and soft-tissue engineering. The range of properties can be furthered by copolymerization with other lactones such as glycolide, lactide, and poly(ethylene oxide) (PEO) or by nanofiller incorporation. Both techniques have been adopted and have shown great potential for biomedical applications [112]. Nanocomposite polymer–clay hydrogels have also been studied extensively by microscopy and scattering techniques [116].

4.9.4 Polypeptide-Based Nanocomposites

Polypeptides as matrices provide an additional array of opportunities in materials design and application in terms of a unique ability to adopt specific secondary, tertiary, and quaternary structures, a feature not available with synthetic polymers. Functionality can also be incorporated by the use of natural and nonnatural amino acids with desired activity at specific sites along the polypeptide backbone. Our group has been investigating the incorporation of polypeptides, particularly, poly(L-lysine) (PLL), as the matrix polypeptide and MMT as the filler. Initial work produced nanocomposites from PLL and MMT that exhibited enhanced mechanical properties and thermal properties comparable to other widely explored biomaterials and engineering thermoplastics [117, 118].

The potential applicability of these materials was further investigated by studying the effect of the molecular weight of the polypeptide on the secondary conformation of the final nanocomposite matrix. PLL is a good choice for such a study because it can display a variety of secondary structures: the random coil, α-helix, or β-sheet in aqueous solution. Moreover, transitions between conformations can be easily achieved using pH, temperature, salt concentration, or alcohol content as an environmental stimulus. Scientists developed responsive hydrogels via self-assembly of short 20-residue peptides. These synthetic peptides are designed to intramolecularly fold under desired aqueous conditions and subsequently intermolecularly self-assemble into supramolecular networks. The network scaffold is comprised of stiff, fibrillar structures that exhibit varied morphology, including twisted or untwisted single fibrils and fibril laminates. Materials like these, in addition to proteins, introduce the possibility of fibril incorporation into nanocomposites as reinforcement with biodegradable matrices [119].

Such unique nanocomposites combine the degradability and strength of the gel matrix with control over the functionality and morphology of the fibrillar fillers. Potential applications of such nanomaterials include drug delivery matrices, tissue engineering scaffolds, and bioengineering materials, to name a few.

4.9.5 Nanocomposites from Other Polymers and Fillers

Nanocomposites from bioactive molecules and clays have also been reported. One such example is smectite nanocomposites that use the ability of smectites to induce specific co-intercalation of purines and pyrimidines. Thus, thymine and uracil are absorbed on MMT appreciably if solutions also contain co-absorbed adenine. Layered double hydroxides (LDHs) present a layered structure comprised of di- and trivalent metal cations [M(II) and M(III)] coordinated by six oxygen anions. The substitution of M(III) cations induces an overall positive charge that is counterbalanced by exchangeable interlayer anions. Thus, LDHs can serve as hosts for polymeric intercalation. An attractive, potential application of LDHs is controlled drug release carriers. Several bioactive compounds like DNA and pharmaceuticals have been incorporated within LDH hosts [120, 121].

Poly(urethane urea) (PUU)-segmented block copolymers are common in ventricular assist devices and total artificial hearts as blood sacs. One of the main disadvantages of PUUs in these devices is their relatively high permeability to air and water vapor, a result of the diffusion through the poly(tetramethylene oxide) soft segments that are present as the majority component of the copolymer. The use of organically modified layered silicates seems particularly attractive among the variety of approaches taken to reduce permeability, while maintaining desired biocompatibility and mechanical properties. Studies on the PUU–clay system have shown increases in the modulus and strength of the nanocomposite. However, in contrast to conventional composite systems, this enhancement does not come at the expense of ductility. In fact, it is reported that the ductility increases with clay loading. Just as nanocomposites used for various biomedical applications have different requirements, nanocomposites used for dental applications have unique necessities. Thermoset methacrylate-based composites are used commonly as dental restorative materials because of a relatively high cure efficiency by free-radical polymerization and excellent aesthetic qualities. However, the demands of the oral environment and the masticatory loads encountered by dental restoratives require further property improvements in these materials.

Specifically, nanocomposites with improved modulus, better efficiency of the free-radical polymerization, low water sorption, improved processability, and low shrinkage are needed [122].

The design of cardiovascular interfaces necessitates a combination of amphiphilicity and antithrombogenicity. Amphiphilicity ensures optimal endothelial cell response at the vascular interface. Thrombogenicity refers to blood clot formation and can lead to early graft occlusion. Using reports of polyhedral oligomeric silsequioxanes (POSS) acting as an amphiphile at the water–air interface, researchers have studied the possibility of using POSS at the vascular interface. The strong intermolecular forces between constituent molecules and neighbors and the robust framework of shorter bond lengths make POSS nanocomposites more resistant to degradation. Initial work has shown that POSS nanocomposites are cytocompatible, making them potentially suitable for tissue engineering [123, 124]. Future efforts in this field are directed at assessing the thrombogenic potential of these nanocomposites, which would be critical in their application as cardiovascular interfaces [123, 124].

4.10 Green Polymer Nanocomposites

Polymer/clay nanocomposites based on traditional thermoplastics such as PPy, PE, nylon, and polyethylene terephthalate (PET) in packaging and structural applications will contribute to the biodegradability (within a defined manner), in addition to the improvement in barrier properties. However, green nanocomposites derived from bioplastics such as PLA polyhydroxyalkanoate (PHA), starch, etc. And with nanomaterials such as layered silicates are an environmentally attractive substitute for conventional polymeric nanocomposites based on petroleum feedstock [4].

Materials are said to be green when they are biodegradable and renewable. The major attractions about green composites are that they are environmentally friendly, fully degradable, and sustainable in every way. At the end of their service life they can be easily disposed of or composted without harming the environment. The challenge of green composites involves obtaining green polymers that are used as matrices for the production of the composites. A polymer is said to be green when it possesses environmentally favorable properties such as renewability and degradability. Biodegradation implies degradation of a polymer in a natural environment that includes changes in the chemical structure, loss of mechanical and structural properties, and change into other compounds that are beneficial to the environment [125, 126].

Green polymers, on the other hand, are those produced using green (or sustainable) chemistry, a term that appeared in the 1990s. According to the International Union of Pure and Applied Chemistry (IUPAC), green chemistry relates to the design of chemical products and processes that reduce or eliminate the use or generation of substances hazardous to humans, animals, plants, and the environment. Thus, green chemistry seeks to reduce and prevent pollution at its source. Natural polymers are usually green [4].

A vast number of biodegradable polymers (e.g., cellulose, chitin, starch, polyhydroxyalkanoates, polylactide, polycaprolactone, collagen, and other polypeptides) have been synthesized or are formed in the natural environment during the growth cycles of two biodegradable polymers Some microorganisms and enzymes capable of degrading such polymers have been identified. A number of natural and other degradable polymers have been used in green composites and some of these are discussed next [127, 128].

4.10.1 Thermoplastic Starch–Based Composites

Thermoplastic starch (TPS) properties also appear to benefit from silica addition. Tang et al. [20] reported that inclusion of dry powder SiO_2 particles in starch polyvinyl alcohol (PVOH) films increased tensile strength at break and improved water barrier properties. Xiong et al. [21] reported improved mechanical properties, transmittance, and water resistance of starch films containing nano-SiO_2 particles [129, 130].

Earlier research on TPS-based composites focused on the use of plasticized starch as a matrix for green nanocomposites. Scientists reported the use of TPS for the production of composites by melt intercalation in a twin-screw extruder. The composites were prepared with regular corn starch plasticized with glycerine and reinforced with hydrated kaolin. The study recorded a significant increase in the tensile strength from 5 to 7.5 MPa for the composite with a 50 phr clay composition. The modulus of elasticity increased from 120 to 290 MPa, while the tensile strain at break decreased from 30% to 14%. The maximum value for the modulus of elasticity observed and the tensile strength corresponded to the maximum quantity of clay that was incorporated in the matrix. The authors noted that an increase in the amount of filler increases the fragility of the composite [131].

TPSs are beginning to show increased commercial use. There are many challenges still to be overcome in order to make commercially viable biodegradable and compostable starch polymers. TPS films generally need to have starch content greater than 70% to have biodegradable or compostable action [18]. The term "composite" refers to matrix polymers that contain dispersed conventional fillers, such as glass, fibers, talc, or clay particles. Polymers with fillers that are nanosized in scale are named nanocomposites. The properties of a composite depend on the volume fraction of the filler and its shape, size, and interfacial adhesion [132, 133].

Starch-based resin 11C is a completely biodegradable and compostable resin based on a blend of TPS and suitable for the manufacturing of film-type products. It can be directly used in the film-blowing process. It does not contain any nondegradable polymer such as PE, PP, PS, and poly(vinyl chloride) (PVC).

Figure 4.4 Thermoplastic starch–based resin 11C [4].

4.10.2 Poly(Lactic Acid)-Based Composites

Researchers prepared nanocomposites using PLA. PLA/organically modified clay (OMLS) blends were made by dissolving the polymer in hot chloroform in the presence of dimethyl distearyl ammonium–modified MMT (2C18MMT). X-ray diffraction results of PLA/MMT show that the silicate layer forming the clay could not be intercalated in the PLA/MMT blends when prepared by the solvent-casting method. The clay existed in the form of tactoids with several stacked silicate monolayers [134].

PLA has caught the attention of polymer scientists and is proving to be a viable alternative biopolymer to petrochemical-based plastics for many applications. PLA is produced from lactic acid, which is itself derived from the fermentation of corn or sugar beet, and due to its biodegradation ability, PLA presents a major advantage to enter in the natural cycle, implying its return to the biomass.

Studies on the thermal, mechanical, and morphological properties of PLA-based composites have been conducted by researchers. Composites were prepared by melt compounding and injection molding. Thermal degradation, thermal transition, morphological properties, and mechanical properties of the composites were evaluated. The tensile modulus of the composites increased from 62.5% to 169.5%, contrary to the earlier studies where no improvement was recorded. More recently, nanocomposites of PLA with a compatibilizer and cellulose fibrils have been developed by scientists. The composites were obtained by solvent-casting methods using *N,N*-dimethylacetamide (DMAC) and characterized PLA reinforced with cellulose nanofibrils resulted in no improvement in tensile strength (30 MPa compared to PLA) and percentage elongation (2.5% compared to ith pure PLA) of the composites [135, 136].

4.10.3 Cellulose-Based Composites

Cellulose from agricultural products has been identified as a source of biopolymers that can replace petroleum polymers. Green nanocomposites have been successfully produced from cellulose acetate (CA), triethyl citrate (TEC) plasticizer, and organically modified clay via melt compounding. All cellulose nanocomposites prepared by partial dissolution using ionic liquids were studied for tendon and ligament applications. Compare the microstructural morphology of undissolved nanofiber networks with those after dissolution. The undissolved network shows layer of nanofibers with homogeneous morphology. The partially dissolved composites showed core–shell morphology with more dissolution on the surface layers [137].

The tensile strength and modulus of cellulosic plastic reinforced with 10 wt% organoclay improved by 75% and 180%, respectively. The thermal stability of the cellulosic plastic also increased. Cellulose was obtained from cotton pulp via acid hydrolysis using a solution of lithium chloride and DMAC. Nanoscaled polypyrrole was used as the second component of the nanocomposites. Nanocomposites were prepared by a polymerization-induced adsorption process.

4.10.4 Plant Oil–Based Composites

Biobased nanocomposites have been developed from epoxidized soybean oil, diglycidyl ether of bisphenol-A (DGEBA), and organically modified montmorillonite (OMM). Soy-based epoxy resin (30/70 wt%) was mixed with different organophilic MMT clay (as reinforcement) concentrations (1, 3, 5, and 7 wt%) to form nanocomposites using a mechanical stirrer. The tensile strengths of 1, 3, 5, and 7 wt% OMM-reinforced soy-based epoxy nanocomposites increased by 2%, 7%, 12.6%, and 23.3%, respectively. The impact strength and flexural properties of the nanocomposites showed significant improvement. A maximum increase of 25% was recorded for flexural strength of the nanocomposites [138].

Nanocomposites reinforced by exfoliated clay were found to exhibit higher storage moduli. Scanning electron microscopy (SEM) images of the composites revealed relatively uniform dispersion of organoclay layers, in addition to a good interface of the material. The improvement of both toughness and impact strength in the case of the exfoliated nanocomposites was attributed to the plasticizing effect of the long aliphatic chains in the soy-based epoxy polymer matrix and the reinforcement effect of the clay nanolayers. The composite produced was not completely from a green source because of the use of organophillic montorillonite clay as reinforcement.

Scientists reported the use of mango puree–based edible films for nanocomposites. Nanocomposites were developed by adding cellulose nanofibers in different concentrations (up to 36 g/100 g) as nanoreinforcement to mango puree–based edible films for packaging applications. The study revealed that mango puree edible films can be used as matrix material for biocomposites as the tensile strength increases from 4.06 MPa to 8.09 MPa and the Young's modulus from 19.85 MPa to 322.05 MPa. This improvement in properties was attributed to the fact that the nanofibers seem to have presented good interactions with the mango puree. They were found to disperse in the matrix without significant agglomeration. The authors suggested that the interactions may have occurred between cellulose and neutral sugar side chains of the mango pectin [139].

Soy-based polyurethane can be used as a matrix for the production of biobased nanocomposites. The polyurethane modified with halloysite nanotube (HNT) nanoparticles was used to manufacture E-glass-reinforced composites using a low-cost vacuum-assisted resin transfer-molding process and characterized for ultimate compressive strength, flexural strength, flexural modulus, and interlaminar shear strength (ILSS). Results showed that interlaminar shear strength increases with an increase in the percentage of HNT used (from 14.28 MPa for 0 wt% HNT to 21.28 MPa for 2.4 wt% HNT). Ultimate compressive strength, flexural modulus, and flexural strength did not show appreciable changes. The 0.8 wt% HNT gave the best combination of properties with a 6%, 28%, and 82% increase in flexural strength, flexural modulus, and ILSS, respectively [140].

4.10.5 Other Biopolymer-Based Composites

Other biopolymers used in nanocomposites synthesized are polyhydroxyl butyrate (PHB) (a naturally occurring polyester produced by numerous bacteria in nature), gelatin (a biopolymer obtained by thermal denaturation of collagen isolated from animal skin and bones with dilute acid), and chitosan (a natural polymer found in exoskeletons of crustaceans and insects and in the cell wall of fungi and microorganisms) [141, 142].

4.11 Applications of Green Polymer Nanocomposites

Nanocomposites have been used in several applications such as mirror housings on various vehicle types, door handles, door panels, trunk liners, instrument panels, parcel shelves, head rests, roofs, upholstery and engine covers, and intake manifolds and timing belt covers. Other applications currently being considered include impellers and blades for vacuum cleaners, power tool housings, mower hoods, and covers for portable electronic equipment such as mobile phones, pagers, etc. Honeywell developed commercial clay–N6 nanocomposite products for drink packaging applications. Starch-based resin 11C is a biodegradable and compostable resin based on a blend of TPS. The basic applications of starch-based resin 11C is to making compostable bags (shopping bags/check-out bags, green bin liners) and produce and meat liners (overwrap packaging, mulch film, breathable film) [143–145].

4.12 Applications of Polymer Nanocomposites

Nanocomposites are of huge interest to biomedical technologies such as controlled drug delivery, tissue engineering, dental applications, and bone replacement/repair. Some biopolymers commonly used in biomedical applications are listed here. Also see Table 4.6.

- HAp–polymer nanocomposites are used as a biocompatible and osteoconductive substitute for bone repair and implantation [99].
- Polypeptidic gelatine is a collagen-based nanocomposite most widely used in wound dressings and pharmaceutical adhesives in clinics [104].
- Polysaccharides like alginate provide a natural polymeric sponge structure that is used in tissue engineering scaffold design [107].
- Composite membranes from HAp nanoparticles and chitosan/collagen sols have also been synthesized to study connective tissue reactions [111].
- Studies of the nucleation of calcium phosphates and bone cell signaling used acidic macromolecules as the nanocomposite matrix. Especially, amino acids like aspartic acid and glutamic acid are used as the matrix protein. Both of these amino acids play an important role in intercellular communication and osteoblast differentiation that increases extracellular mineralization [112].

- Aliphatic polyesters are used as a major choice for materials in degradable drug delivery systems [113].
- Polyesters are used in biomedical fields; PLLA is the most widely used one [114]. PLLA has general applications in drug delivery devices, scaffolds, sutures, prosthetics, bone screws, and vascular grafts for temporary internal fixation [115].
- MMT in a PLLA solution is used to improve mechanical and biodegradation characteristics and ideal pore sizes (microns) for blood vessel invasion, cell growth, and nanosized pores for nutrient/waste transfer [146].
- The use of organically modified layered silicates is predominantly attractive for biocompatibility and mechanical properties [124].
- Quantum dots (QDs) can be treated with drug moieties, for instance, nonsteroidal anti-inflammatory drugs, in order to specifically target certain organ or cell organelles [114].
- The nanoparticle size plays a significant role in determining the magnetic response of a material and hence influences its biomedical activity [147].
- Nanocomposites can be used as an agent in nanomedicine. For example, one of the most promising applications is bimodal anticancer therapy encompassing photodynamic and hyperthermic capabilities [148].
- Applications of magnetic fluorescent nanoparticles are in cell tracking, diagnostics, cytometry, magnetic separation, and treatment of various diseases [148].
- Nanocomposites are used in chemical sensors, electroluminescent devices, electrocatalysis, batteries, smart windows, and memory devices.
- Another potential application includes photovoltaic (PV) cells and photodiodes, supercapacitors, printable conductors, light-emitting diodes (LEDs), and field-effect transistors [149].
- The electrical conductivity of CNTs in insulating polymers has also been a topic of considerable interest. The potential applications include electromagnetic interference shielding, transparent conductive coatings, electrostatic dissipation, supercapacitors, electromechanical actuators, and various electrode applications [150, 151].

- Conjugated polymers with various nanoscale filler inclusions have been investigated for sensor applications, including gas sensors, biosensors, and chemical sensors. The nanofillers employed include metal oxide nanowires, CNTs, and nanoscale gold, silver, nickel, copper, platinum, and palladium particles [152].
- The emerging interest in graphene for electronics applications parallels that which occurred with CNT discovery. Graphene, in sheet form, may offer promise in replacing silicon as Moore's law limits are achieved [153].
- Polymer-based solar cells have the capability of being used to make cheap, large, flexible panels. The only downside is substantially low efficiency compared to commercial solar cells. Cadmium chalcogenides such as CdS nanoparticles in polymers have been used to make solar cells [154].
- Polymer semiconductor nanocomposites offer the promise of a new generation of hybrid materials with numerous possibilities of applications such as in optical displays, catalysis, PV cells, gas sensors, electrical devices, mechanics, photoconductors, and superconductor devices [155, 156].

Table 4.6 Potential applications of polymer nanocomposite systems

Nanocomposites	Applications
Polycaprolactone/SiO_2	Bone-bioerodible for skeletal tissue repair
Polyimide/SiO_2	Microelectronics
PMMA/SiO_2	Dental application, optical devices
Polyethylacrylate/SiO_2	Catalysis support, stationary phase for chromatography
Poly(*p*-phenylene vinylene)/SiO_2	Nonlinear optical material for optical waveguides
Poly(amide-imide)/TiO_2	Composite membranes for gas separation applications
Poly(3,4-ethylene-dioxythiophene)/V_2O_5	Cathode materials for rechargeable lithium batteries
PC/SiO_2	Abrasion-resistant coating
Shape memory polymers/SiC	Medical devices for gripping or releasing therapeutics within blood vessels

Nanocomposites	Applications
N6/ cyclopentyladenosine (LS)	Automotive timing belt: Toyota
PEO/LS	Airplane interiors, fuel tanks, components in electrical and electronic parts, brakes, and tires
PLA/LS	Lithium battery development
PET/clay	Food-packaging applications, for example, packaging for processed meats, cheese, confectionery, cereals and boil-in-the-bag foods, fruit juice and dairy products, beer, and carbonated drinks bottles
Thermoplastic olefin/ clay	Beverage container applications
Polyimide/clay	Automotive step assists: GM Safari and Astra Vans
Epoxy/MMT	Materials for electronics
SPEEK/laponite	Direct methanol fuel cells

4.13 Conclusion

Nanocomposites are suitable materials to meet the emerging demands arising from scientific and technologic advances. Processing methods for different types of nanocomposites (ceramic matrix nanocomposites [CMNCs], metal matrix nanocomposites [MMNCs], and PMNCs) are available, but some of these pose challenges, thus giving opportunities to researchers to overcome the problems being encountered with nanosized materials. They offer improved performance over monolithic and microcomposite counterparts and are consequently suitable candidates to overcome the limitations of many currently existing materials and devices. A number of applications already exist, while many potentials are possible for these materials, which opens new vistas for the future. In view of their unique properties such as very high mechanical properties even at low loading of reinforcements, gas barrier, and flame-related properties, many potential applications and hence the market for these materials have been projected in various sectors. Thus all three types of nanocomposites provide opportunities and rewards, creating new worldwide interest in these new materials. Nanocomposites are promising materials to replace conventional metal/alloys and their composites with nanosized reinforcements. Judicious selection of a processing method and standardization of parameters are absolutely essential in producing high-quality nanocomposites. Nanocomposites produced usually have high ductility and high toughness.

References

1. Twardowski, T. E. (2007). *Introduction to Nanocomposite Materials: Properties, Processing, Characterization* (DEStech Publications, Lancaster).
2. Di Ventra, M., Evoy, S. and Heflin, J. R. (2004). *Introduction to Nanoscale Science and Technology* (Springer-Verlag US).
3. Camargo, P. H. C., Satyanarayana, K. G. and Wypych, F. (2009). Nanocomposites: synthesis, structure, properties and new application opportunities, *Mater. Res.*, **12**(1), pp. 1–39.
4. Modi, V. K., Shrives, Y., Sharma, C., Sen, P. K. and Bohidar, S. K. (2014). Review on green polymer nanocomposite and their applications, *Int. J. Innovative Res. Sci. Eng. Technol.*, **3**(11), pp. 17651–17656.
5. Cammarata, R. (2006). *Introduction to Nano Scale Science and Technology* (Springer, USA).
6. Okpala, C. (2013). Nanocomposites: an overview, *Int. J. Eng. Res. Dev.*, **8**(11), pp. 17–23.
7. Zapata, P., Quijada, R., Retuer, J. and Moncada, E. (2008). Preparation of nanocomposites by in situ polimerization, *J. Chil. Chem. Soc.*, **53**(1). Available online: http://www.scielo.cl/scielo.php?pid=S0717-970720 08000 100006&script=sci-arttext (accessed on August 10, 2014).
8. Asmatulu, R., Khan, W. S., Reddy, R. J. and Ceylan. M. (2015). Synthesis and analysis of injection-molded nanocomposites of recycled high-density polyethylene incorporated with graphene nanoflakes, *Polym. Compos.*, **36**(9), pp. 1565–1573.
9. Paul, D. R. and Robeson, L. M. (2008). Polymer nanotechnology: nanocomposites, *Polymer*, **49**, pp. 3187–3204.
10. Choi, S.-M. and Awaji, H. (2005). Nanocomposites–a new material design concept, *Sci. Technol. Adv. Mater.*, **6**, pp. 2–10.
11. Choi, S. M., Honda, S., Nishikawa, T., Awaji, H., Gnanm, F. D., Vishita, K. and Kuroyama, T. (2004). Design concept of strengthing and toughening

mechanism in nanocomposites–ceremic metal nanocomposites, *J. Ceram. Soc. Jpn.*, **112**, pp. S912–S915.

12. Nanocomposites: properties and applications, The International Nanoscience Community, Posted by Nano on January 4, 2008 in Polymer Nanocomposite Group.
13. Gangopadhyay, R. and Amitabha, D. (2000). Conducting polymer nanocomposites: a brief overview, *Chem. Mater.*, **12**(7), pp. 608–622.
14. Azioune, A., Peck, K., Soudi, B., Chehimi, M. M., McCarthy, G. P. and Armes, S. P. (1999). Adsorption of human serum albumin onto polypyrrole powder and polypyrrole-silica nanocomposites, *Synth. Met.*, **102**(1–3), pp. 1419–1420.
15. Ogasawara, T., Ishida, Y., Ishikawa, T. and Yokota, R. (2004). Characterization of multi-walled carbon nanotube/phenylethynyl terminated polyimide composites, *Composites Part A*, **35**(1), pp. 67–74.
16. Bafna, A., Beaucage, G., Mirabella, F. and Mehta, S. (2002). Shear induced orientation and associated property enhancement in polymer/clay nanocomposites, *Proceedings Nanocomposites*, San Diego, USA.
17. Jimenez, G., Ogata, N., Kawai, H. and Ogihara, T. (1997). Structure and thermal/mechanical properties of poly (ε-caprolactone) - clay blend, *J. Appl. Polym. Sci.*, **64**, pp. 2211–2220.
18. Sandler, J., Shaffer, M. S. P., Prasse, T., Bauhofer, W., Schulte, K. and Windle, A. H. (1999). Development of a dispersion process for carbon nanotubes in an epoxy matrix and the resulting electrical properties, *Polymer*, **40**(21), pp. 5967–5971.
19. Roslaniec, Z., Broza, G. and Schulte, K. (2003). Nanocomposites based on multiblock polyester elastomers (PEE) and carbon nanotubes (CNT), *Compos. Interfaces*, **10**(1), pp. 95–102.
20. Fernando, W. and Satyanarayana, K. G. (2005). Functionalization of single layers and nanofibers: a new strategy to produce polymer nanaocomposites with optimized properties, *J. Colloid Interface Sci.*, **285**(1), pp. 532–543.
21. Tyagi, M. and Tyagi, D. (2014). Polymer nanocomposites and their applications in electronics industry, *Int. J. Electron. Electr. Eng.*, **7**, pp. 603–608.
22. Sambarkar, P. P., Patwekar, S. L. and Dudhgaonkar, B. M. (2012). Polymer nanocomposites: an overview, *Int. J. Pharm. Pharm. Sci.*, **4**(2), pp. 60–65.

23. Winey, K. I. and Vaia, R. A. (2007). Polymer nanocomposites, *MRS Bull.*, **32**, pp. 314–322.
24. Jordon, J., Jacob, K. I., Tannenbaum, R., Sharaf, M. A. and Jasiuk, I. (2005). Experimental trends in polymer nanocomposites: a review, *Mater. Sci. Eng. A*, **393**, pp. 1–11.
25. Mai, Y.-W. and Yu, Z.-Z. (2006). *Polymer Nanocomposites* (Woodhead, Cambridge, UK).
26. Thostenson, E. T., Li, C. and Chou, T.-W. (2005). Nanocomposites in context, *Compos. Sci. Technol.*, **65**, pp. 491–516.
27. Kamigaito, O. (1991). What can be improved by nanometer composites?, *J. Jpn. Soc. Powder Metal.*, **38**, pp. 315–321.
28. Alexandre, M. and Dubois, P. (2000). Polymer-layered silicate nanocomposites: preparation, properties and uses of a new class of materials, *Mater. Sci. Eng., R*, **28**(1–2), pp. 1–63.
29. Zhao, X., Urano, K. and Ogasawara, S. (1989). Adsorption of polyethylene glycol from aqueous solutions on monmorillonite clays, *Colloid. Polym. Sci.*, **267**(10), pp. 899–906.
30. Usuki, A., Kojima, Y., Kawasumi, M., Okada, A., Fukushima, Y., Kurauchi, T. and Kamigaito, O. (1993). Synthesis of nylon-6-clay hybrid, *J. Mater. Res.*, **8**(5), pp. 1179–1183.
31. Messersmith, P. B. and Giannelis, E. P. (1994). Synthesis and characterization of layered silicate-epoxy nanocomposites, *Chem. Mater.*, **6**(10), pp. 1719–1725.
32. Vaia, R. A. and Giannelis, E. P. (1997). Lattice of polymer melt intercalation in organically modified layered silicates, *Macromolecules*, **30**(25), pp. 7990–7999.
33. Vaia, R. A. and Giannelis, E. P. (1997). Polymer melt intercalation in organically modified layeredsilicates: model predictions and experiment, *Macromolecules*, **30**(25), pp. 8000–8009.
34. Tomasko, D. L., Han, X., Liu, D. H. and Gao, W. (2003). Supercritical fluid applications in polymer nanocomposites, *Curr. Opin. Solid State Mater. Sci.*, **7**(4–5), pp. 407–412.
35. Carrado, K. A. and Xu, L. Q. (1998). In situ synthesis of polymer-clay nanocomposites from silicate gels, *Chem. Mater.*, **10**(5), pp. 1440–1445.
36. Mbhele, Z. H., Salemane, M. G., Van Sittert, C. G. C. E., Nedeljkovic, J. M., Djokovic, V. and Luyt, A. S. (2003). Fabrication and characterization of silver-polyvinyl alcohol nanocomposites, *Chem. Mater.*, **15**(26), pp. 5019–5024.

37. Xu, X., Yin, Y., Ge, X., Wu, H. and Zhang, Z. (1998). Γ-radiation synthesis of poly (acrylic acid)-metal nanocomposites, *Mater. Lett.*, **37**(6), pp. 354–358.

38. Liu, J., Gao, Y., Wang, F., Li, D. and Xu, J. (2002). Preparation and characteristic of a new class of silica/polyimide nanocomposites, *J. Mater. Sci.*, **37**(14), pp. 3085–3088.

39. Avadhani, C. V. and Chujo, Y. (1997). Polyimide-silica gel hybrids containing metal salts: preparation via the sol-gel reaction, *Appl. Organomet. Chem.*, **11**(2), pp. 153–161.

40. Kickelbick, G. (2003). Concepts for the incorporation of inorganic building blocks into organic polymers on a nanoscale, *Prog. Polym. Sci.*, **28**(1), pp. 83–114.

41. Lu, H., Hu, Y., Li, M., Chen, Z. and Fan, W. (2006). Structure characteristics and thermal properties of silane-grafted-polyethylene/clay nanocomposite prepared by reactive extrusion, *Compos. Sci. Technol.*, **66**, pp. 3035–3039.

42. Schmidt, G. and Malwitz, M. M. (2003). Properties of polymer–nanoparticle composites, *Curr. Opin. Colloid Interface Sci.*, **8**, pp. 103–108.

43. Zeng, Z., Yu, J. and Guo, Z. X. (2005). Preparation of polymer/silica composite nanoparticles bearing carboxyl groups on the surface via emulsifier-free emulsion copolymerization, *J. Polym. Sci.*, **43**, pp. 2826–2835.

44. Vaia, R. A. and Wagner, H. D. (2004). Framework for nanocomposites, *Mater. Today* **7**, pp. 32–37.

45. Pavlidou, S. and Papaspyrides, C. D. (2008). A review on polymer–layered silicate nanocomposites, *Prog. Polym. Sci.*, **33**, pp. 1119–1198.

46. Sailaja, A. K., Amareshwar, P. and Chakravarty, P. (2011). Different techniques used for the preparation of nanoparticles using natural polymers and their application, *Int. J. Pharm. Pharm. Sci.* **3**(Suppl 2), pp. 45–50.

47. Ciprari, D., Jacob, K. and Tannenbaum, R. (2006). Characterization of polymer nanocomposite interphase and its impact on mechanical properties, *Macromolecules*, **39**, pp. 6565–6573.

48. Miranda-Trevino, J. C. and Coles, C. A. (2003). Kaolinite properties, structure and influence of metal retention on pH, *Appl. Clay Sci.*, **23**, pp. 133–139.

49. Kim, J., Wainaina, J. and Na, J. S. (2011). Synthesis of amphiphilic silica/polymer composite nanoparticles as waterdispersible nano-absorbent for hydrophobic pollutants, *J. Ind. Eng. Chem.*, **17**, pp. 681–690.
50. Kojima, Y., Usuki, A., Kawasumi, M., Kurauchi, T. and Kamigaito, O. (1993). Onepot synthesis of nylon 6 clay hybrid, *J. Polym. Sci., Part A: Polym. Chem.*, **31**, pp. 1755–1758.
51. Kojima, Y., Usuki, A., Kawasumi, M., Okada, A., Fukushima, Y., Kurauchi, T. and Kamigaito, O. (1993). Mechanical properties of nylon 6–clay hybrid, *J. Mater. Res.*, **8**, pp. 1185–1189.
52. Blumstein, A. (1965). Polymerization of adsorbed monolayers: II. Thermal degradation of the inserted polymers, *J. Polym. Sci., Part A: Polym. Chem.*, **3**, pp. 2665–2673.
53. Vaia, R. A., Vasudevan, S., Krawiec, W., Scanlon, L. G. and Giannelis, E. P. (1995). New polymer electrolyte nanocomposites: melt intercalation of poly (ethylene oxide) in mica-type silicates, *Adv. Mater.*, **7**, pp. 154–156.
54. Lira-Cantu, M. and Gomez-Romero, P. (1998). Electrochemical and chemical synthesis of the hybrid organic-inorganic electroactive material formed by phosphomolybdate and polyaniline. Application as cation-insertion electrodes, *Chem. Mater.*, **10**, pp. 698–704.
55. (a) Hata, K., Futaba, D. N., Mizuno, K., Namai, T., Yamura, M. and Iijima, S. (2004). Water-assisted highly efficient synthesis of impurity-free single-wall carbon nanotubes, *Science*, **306**, pp. 1362–1364. (b) Ji, L. and Zhang, X. (2008). Ultrafine polyacrylonitrile/silica composite fibers via electrospinning, *Mater. Lett.*, **62**(14), pp. 2161–2164.
56. Gavillet, J., Loiseau, A., Journet, C., Willaime, F., Ducatelle, F. and Charlier, J. C. (2001). Root-growth mechanism for single-wall carbon nanotubes, *Phys. Rev. Lett.*, **87**, p. 275504.
57. Baughman, R. H., Zakhidov, A. A. and de Heer, W. A. (2002). Carbon nanotubes: the route toward applications, *Science*, **297**, pp. 787–792.
58. Qian, D., Dickey, E. C., Andrews, R. and Rantell, T. (2000). Load transfer and deformation mechanisms in carbon nanotube-polystyrene composites, *Appl. Phys. Lett.*, **76**(20), pp. 2868–2870.
59. Kubayashi, K. and Hayashi, S. (1992). Woven fabric made of shape memory polymers, United States, Patent 5, 128.
60. Tang, W., Santare, M. H. and Advani, S. G. (2003). Melt processing and mechanical property characterization of multi-walled carbon nanotube/high density polyethylene (MWNT/HDPE) composite films, *Carbon*, **41**(14), pp. 2779–2785.

61. Andrews, R., Jacques, D., Minot, M. and Rantell, T. (2002). Fabrication of carbon multiwall nanotube/polymer composites by shear mixing, *Macromol. Mater. Eng.*, **287**(6), pp. 395–403.

62. Liu, T. X., Phang, I. Y., Shen, L., Chow, S. Y. and Zhange, W. D. (2004). Morphology and mechanical properties of multiwalled carbon nanotubes reinforced nylon-6 composites, *Macromolecules*, **37**(19), pp. 7214–7222.

63. Chen, W., Zhang, J. and Cai, W. (2003). Sonochemical preparation of Au, Ag, Pd/SiO_2 mesoporous nanocomposites, *Scr. Mater.*, **48**(2), pp. 1061–1066.

64. Xia, H. S., Wang, Q., Li, K. S. and Hu, G. H. (2004). Preparation of polypropylene/carbon nanotube composite powder with a solid-state mechanochemical pulverization process, *J. Appl. Polym. Sci.*, **93**(1), pp. 378–386.

65. Ramamurthy, P. C., Malshe, A. M., Harrell, W. R., Gregory, R. V., McGuire, K. and Rao, A. M. (2004). Polyaniline/signle-walled carbon tube composite electronic devises, *Solid State Electrochem.*, **48**(10–11), pp. 2019–2024.

66. Yu, M.-F., Files, B. S., Arepalli, S. and Ruoff, R. S. (2000). Tensile loading of ropes of single wall carbon naotubes and their mechanical properties, *Phys. Rev. Lett.*, **84**(24), pp. 5552–5555.

67. Carbon nanotechnologies, Inc., http://www.cnanotech.com.

68. Hone, J., Whitney, M., Piskoti, C. and Zettl, A. (1999). Thermal conductivity of single-walled carbon nanotubes, *Phys. Rev. B*, **59**, p. R2514(R); Thess, A., Lee, R., Nikolaev, P., et al. (1996). Crystalline ropes of metallic carbon nanotubes, *Science*, **273**, pp. 483–487.

69. Mehrez, T. J., Guo, H., Wang, J. and Roland, C. (2000). Carbon nanotube based magnetic tunnel junctions, *Phys. Rev. Lett.*, **84**(11), pp. 2682–2685.

70. Tombler, T. W., Zhou, C., Alexseyev, L., Kong, J., Dai, H., Liu, L., et al. (2000). Reversible electromechanical characteristics of carbon nanotubes under local-probe manipulation, *Nature*, **405**, pp. 769–772.

71. Yamamoto, K., Akita, S. and Nakayama, Y. (1998). Orientation and purification of carbon nanotubes using ac electrophoresis, *J. Phys. D: Appl. Phys.*, **31**(8), pp. 34–36.

72. Rong, M. Z., Zhang, M. Q., Shi, G., Ji, Q. L., Wetzel, B. and Friedrich, K. (2003). Graft polymerization onto inorganic nanoparticles and its effect on tribological performance improvement of polymer composites, *Tribol. Int.*, **36**, pp. 697–707.

73. Cong, H., Radosz, M., Towler, B. F. and Shen, Y. (2007). Polymer-inorganic nanocomposite membranes for gas separation, *Sep. Purif. Technol.*, **55**, pp. 281–291.

74. Okamoto, M., Morita, S., Kim, Y. H., Kotaka, T. and Tateyama, H. (2000). Synthesis and structure of smectic clay/poly(methyl methacrylate) and clay/polystyrene nanocomposites via in situ intercalative polymerization, *Polymer*, **41**, pp. 3887–3890.

75. Kawasumi, M., Hasegawa, N., Kato, M., Usuki, A. and Okada, A. (1997). Preparation and mechanical properties of polypropylene-clay hybrids, *Macromolecules*, **30**, pp. 6333–6338.

76. Gyoo, P., Venkataramani, S. and Kim, S. (2006). Morphology, thermal, and mechanical properties of polyamide 66/clay nanocomposites with epoxy-modified organoclay, *J. Appl. Polym. Sci.*, **101**, pp. 1711–1722.

77. Tomasko, D. L., Han, X., Liu, D. and Gao, W. (2003). Supercritical fluid applications in polymer nanocomposites, *Curr. Opin. Solid State Mater. Sci.*, **7**, pp. 407–412.

78. Park, S. S., Bernet, N., De La Roche, S. and Hahn, H. T. (2003). Processing of iron oxide-epoxy vinyl ester nanocomposites, *J. Compos. Mater.*, **37**, pp. 465–476.

79. Avadhani, C. V. and Chujo, Y. (1997). Polyimide-silica gel hybrids containing metal salts: Preparation via the sol-gel reaction, *Appl. Organometal. Chem.*, **11**, pp. 153–161.

80. Mathur, S., Veith, M., Shen, H., Hufner, S. and Jilavi, M. (2002). Structural and optical properties of NdAl03 nanocrystals embedded in an Al_2O_3 matrix, *Chem. Mater.*, **14**(2), pp. 568–582.

81. Livage, J. (1997). Sol-gel processes, *Curr. Opin. Solid State Mater. Sci.*, **2**(2), pp. 132–136.

82. Fischer, H. (2003). Polymer nanocomposites: from fundamental research to specific applications, *Mater. Sci. Eng., C*, **23**(6–8), pp. 763–772.

83. Ray, S. S. and Okamoto, M. (2003). Polymer - layered silicate nanocomposites: a review from preparation to processing, *Prog. Polym. Sci.*, **28**(11), pp. 1539–1641.

84. Evora, V. M. F. and Shukla, A. (2003). Fabrication, characterization, and dynamic behavior of polyester/TiO_2 nanocomposites, *Mater. Sci. Eng.*, **361**(1–2), pp. 358–366.

85. Han, K. and Yu, M. (2006). Study of the preparation and properties of UV-blocking fabrics of a PET/TiO_2 nanocomposite prepared by in situ polycondensation, *J. Appl. Polym. Sci.*, **100**(2), pp. 1588–1593.

86. Sarwar, M. I., Zulfiqar, S. and Ahmad, Z. (2008). Polyamide-silica nanocomposites: mechanical, morphological and thermomechanical investigations, *Polym. Int.*, **57**, pp. 292–296.

87. Garnett, J. (1906). Colours in metal glasses, in metallic films, and in metallic solutions, *Philos. Trans. R. Soc. London, Ser. A*, pp. 237–288.

88. Biswas, A., Aktas, O. C., Kanzow, J., Saeed, U., Strunskus, T., Zaporojtchenko, V., et al. (2004). Polymer-metal optical nanocomposites with tunable particle plasmon resonance prepared by vapor phase co-deposition, *Mater. Lett.*, **58**, pp. 1530–1534.

89. Jeon, I.-Y. and Baek, J.-B. (2010). Nanocomposites derived from polymers and inorganic nanoparticles, *Materials*, **3**, pp. 3654–3674.

90. Su, S. J. and Kuramoto, N. (2000). Processable polyaniline-titanium dioxide nanocomposites effect of titanium dioxide on the conductivity, *Synth. Met.*, **114**, pp. 147–153.

91. Yu, S., Hing, P. and Hu, X. (2002). Thermal conductivity of polystyrene-aluminum nitride composite, *Composites Part A*, **33**, pp. 289–292.

92. Ruiz-Hitzky, E., Darder, M. and Aranda, P. (2005). Functional biopolymer nanocomposites based on layered solids, *J. Mater. Chem.*, **15**, p. 3650.

93. Alivisatos, A. P. (1996). Semiconductor clusters, nanocrystals, and quantum dots, *Science*, **271**, p. 933.

94. Daniels, A. U., Chang, M. K. O. and Andriano, K. P. (1990). Mechanical properties of biodegradable polymers and composites proposed for internal fixation of bone, *J. Appl. Biomater.*, **1**(1), p. 57.

95. Bradley, G. W., et al. (1979). Effects of flexural rigidity of plates on bone healing, *J. Bone Joint Surg.*, **61**(6), p. 866.

96. Terjesen, T. and Apalset, K. (1988). The influence of different degrees of rigidity of fixation plates on experimental bone healing, *J. Orthop. Res.*, **6**, p. 293.

97. Gillett, N., Brown, S. A., Dumbleton, J. H. and Pool, R. P. (1985). The use of short carbon fibre reinforced thermoplastic plates for fracture fixation, *Biomaterials*, **6**(2), p. 113.

98. Uchida, A., et al. (1990). The use of calcium hydroxyapatite ceramic in bone tumour surgery, *J. Bone Joint Surg. Br.*, **72B**(2), p. 298.

99. Cooke, F. W. (1992). Ceramics in orthopedic surgery, *Clin. Orthop. Relat. Res.*, **276**, p. 135.

100. Yamaguchi, I., et al. (2001). Preparation and microstructure analysis of chitosan/hydroxyapatite nanocomposites, *J. Biomed. Mater. Res.*, **55**(1), p. 20.

101. Ogawa, K., et al. (1984). A new polymorph of chitosan, *Macromolecules*, **17**, p. 973.

102. Okuyama, K., et al. (1999). Structural study of anhydrous tendon chitosan obtained via chitosan/acetic acid complex, *Int. J. Biol. Macromol.*, **26**(4), p. 285.

103. (a) Kim, H.-W., Kim, H.-E. and Salih, V. (2005). Stimulation of osteoblast responses to biomimetic nanocomposites of gelatin–hydroxyapatite for tissue engineering scaffolds, *Biomaterials*, **26**(25), p. 5221. (b) Sobhana, S. S. Liji, Sundaraseelan, J., Sekar, S., Sastry, T. P. and Mandal, A. B. (2008). Gelatin–chitosan composite capped gold nanoparticles: a matrix for the growth of hydroxyapatite, *J. Nanopart. Res.*, **11**(2), pp. 333–340.

104. Cenni, E., et al. (2000). Biocompatibility and performance in vitro of a hemostatic gelatin sponge, *J. Biomater. Sci. Polym. Ed.*, **11**, p. 685.

105. Mann, S. and Ozin, G. A. (1996). Synthesis of inorganic materials with complex form, *Nature*, **382**, p. 313.

106. Zerwekh, J. E., et al. (1992). Fibrillar collagen-biphasic calcium phosphate composite as a bone graft substitute for spinal fusion, *J. Orthop. Res.*, **10**(4), p. 562.

107. Lin, H.-R. and Yeh, Y.-J. (2004). Porous alginate/hydroxyapatite composite scaffolds for bone tissue engineering: Preparation, characterization, and in vitro studies, *J. Biomed. Mater. Res. Part B: Appl. Biomater.*, **71B**(1), p. 52.

108. Ito, M., Hidaka, Y., Nakajima, M., Yagasaki, H. and Kafrawy, A. H. (1999). Effect of hydroxyapatite content on physical properties and connective tissue reactions to a chitosan-hydroxyapatite composite membrane, *J. Biomed. Mater. Res.*, **45**(3), p. 204.

109. Du, C., et al. (1998). Tissue response to nano-hydroxyapatite/collagen composite implants in marrow cavity, *J. Biomed. Mater. Res.*, **42**(4), p. 540.

110. Boanini, E., et al. (2006). Nanocomposites of hydroxyapatite with aspartic acid and glutamic acid and their interaction with osteoblast-like cells, *Biomaterials*, **27**(25), p. 4428.

111. Edlund, U. and Albertsson, A.-C. (2002). Degradable polymer microspheres for controlled drug delivery, *Adv. Polym. Sci.*, **157**, p. 67.

112. Kronenthal, R. L. (1975). Biodegradable polymers in medicine and surgery, in *Polymer Medicine and Surgery* (Plenum Press, New York), p. 119.

113. Tsuji, H. and Ikada, Y. (2000). Properties and morphology of poly(l-lactide) 4. Effects of structural parameters on long-term hydrolysis of poly(l-lactide) in phosphate-buffered solution, *Polym. Degrad. Stab.*, **67**, p. 179.

114. Bleach, N. C., et al. (2002). Effect of filler content on mechanical and dynamic mechanical properties of particulate biphasic calcium phosphate--polylactide composites, *Biomaterials*, **23**(7), p. 1579.

115. Alexander, H., Langrana, N., Massengill, J. and Weiss, A. (1981). Development of new methods for phalangeal fracture fixation, *J. Biomech.*, **14**(6), p. 377.

116. Loizou, E., et al. (2005). Large-scale structures in nanocomposite hydrogels, *Macromolecules*, **38**(6), p. 2047.

117. Hule, R. A. and Pochan, D. J. (2007). Poly(L-lysine) and clay nanocomposite with desired matrix secondary structure: effects of polypeptide molecular weight, *J. Polym. Sci., Part B: Polym. Phys.*, **45**, **p.** 239.

118. Krikorian, V., et al. (2002). Polypeptide-based nanocomposite: Structure and properties of poly(L-lysine)/Na+-montmorillonite, *J. Polym. Sci., Part B: Polym. Phys.*, **40**, p. 2579.

119. Ozbas, B., et al. (2004). Salt-triggered peptide folding and consequent self-assembly into hydrogels with tunable modulus, *Macromolecules*, **37**(19), p. 7331.

120. Lailach, G. E. and Brindley, G. W. (1969). Specific co-absorption of purines and pyrimidines by montmorillonite (Clay-organic studies XV), *Clays Clay Miner.*, **17**, p. 95.

121. Wei, M., et al. (2005). Layered solids as a "molecular container" for pharmaceutical agents: L-tyrosine-intercalated layered double hydroxides, *J. Mater. Chem.*, **15**, p. 1197.

122. Xu, R., Manias, E., Snyder, A. J. and Runt, J. (2001). New biomedical poly(urethane urea)–layered silicate nanocomposites, *Macromolecules*, **34**(2), p. 337.

123. Wilson, K. S., Zhang, K. and Antonucci, J. M. (2005). Systematic variation of interfacial phase reactivity in dental nanocomposites, *Biomaterials*, **26**(25), p. 5095.

124. Kannan, R. Y., Salacinski, H. J., Butler, P. E. and Seifalian, A. M. (2005). Polyhedral oligomeric silsesquioxane nanocomposites: the next generation material for biomedical applications, *Acc. Chem. Res.*, **38**(11), p. 879.

125. Guan, J. and Hanna, M. A. (2006). Selected morphological and functional properties of extruded acetylated starch-cellulose foams, *Bio. Resour. Technol.*, **97**, pp. 1716– 1726.

126. Jamshidian, M., Tehrany, E. A., Imran, M., Jacquot, M. and Desobry, S. (2010). Poly-lactic acid: production, applications, nanocomposite, and release studies, *Compr. Rev. Food Sci. Food Saf.*, **9**, pp. 552–571.

127. Kaplan, D. L., Mayer, J. M., Ball, D., McCassie, J., Allen, A. L. and Stenhouse, P. (1993). Fundamentals of bio-degradable polymers, in *Biodegradable Polymers and Packaging*, Ching, C., Kaplan, D. L. and Thomas, E. L., eds. (Technomic Pub Co, Lancaster), pp. 1–42.

128. Chandra, R. and Rustgi, R. (1998). Biodegradable polymers, *Prog. Polym. Sci.*, **23**(7), pp. 1273–1335.

129. Tang, S., Zou, P., Xiong, H. and Tang, H. (2008). Effect of nano-SiO_2 on the performance of starch/polyvinyl alcohol blend films, *Carbohydr. Polym.*, **72**, pp. 521–526.

130. Xiong, H., Tang, S., Tang, H. and Zou, P. (2008). The structure and properties of a starch-based biodegradable film, *Carbohydr. Polym.*, **71**, pp. 263–268.

131. Carvalho, A. J. F., Curvelo, A. A. S. and Agnelli, J. A. M. A. (2001). First insight on composites of thermoplastic starch and kaolin, *Carbohyd. Polym.*, **45**, pp. 189–194.

132. Environment Australia (2002). *Biodegradable Plastics: Developments and Environmental Impacts* (Nolan-Itu).

133. Billmeyer, F. W. J. (1984). *Textbook of Polymer Science*, 3rd ed. (John Wiley & Sons).

134. Ogata, N., Jimenez, G., Kawai, H. and Ogihara, T. (1997). Structure and thermal/mechanical properties of poly (L-lactide)-clay blend, *J. Polym. Sci., Part B: Polym. Phys.*, **35**, pp. 389–396.

135. Lee, S., Kang, I., Doh, G., Yoon, H., Park, B. and Wu, Q. (2008). Thermal and mechanical properties of wood flour/talc-filled polylactic acid composites: effect of filler content and coupling treatment, *J. Thermoplast. Compos. Mater.*, **21**, pp. 209– 223.

136. Qu, P., Gao, Y., Wu, G. and Zhang, L. (2010). Nanocomposites of poly (lactic acid) reinforced with cellulose nanofibrils, *BioResources*, **5**, pp. 1811–1823.

137. Misra, M., Park, H., Mohanty, A. K. and Drzal, L. T. (2004). Injection molded "Green" nanocomposite materials from renewable resources, *Global Plastics Environmental Conference*, Detroit, MI, USA.

138. Sithique, M. A. and Alagar, M. (2010). Preparation and properties of bio-based nanocomposites from epoxidized soy bean oil and layered silicate, *Malaysian Polym. J.*, **5**, pp. 151–161.

139. Azeredo, H. M. C., Mattoso, L. H. C., Wood, D., Williams, T. G., Avena-Bustillos, R. J. and Mchugh, T. H. (2009). Nanocomposite edible films from mango puree reinforced with cellulose nanofibers, *J. Food Sci.*, **74**, pp. 31–35.

140. Tate, J. S., Akinola, A. T. and Kabakov, D. (2010). Bio-based nanocomposites: an alternative to traditional composites, *J. Technol. Stud.*, **1**, pp. 25–32.

141. Nunes, M. R. S., Silva, R. C., Silva, J. G., Jr., Tonholo, J. and Ribeiro, A. S. (2009). Preparation and morphological characterization of chitosan/clay nanocomposites, *Proceedings of the 11th International Conference on Advanced Materials*, Rio de Jenero, Brazil, pp. 20–25.

142. Maiti, P., Batt, C. A. and Giannelis, E. P. (2003). Renewable plastics: synthesis and properties of PHB nanocomposites, *Polym. Mater. Sci. Eng.*, **88**, pp. 58–59.

143. Gao, F. (2004). Clay/polymer composites: the story, *Mater. Today*, **7**(11), pp. 50–55. (Reproduced in part with permission from Elsevier 2006.)

144. Hay, J. N. and Shaw, S. J. *Nanocomposites: properties and applications*. Available online: http://www.azom.com/Details.asp?ArticleID=921 (accessed on August 15, 2010).

145. Pandey, J. K., Chu, W. S., Lee, C. S. and Ahn, S. H. (2007). Preparation characterization and performance evaluation of nanocomposites from natural fiber reinforced biodegradable polymer matrix for automotive applications, *International Symposium on Polymers and the Environment: Emerging Technology and Science, Bioenvironmental Polymer Society (BEPS)*, Vancouver, WA, USA.

146. Lee, Y. H., et al. (2005). Electrospun dual-porosity structure and biodegradation morphology of montmorillonite reinforced PLLA nanocomposite scaffolds, *Biomaterials*, **26**(16), p. 3165.

147. Byrne, S. J., le Bon, B., Corr, S. A., Stefanko, M., O'Connor, C., Gun'ko, Y. K., Rakovich, Y. P., Donegan, J. F., Williams, Y., Volkov, Y. and Evans, P. (2007). Synthesis, characterisation, and biological studies of CdTe quantum dot-naproxen conjugates, *ChemMedChem*, **2**, p. 183.

148. Tartaj, P., Morales, M. D., Veintemillas-Verdaguer, S., Gonzalez-Carreno, T. and Serna, C. J. (2003). The preparation of magnetic nanoparticles for applications in biomedicine, *J. Phys. D: Appl. Phys.*, **36**, p. R182.

149. Corr, S. A., Rakovich, Y. P. and Gun'ko, Y. K. (2008). Multifunctional magnetic-fluorescent nanocompositesfor biomedical applications, *Nanoscale Res. Lett.*, **3**, pp. 87–104.

150. Baibarac, M. and Gómez-Romero, P. (2006). Nanocomposites based on conducting polymers and carbon nanotubes: from fancy materials to functional applications, *J. Nanosci. Nanotechnol.*, **6**, pp. 289–302.

151. Baughman, R. H., Zakhidov, A. A. and de Heer, W. A. (2002). Carbon nanotubes: the route toward applications, *Science*, **297**, pp. 787–792.

152. Moniruzzaman, M. and Winey, K. I. (2006). Polymer nanocomposites containing carbon nanotubes, *Macromolecules*, **39**, pp. 5194–5205.

153. Sandler, J. K. W., Kirk, J. E., Kinloch, I. A., Shaffer, M. S. P. and Windle, A. H. (2003). Ultra-low electrical percolation threshold in carbon-nanotube-epoxy composites, *Polymer*, **44**, pp. 5893–5899.

154. Van Noorden, R. (2006). Moving towards a graphene world, *Nature*, **442**, pp. 228–229.

155. Kothurkar, N. K. (2004). *Solid State, Transparent, Cadmium Sulfide-Polymer Nanocomposite*, Dissertation, University of Florida.

156. Hule, R. A. and Pochan, D. J. (2007). Polymer nanocomposites for biomedical applications, *MRS Bull.*, **32**, pp. 354–358.

Chapter 5

Ion-Exchange Nanocomposites: Avant garde Materials for Electrodialysis

Shaswat Barua,[a] Swagata Baruah,[b] and Rocktotpal Konwarh[c]
[a]*Department of Chemistry, School of Basic Sciences, Assam Kaziranga University, Jorhat 785006, Assam, India*
[b]*Applied Organic Chemistry Group, Chemical Science and Technology Division, CSIR-NEIST, Jorhat 785006, Assam, India*
[c]*Department of Biotechnology, College of Biological and Chemical Engineering, Addis Ababa Science and Technology University, Addis Ababa, P.O. Box 16417, Ethiopia*
rock1311@gmail.com

With a wide gamut of use, nanocomposites have etched an indelible impression on the canvas of materials science and other domains. The most recent innovative thrust has been directed toward utilization of ion-exchange nanocomposites for electrodialysis. The chapter begins with a note on the design criteria and hallmark attributes of electrodialysis membranes. The major focus is on the fabrication of various ion-exchange nanocomposites using nanomaterials of different dimensions, their physicochemical attributes, the advantages, and the associated challenges. Among a plethora of

Hybrid Nanocomposites: Fundamentals, Synthesis, and Applications
Edited by Kaushik Pal

ISBN 978-981-4800-34-1 (Hardcover), 978-0-429-00096-6 (eBook)
www.panstanford.com

applications, we have specifically stressed upon their biomedical relevance and their applications in the domain of water desalination. The practical pertinence of these materials has been put forth vis-à-vis the challenges and the scope for improvement. Thus, the chapter will be a portal to the current research on electrodialysis membranes, fabricated using different polymeric nanocomposites. This will act as a catalyst in the niche of materials science for the seeding of novel ideas toward the development of commercially viable, industrially and biomedically relevant electrodialysis membranes.

5.1 Introduction

Electrodialysis is a process that uses semipermeable membranes to transport ions under the influence of an electric potential. It has a wide range of applications, including desalination of seawater, wastewater treatment, food processing, agriculture, artificial kidneys, glycerin purification, etc. [1]. Ion-exchange membranes are the prime components of these processes, which restrict the transport of co-ions and allow the passage of counterions. Transport of ions is carried out with the help of electrodialysis cells. In a typical electrodialysis cell, anion- and cation-exchange membranes are placed between two electrodes. These constitute the dilute and concentrated compartments. The negatively charged ions move toward the anode, while the positively charged ones migrate toward the cathode under the influence of an applied electric potential. This process is generally used to desalinate water. Suppose, a sample of water containing a very high concentration of a salt, say sodium chloride, is desalinated by electrodialysis; then the individual ions (Na^+ and Cl^-) start migrating toward the opposite electrodes. The ion-exchange membranes used in a typical cell, thus, provide a barrier for the ions to pass through, which results in the formation of two streams, that is, a concentrated and a dilute stream. With operation time, the ion concentration increases in the concentrated stream, while a corresponding decrease is observed in the other stream. The function of an electrodialysis cell is schematically presented in Fig. 5.1.

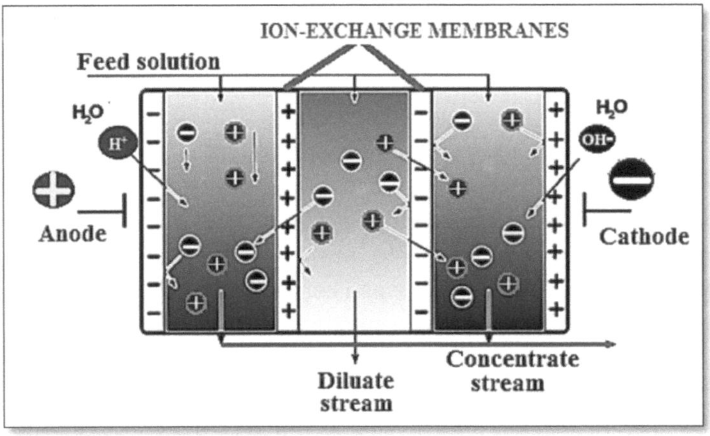

Figure 5.1 Principle of electrodialysis.

The efficiency of the membrane materials dictates the efficiency of the ion-exchange process. The main function of a membrane is to act as an interphase between the two compartments of different phases, thereby regulating the transport of ions. A plethora of reports has attested the commercial potential and viability of different ion-exchange membranes since the past five decades [2–4]. The crucial criteria for selection of a membrane material include operational cost, environmental safety, and efficacy of ion exchange. The most commonly used ion-exchange materials in such membranes are ion-exchange polymers. Amongst them, polystyrene-based composites have been extensively used [5]. Reports are also available on the application of cellulose acetate, poly(4-vinylpyridine), polysulfone, sulfonated polyphosphazene, and poly(benzylsulfonic acid)siloxane (PBSS) for the purpose [6, 7]. However, the use of pristine polymer-based membranes is intertwined with numerous issues like poor durability, inadequate mechanical performance, inefficient ion-exchange capacity (IEC), and shorter service life [8]. As a prospective solution, scientists have resorted to the approach of polymeric blending to achieve advancement in properties [9]. Further, with the outbreak of the *nanotechnological revolution*, researchers have explored the possibility of using nanomaterials in ion-exchange membranes. Nanomaterials are known for their unique shape-size-surface-chemistry accord. Polymeric nanocomposites have geared up research across diverse realms, including the biomedical world, catalysis, biomolecule immobilization, etc. [10–18]. It is of pertinence to note that various nanocomposites have catalyzed the novel fabrication of tunable membranes for multifaceted applications. In this chapter, the progress of ion-exchange materials vis-à-vis nanocomposites has been highlighted in the context of electrodialysis. Herein, the critical parameters of electrodialysis membranes are described with special emphasis on the membrane matrix selection and the choice of nanomaterials. Furthermore, an effort has been made to present a comprehensive discussion on the development of ion-exchange membranes in recent years. Thus, this chapter is expected to be of high interest to readers from diverse backgrounds for retrieving detailed information on the current status of various ion-exchange nanocomposites.

5.2 Hallmarks of an Electrodialysis Membrane

As described by Kariduraganavar et al., the critical aspects of an ion-exchange membrane include retrieval and enhancement of important ions and exclusion of undesired ions, especially toxic metals [19]. Primarily, an ion-exchange membrane separates cations from anions and vice versa.

Such membranes are fabricated as polymeric foils with ion-exchange ability. Different synthetic fibers are impregnated to such matrices to attain adequate mechanical performance. Both cation- and anion-exchange membranes should satisfy the following requisites for efficient functioning: (i) electrical conductivity or low resistivity, (ii) strength as well as flexibility to form stacks, (iii) hydrophobicity to avoid degradation in aqueous solutions, (iv) resistance to small changes in pH and temperature, (v) prolonged service life, and (vi) antifouling or fouling resistance [20].

Another important parameter of such a membrane is the ion transport number. This is also simply termed as "transport number" or "transference number." It is defined as the fraction of total current carried by an ion "a" in an electrolyte. The transport number (T_a) can be expressed as follows:

$$T_a = \frac{I_a}{I_{tot}}, \tag{5.1}$$

where I_a is the fraction of current carried by the ion "a" and I_{tot} is the total current.

Good permeation or a high transport number implies efficient transportation of a particular kind of ion. Thus, a membrane selective to monovalent ions can resist or restrict the flow of divalent or trivalent ions. The membranes that can permit the passage of only counterions and block the co-ions completely are termed as "ideal permselective membranes" (Fig. 5.2) [21]. Permselectivity refers to the preferential permeation of certain ionic species through ion-exchange membranes.

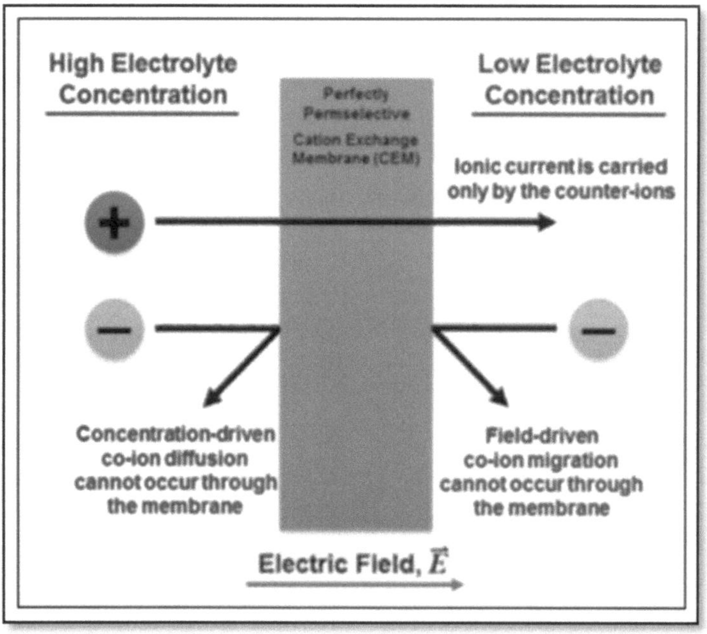

Figure 5.2 Ion transport through a perfectly permselective cation-exchange membrane. Reproduced from Ref. [21] with permission of the PCCP Owner Societies.

However, a trade-off in properties cannot be overruled while fabricating membranes with particular features. For example, mechanical strength may be increased by the increment of crosslinking density in a polymer, which, on the other hand, decreases the conductivity of the matrix and, consequently, the IEC.

Therefore, it was well understood by the scientific fraternity that polymers alone might not act as the best material for ion-exchange membranes. Polymers generally aid in the separation of ions via the solution diffusion method, where penetrants diffuse through the membrane [22]. Subsequent downstream desorption results in efficient ion exchange. However, in the case of electrodialysis, electrical properties of the membranes play a major role. Though a few polymers like poly(pyrrole), polycarbazoles, polyaniline, etc., show good conductivity, they do not suffice the requirements to be used in electrodialysis without forming composites [23]. Again, leaching of nanomaterials is a serious issue, which raises the concern of toxicity, particularly in water desalination and biomedical membranes [24]. Conversely, nanomaterials impart unique attributes like selectivity, conductivity, antifouling, and magnetic properties to the composite system, while polymers regulate their activity, besides rendering excellent stability and durability.

The size-sieving mechanism underlines separation in simple filtration that involves pore flow. However, attributed to polymer surface charge, low to moderate rejection of higher-valent ions and low rejection of monovalent ions by pore flow membranes are a common observation in nanofiltration (NF) membranes, where pore size is considerably small. This had led to the concept of utilizing composite materials in ion-exchange membranes. With the emergence of nanotechnology, nanocomposites have taken over the domain of composite science [25, 26].

Delving into the mechanism of ion exchange through nanocomposite membranes, Tong and coworkers found that as the nanomaterial loading increases, permselectivity and swelling of cation-exchange membranes increase initially and then decrease. The IEC and ionic resistance of such membranes exhibit a similar nature irrespective of the nanomaterial used. The study also revealed that a change in the membrane microstructure dictates the membrane properties, as evident from modeling analysis (Fig. 5.3) [27].

In subsequent sections, the role of nanocomposites in electrodialysis membranes will be discussed thoroughly.

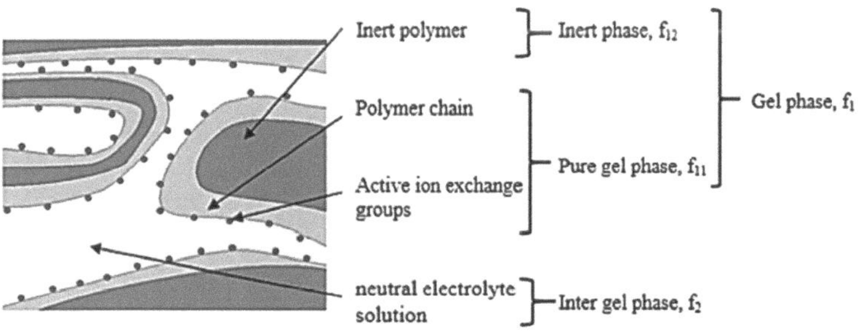

Figure 5.3 Microphases of ion-exchange membrane based on a three-phase model. Reprinted with permission from Ref. [27]. Copyright (2017) American Chemical Society.

5.3 Nanocomposites in Electrodialysis Membranes

Significant efforts have been put forth by various researchers to address the practical challenges of using pristine polymeric membranes by preparing appropriate composites. Composites are defined as materials containing two or more phases in either micro- or macroscales. Useful characteristics of the materials aid the composite system individually [28]. Rapid progress in various nanocomposites has added new hues to the canvas of materials science. Excellent barrier properties and reinforcement of mechanical attributes are the two prime characteristics of nanocomposites [29]. Nanocomposites can be defined as "multiphase material, which has one, two or all the dimensions within 100 nm range." Polymer nanocomposites may be broadly classified into two types: (i) intercalated and (ii) exfoliated nanocomposites. In the case of the intercalated nanocomposites, nanomaterials penetrate the polymer matrix in an organized crystallographic mode. In exfoliated nanocomposites, nanomaterials do not retain their identical orientation and get dispersed within the polymer [28].

Nanomaterials are often categorized as 0D, 1D, 2D, and 3D ones from a broader perspective. Apart from the mixing strategies, interfacial interaction between the nanomaterials and polymers dictates the type of nanocomposites formed. Thus, the electrical conductivity, mechanical strength, hydrophobicity, antifouling properties, etc., of an electrodialysis membrane can be tailored by varying the polymeric structures and the composite recipe. The utilization of polymer nanocomposites for the fabrication of electrodialysis membranes is discussed in subsequent sections in terms of the nanomaterials used.

5.3.1 Application of 0D Nanomaterials in Electrodialysis Membranes

Zero-dimensional nanomaterials represent the class where all the three dimensions of the material are in the nanoregime. Metal nanoparticles (MNPs), nanoclusters, and quantum dots are the representative members of this category. Impregnation of inorganic nanoparticles into polymeric matrices imparts many interesting attributes to the nanocomposites, which include increment of mechanical and thermal stability. Furthermore, this assists in fabricating high-temperature- and chemical-resistant materials capable of withstanding severe oxidizing environments. Synergistic effects of the organic-inorganic components also improve the permselectivity [30].

Amongst MNPs, silver has been widely explored in membranes due to its significant electrical and antibacterial properties [31]. In 2010, Hosseini and coworkers investigated the electrical conductance, transport number, selectivity, and membrane potential of the electrodialysis membranes, the design of which involved impregnation of silver nanoparticles (Ag NPs) within a modified polymer matrix, composed of poly(vinyl chloride) (PVC) and styrene-butadiene-rubber, through plasma treatment [31]. They investigated the morphology of the modified and unmodified membranes by scanning optical microscopy (SOM; Fig. 5.4a,b). Plot of the afore-stated parameters of a representative 40-nm-thick membrane with respect to solution concentration is presented in Fig. 5.4c [31].

On the other hand, Zendehnam et al. fabricated an anion-exchange membrane composed of acrylonitrile butadiene styrene/activated carbon/silver nanoparticles [32]. They observed an increased selectivity of the membrane with an increase in the thickness of the deposited Ag NP layer, along with significant antimicrobial action against *E. coli* (Fig. 5.5a). A silver/PVC nanocomposite was also used by Zarrinkhameh et al. for fabricating an antifouling dialysis membrane [33]. Another work focused on the use of copper nanoparticle–embedded multiwalled carbon nanotubes (MWCNTs) in a PVC-based electrodialysis membrane [34]. The researchers found an increment in the membrane potential, electrical conductivity, and transport number on increasing the loading of the nanomaterials. A recent report showcased the fabrication of a bipolar electrodialysis membrane composed of polyethylene, polypropylene, and sulfonated poly(2,6-dimethyl-1,4-phenylene oxide), embedded with iron oxide nanoparticles [35]. The electrodialysis membrane is pictorially presented in Fig. 5.5b. The stability of the fabricated membranes was investigated by measuring the Cl^- and SO_4^{2-} concentrations in specific time intervals. This investigation was instrumental in understanding the time-dependent electrodialysis efficacy of the membranes.

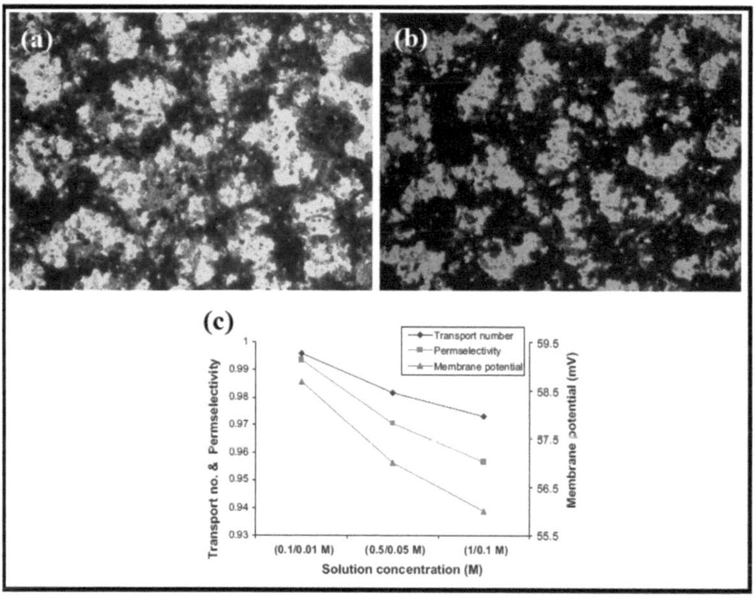

Figure 5.4 SOM images of (a) unmodified and (b) modified PVC and styrene-butadiene-rubber, and (c) effect of NaCl concentration on selectivity, membrane potential, and transport number of silver nanoparticle–impregnated PVC and styrene-butadiene-rubber membrane. Reprinted from Ref. [31], Copyright (2010), with permission from Elsevier.

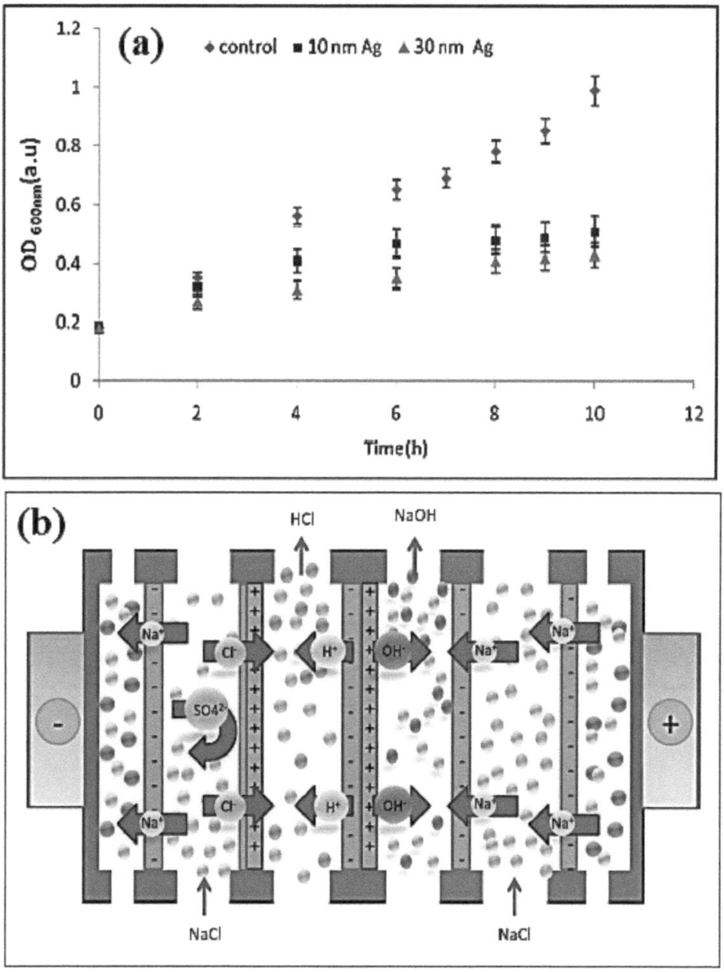

Figure 5.5 (a) Inhibition of the growth of *E. coli* with time by acrylonitrile butadiene styrene/activated carbon/silver nanocomposite membrane. Reprinted from Ref. [32], Copyright (2013), with permission from Elsevier. (b) Electrodialysis bipolar membranes. Reprinted from Ref. [35], Copyright (2017), with permission from Elsevier.

In yet another endeavor, Fe_3O_4 nanoparticle–impregnated polyaniline membranes were synthesized by Daraei et al. The membranes exhibited tremendous potential for copper ion removal from water [36]. Similarly, an anion-selective membrane was fabricated by incorporating magnetic iron/nickel oxide nanoparticles in PVC, which showed greater permeability for bivalent ions as compared to the monovalent counterparts [37]. Increase in the concentration of Fe_2NiO_4 nanoparticles improved the IEC (plausibly due to the better interaction of the ions with the surface of membrane, thereby facilitating the ion transport) (Fig. 5.6a).

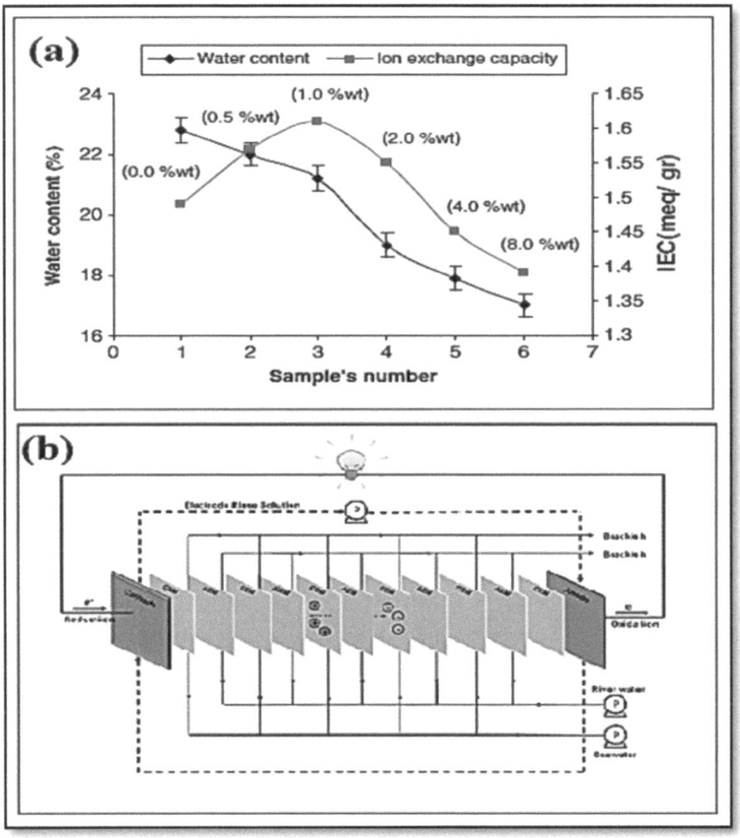

Figure 5.6 (a) Effect of Fe_2NiO_4 nanoparticle loading onto PVC and the IEC of the membranes. Reprinted from Ref. [37], Copyright (2012), with permission from Elsevier. (b) Stack of reverse electrodialysis membranes. Reprinted from Ref. [38], Copyright (2014), with permission from Elsevier.

Functionalized iron oxide nanoparticle–decorated sulfonated poly(2,6-dimethyl-1,4-phenylene oxide) membranes for reverse electrodialysis (RED) have been recently reported [38]. A typical RED device is commonly made up of alternating series of anion- and cation-exchange membranes, as shown in Fig. 5.6b.

Another exceptional class of 0D nanomaterials are quantum dots. However, very few reports can be retrieved on the use of quantum-dot-based polymeric nanocomposites in electrodialysis membranes.

5.3.2 Application of 1D Nanomaterials in Electrodialysis Membranes

This class of nanomaterials is specified by members with two dimensions in the nanometer range and the third in the micrometer scale. The various remarkable morphologies include nanotubes, nanofibers, nanorods, nanowires, etc.

The most widely used 1D nanomaterials are carbon nanotubes (CNT), which are endowed with numerous hallmark features for a number of advanced applications. Excellent mechanical and electrical properties have endorsed CNTs for application in electrodialysis. CNT architecture comprises sp^2-hybridized carbon atoms rolled in the shape of tubes. CNTs are classified as single-walled (SW), double-walled (DW), and multiwalled (MW) CNTs, depending upon the number of graphitic layers, present in the structure. Different techniques have been reported for synthesizing CNTs for the past two decades.

Li et al. reported a novel membrane device by the unison of carbon nanofibers (CNFs) and a CNT composite [39]. They described the fabrication of the membrane by low-pressure vapor deposition methods. In this work, 49.2% improvement in the desalination performance was documented for the fabricated membrane as compared to the conventional capacitive deionization. The unique setup is schematically depicted in Fig. 5.7.

MNP-embedded CNT hybrids were further investigated by a few research groups. This provides a strong matrix when incorporated in polymers and confers unique attributes like excellent mechanical performance and infection-resistant surfaces. In this quest, Arai et al. reported the fabrication of a copper/SWCNT composite film [40]. Some beaded structures of the composites were identified using scanning electron microscopy (SEM) (Fig. 5.8). The most important feature of a nanohybrid or composite depends on the interaction amongst the nanomaterials. The uniform decoration of the silver nanoparticles onto the surface of SWCNTs is depicted in Fig. 5.8a–c. A similar complementary benefit of each material has been demonstrated in recent works of other research groups.

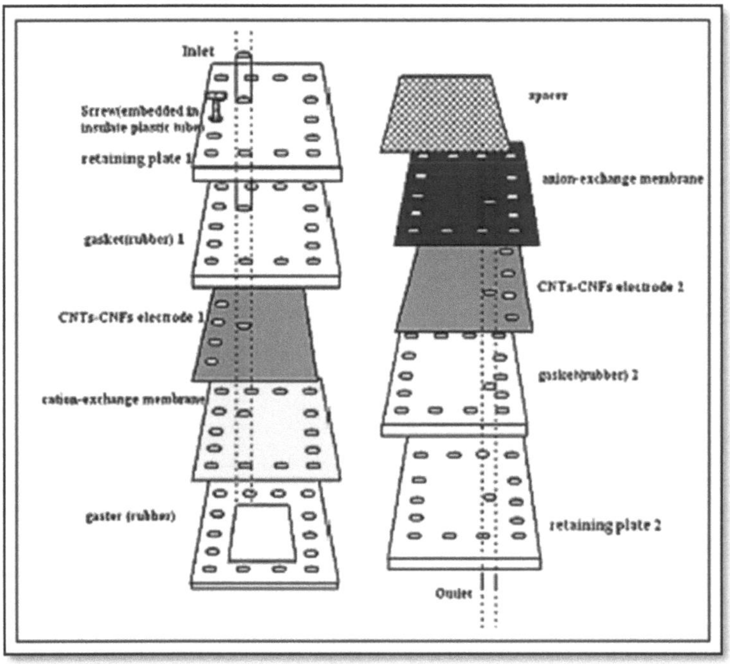

Figure 5.7 CNT–CNF composite membrane capacitive deionization device. Reprinted from Ref. [39], Copyright (2008), with permission from Elsevier.

The most widely preferred material in membrane technology is cellulose. Hosseini and coworkers also reported the preparation of a PVC/cellulose acetate cation-exchange electrodialysis membrane [41]. They observed that the increase in cellulose acetate content improved the IEC from 1.45 to 1.57 meq/g. Due to the hydrophilic characteristic of the cellulose matrix, ion transportation was facilitated, resulting in the efficient IEC of the membrane. Figure 5.9 illustrates the IEC of this nanocomposite membrane with different PVC:cellulose acetate ratios [41].

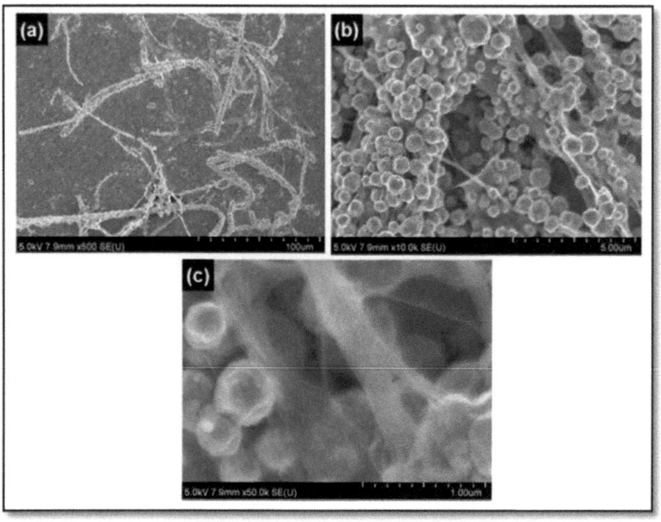

Figure 5.8 SEM micrographs of copper/SWCNT composite in different magnifications. Reproduced from Ref. [40], published under the terms of CC BY NC ND.

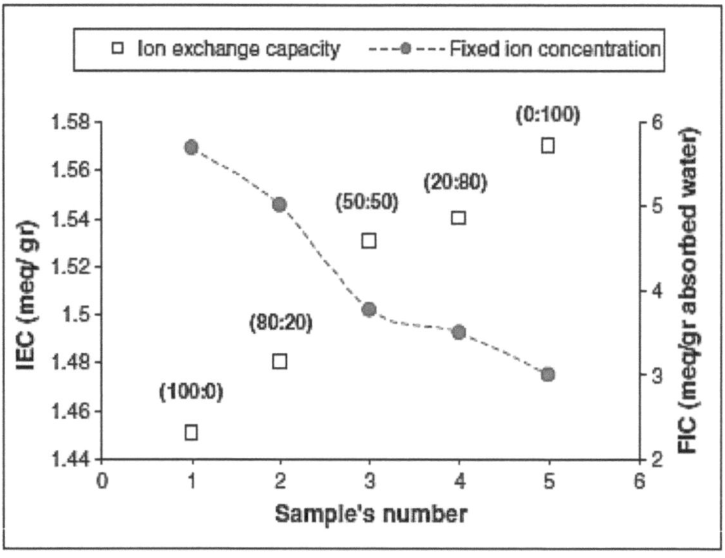

Figure 5.9 IEC and fixed ion concentration (FIC) of membranes with different PVC:cellulose acetate ratios. Reprinted from Ref. [41], Copyright (2012), with permission from Elsevier.

Nanocrystalline and nanofibrillar cellulose are also used in electrodialysis membranes. Ferraz and coworkers designed a conducting nanocomposites based on nanocellulose and polypyrrole, which showed encouraging results for blood purification [42]. This group also investigated the hemodialysis efficiency of a porous nanocellulose/polypyrrole membrane [43]. They witnessed a faster diffusion of uremic toxins across the nanocomposite membrane as compared to a bare cellulose membrane.

A novel nanocellulose membrane impregnated with silver nanoparticles was designed by Suman et al. for the effective removal of microbes, dyes, and metal contaminants from water [44].

5.3.3 Application of 2D Nanomaterials in Electrodialysis Membranes

Two-dimensional nanomaterials include hydrophobic and hydrophilic nanoclay, graphene, graphene oxide (GO), reduced graphene oxide (rGO), etc. Nanoclays are constituted of tetrahedrally bound silicon atoms with $Al(OH)_3$ or $Mg(OH)_2$. They possess a very high surface-to-volume ratio and layers are bound by weak physical forces to form galleries. A range of nanoclay materials are known till date. This includes bentonite, illite, nacrite, sapiolite, antigorite, talc, montmorillonite, kenyaite, beidellite, belinite, kolinite, brucite, halloysite, octosilicate, etc. Due to the unique structural features, nanoclays exhibit an excellent IEC. Large surface functionalities of nanoclays form excellent composites with polymers. Electrostatic forces play a major role in such interactions [45]. In a previous study, hydrophilic bentonite-supported MNPs were reported to exhibit >90% efficiency in the removal of Pd, Cr, and Cu ions from wastewater [46].

Two-dimensional graphene-based water desalination membranes were fabricated by Konatham et al. [47]. They studied the ion and water translocation across these membranes with the help of potential energy surfaces. Sulfonated polyethersulfone (SPES) and GO nanocomposite–based membranes were designed by Gahlot and coworkers [48]. The researchers observed a complementary effect on the transport properties of ions when GO was incorporated [48]. The d-spacing of the graphitic layers was enhanced upto 7.81 Å upon formation of GO (Fig. 5.10).

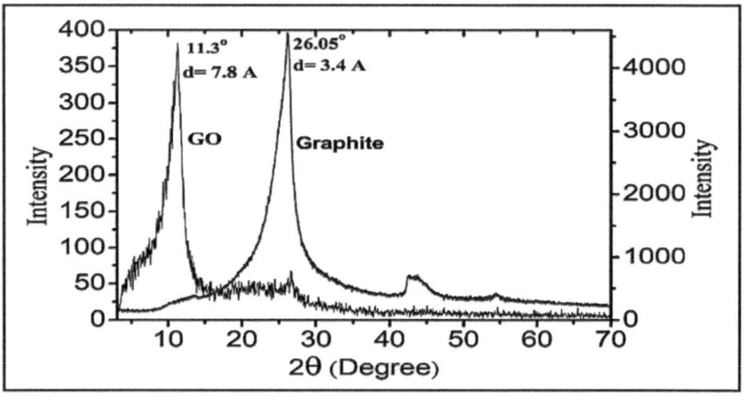

Figure 5.10 X-ray diffraction (XRD) patterns for graphite and GO. Reproduced from Ref. [48] with permission of The Royal Society of Chemistry.

Due to the high cost of graphene, rGO is preferred in most cases [49]. rGO is known to exhibit the outstanding characteristics of graphene [50]. The conventional synthesis of rGO involves direct reduction of GO using reducing agents such as hydrazine, hydroquinone, sodium borohydride, etc. [51–53]. Such reagents if present in trace amounts may cause severe damage to biological systems. In this milieu, efforts have been made to use environment-friendly benign materials, such as aqueous extracts of leaves and fruits [54]. Thakur and Karak have reviewed the reports on alternative methods and nature-based reagents for the reduction of GO [55]. They had proposed various mechanisms for the reduction of GO by vitamins, carbohydrates, and amino acids. They had particularly focused upon the reduction-based strategy, using various plant leaf extracts, such as tea, *Colocasia esculenta*, *Mesuaferrea* L., *Citrus sinensis*, *Ginkgo biloba*, *Rosa damascene*, etc. [55]. Further, rGO and MNP-based hybrid materials have been synthesized by simultaneous green reductive techniques. In this regard, silver, copper/copper oxide, and iron oxide nanoparticles decorated rGO have been reported [54, 55].

Such nanohybrids combine the properties of both the nanomaterials and confer fascinating attributes like stability, antimicrobial property, photoactivity, and electrical activity. Thus, these kinds of hybrid nanomaterials with graphene-based structures are envisaged to be befitting for the fabrication of electrodialysis membranes. Furthermore, the retention of functionalities of rGO and its high surface energy make it compatible with various polymeric matrices. rGO/TiO_2-embedded m-phenylenediamine membranes with antifouling properties were prepared by Safarpour and coworkers for desalination of water [56]. They observed superior water flux and salt rejection for the composite membranes as compared to bare membranes (Fig. 5.11).

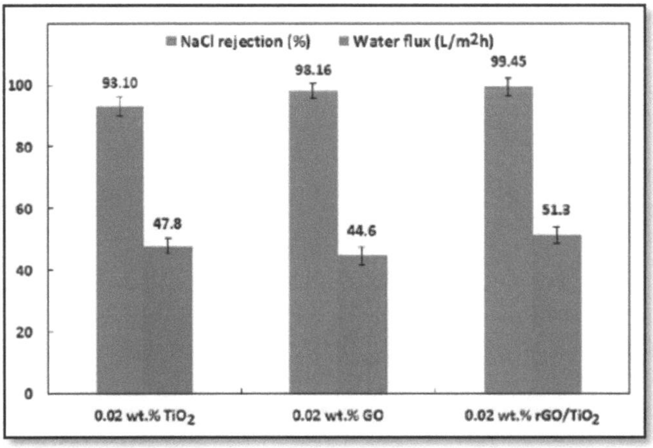

Figure 5.11 Water flux and salt rejection of rGO/TiO_2-embedded m-phenylenediamine membranes. Reprinted from Ref. [56], Copyright (2015), with permission from Elsevier.

Fouling is one of the pertinent problems associated with desalination membranes [57]. Due to the presence of organic and inorganic impurities and biogenic materials, surface fouling occurs in the membranes [58]. Biological impurities such as pathogenic microbes form a biofilm, which results in the retardation of the desalination process, consumes high energy, and hampers membrane durability [59]. Unison of MNPs and graphene-based materials is expected to impart excellent toughness as well as fouling resistance capacity to the membranes. Moreover, their tunable electrical conductance would aid in the fabrication of high-performance electrodialysis membranes.

5.3.4 Application of 3D Nanomaterials in Electrodialysis Membranes

Three-dimensional nanomaterials include polycrystals, mutilayered materials, etc. Box-shaped graphene (BSG) nanoclusters are also included in this category. Cleavage of pyrolytic carbon/pyrolytic graphite yields BSG nanoclusters, which are multilayered systems of nanochannels having a quadrangular cross section [60]. However, research on such 3D nanostructures, particularly in the fabrication of nanocomposite membranes for application in electrodialysis, is still incipient.

5.4 Applications

As explained in the preceding sections, nanocomposites have found immense use in ion-exchange membrane technology. A nonexhaustive list of polymers and nanomaterials that have been utilized in the fabrication of electrodialysis membranes is presented in Table 5.1.

Table 5.1 Representative nanocomposite-based electrodialysis membranes

S. no.	Nanomaterial	Polymer	Application	References
1	Silver	PVC/Styrene-butadiene-rubber	Cation-exchange membrane	[31]
2	Silver	Acrylonitrile-butadiene-styrene	Heterogeneous anion-exchange membrane	[32]
3	Silver	PVC	Cation-exchange membrane	[33]
4	Iron oxide	Polyethylene, polypropylene, sulfonated poly(2,6-dimethyl-1,4-phenylene oxide)	Bipolar electrodialysis membrane	[35]

S. no.	Nanomaterial	Polymer	Application	References
5	Iron oxide	Polyethersulfone and polyaniline	Membrane for Cu removal from water	[36]
6	Magnetic iron/ nickel oxide nanoparticles	PVC/styrene-butadiene-rubber	Heterogeneous cation (monovalent)-exchange membrane	[37]
7	Iron oxide	Poly(2,6-dimethyl-1,4-phenylene oxide)	RED membrane	[38]
8	Cellulose	PVC	Cation-exchange membrane	[41]
9	Nanocellulose	Polypyrrole	Blood purification membrane	[42]
10	*Cladophora* cellulose	Polypyrrole	Hemodialysis membrane	[43]
11	Nanocellulose and silver	-	Dye, heavy metal, and microbial removal	[44]
12	Bentonite clay and iron	-	Removal of Cr(VI) ions	[46]
13	GO	SPES	Desalination of water	[47]
14	rGO/TiO_2	Poly m-phenylenediamine	Desalination of water	[56]

In the subsequent section, we have highlighted the applications of such nanocomposites under two broad headings.

5.4.1 Water Desalination

Water desalination is a prime concern in the context of water recovery and treatment processes. Recent investigations have established the prospects of nanocomposite-based ion-exchange membranes for this purpose. Hosseini et al. fabricated such a membrane by impregnating TiO_2 nanoparticles within a PVC matrix [61]. They observed that the permeability of sodium ions can be increased by increasing TiO_2 loading. In another of their significant attempts, better selectivity, transport number, and electrical conductance on incorporation of zeolite nanoparticles up to 8 wt% were reported [62]. On the other hand, CNT-based membranes have been fabricated with different polymeric systems, exhibiting excellent potential for water desalination. Polyethersulfone, polydimethylsiloxane, polystyrene, polyvinylidene fluoride, etc., are some of the polymers that have been used for designing CNT-based nanocomposite membranes with a wide range of selectivity (44.9% to 96.1%) and permeability (1.2 to 19.1 L m^{-2} h^{-1} bar^{-1}) [63–66]. It is pertinent to note that Gethard and coworkers demonstrated a mechanism, establishing the role of CNTs in the desalination process. They demonstrated that immobilization of CNTs in the membrane pores altered the interaction between membrane and water, thereby affecting the permeability and selectivity of the membrane (Fig. 5.12) [66].

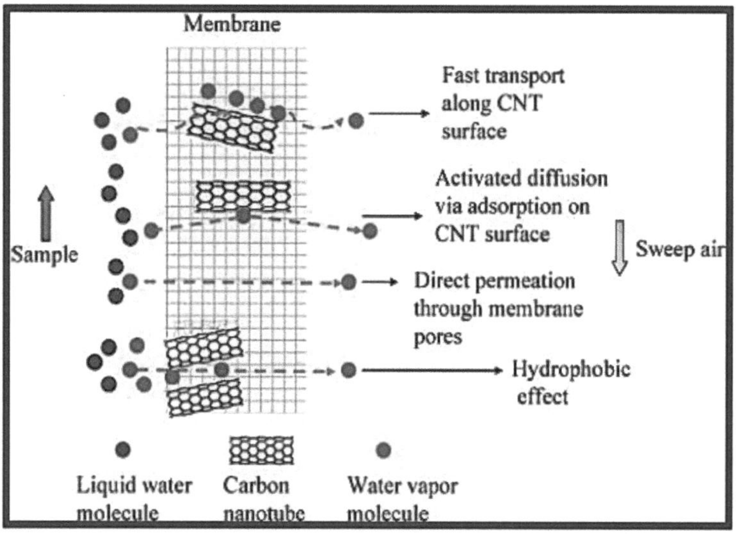

Figure 5.12 Mechanism of water desalination in the presence of CNTs. Reprinted with permission from Ref. [66]. Copyright (2011) American Chemical Society.

The use of carbon-based nanomaterials, such as CNTs, graphene, rGO, or MNPs like silver, copper/copper oxide, zinc oxide, titanium dioxide, and iron oxide, further confers antifouling attributes to the membrane, thereby enhancing the longevity and performance [67–69]. Again, some inorganic materials, including alumina, zirconia, and silica, have been employed to improve the quality of the membranes [70–73]. However, due to various issues related to toxicity and safety regulations, the hunt for an ideal electrodialysis membrane for water desalination continues.

5.4.2 Biomedical Applications

An ideal biomedically applicable membrane should satisfy the requisites of biocompatibility and inertness without compromising its IEC. Thus, selection of the polymer and the nanomaterial is very crucial in this regard. Most of the biocompatible polymers, including chitosan, polyethylene glycol, chitin, polylactic acid, or polyglycolic acid, hardly contribute to the overall mechanical performance of the nanocomposite. Resorting to the use of cellulose, the most abundant organic polymer, may be a prudent step for fabricating tough, durable, biocompatible biomembranes [74]. However, this necessitates the incorporation of electroactive nanomaterials for efficient electrodialysis. Thus, there is broad scope for the development of electrodialysis membrane technology for biomedical utility.

A few reports stand in testimony to the use of nanocomposites for fabrication of biomedically relevant electroactive membranes. To cite for evidence, heparinized composites of nanocellulose and polypyrrole were fabricated as hemodialysis membranes with excellent IEC and hemocompatibility [42]. A similar study illustrated that an electroactive polypyrrole/*Cladophora* cellulose membrane could efficiently remove small water-soluble uremic toxins as well as large pathogenic proteins [43].

Kaleekkal and coworkers reported the fabrication of a GO/poly(vinylpyrrolidone)-incorporated polyetherimide nanocomposite membrane with "reduced protein adsorption, suppressed platelet adhesion and lower complement activation" [75]. The structural and functional attributes of the fabricated membrane are schematically presented in Fig. 5.13.

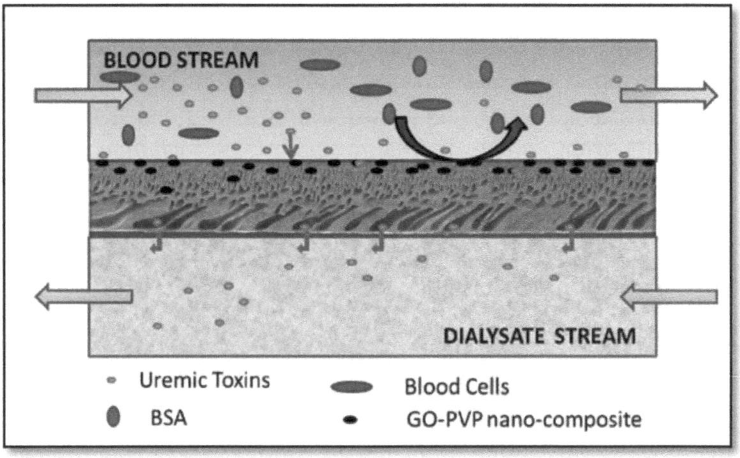

Figure 5.13 GO/poly(vinylpyrrolidone)-incorporated polyetherimide nanocomposite membrane for blood purification. Reprinted with permission from Ref. [75]. Copyright (2015) American Chemical Society.

5.5 Future Perspective

Successful application of membranes of electrodialysis is intertwined with the challenges of developing efficient, cost-effective materials and technological strategies to obtain the desired qualities. In this context, novel biocompatible polymers may be designed with multifunctional moieties. Tailoring the structure of the polymers is expected to confer the requisite attributes such as mechanical toughness, biocompatibility, and electrical property. Furthermore, this may be projected as an avenue to establish an optimum balance between permeability and selectivity [76, 77]. Such polymers will also provide a compatible multifunctional surface for nanomaterial impregnation. Depending on the target application, the available choices for nanomaterials are generally diverse. Recently, transition metal dichalcogenides (TMDs) have attracted significant attention because of their unique electronic and optical properties. Researchers have shown the potential of TMDs in biosensors, solar cells, and supercapacitors. It is of relevance to note that Azamat and Khataee have reported the potential of a MoS_2 membrane in heavy metal separation from water [78]. Thus, there is an immense possibility of utilizing TMDs (MoS_2, WS_2, etc.) in the fabrication of polymer-based ion-exchange membranes in the near future.

5.6 Conclusion

Continuous endeavors have been underway for obtaining high-performance membranes for electrodialysis. However, the desirable mechanical toughness and fouling-resistant capacity, two of the prerequisites, have not been achieved as per requirements. As evident from Table 5.1, most of the researchers have utilized 0D/metal-based nanoparticles in membrane fabrication. This novel strategy is dictated by the objective of improving the antimicrobial efficacy and electrical properties of the materials. However, an appreciable mechanical toughness has not been registered in such endeavors. In recent times, mainly CNTs and graphene-based materials have received copious attention in this regard. Excellent mechanical attributes of carbon-based nanomaterials and their inherent electrical conductivity provide avenues to overcome the drawbacks of prevailing materials. Consequently, metal/carbon combinations have been used to fabricate hybrid nanomaterials. However, such materials have not been delved into thoroughly. The second concern is the polymeric matrix used for membrane fabrication. The desired qualities of the polymer include toughness, biocompatibility, and durability. Though the various polymers employed so far have been considerably successful in conferring the requisite properties, there is tremendous scope for improvement. Molecular tinkering of the functionalities of the polymer and the nanomaterials is expected to provide an avenue to fabricate unique materials with tailored properties. The use of hyperbranched polymers or dendrimers may be a prudent step. Fabrication procedures also involve less effort and time for such polymeric systems. Thus, nanocomposites need to be investigated more critically for electrodialysis membrane technology. From the perspective of biomedical applications, the in vitro and in vivo biocompatibility assessments are pre-requisites for forwarding such materials to the market.

Acknowledgments

The authors extend their gratitude to Prof. Niranjan Karak (Department of Chemical Sciences, Tezpur University, Assam, India) for his never-ceasing support and guidance.

References

1. Xu, T. and Huang, C. (2008). Electrodialysis-based separation technologies: a critical review, *Am. Inst. Chem. Eng. J.*, **54**, pp. 3147–3159.
2. Yee, R. S. L., Rozendal, R. A., Zhang, K. and Ladewig, B. P. (2012). Cost effective cation exchange membranes: a review, *Chem. Eng. Res. Des.*, **90**, pp. 950–959.
3. Savadogo, O. (2004). Emerging membranes for electrochemical systems. Part II: high temperature composite membranes for polymer electrolyte fuel cell (PEFC) applications, *J. Power Sources*, **127**, pp. 135–161.
4. Smitha, B., Sridhar, S. and Khan, A. A. (2005). Solid polymer electrolyte membranes for fuel cell applications: a review, *J. Membr. Sci.*, **259**, pp. 10–26.
5. Vinodh, R., Ilakkiya, A., Elamathi, S. and Sangeetha, D. (2010). A novel anion exchange membrane from polystyrene (ethylene butylene) polystyrene: synthesis and characterization, *Mater. Sci. Eng. B*, **167**, pp. 43–50.
6. Irvine, G. J., Rajesh, S., Georgiadis, M. and Phillip, W. A. (2013). Ion selective permeation through cellulose acetate membranes in forward osmosis, *Environ. Sci. Technol.*, **47**, pp. 13745–13753.
7. Aminabhavi, T. M., Kulkarni, E. V. and Kariduraganavar, M. Y. (2004). Ion exchange membranes, methods and processes for production thereof and uses in specific applications, US Patent 6,814,865.
8. Strathmann, H., Grabowski, A. and Eigenberger, G. (2013). Ion-exchange membranes in the chemical process industry, *Ind. Eng. Chem. Res.*, **52**, pp. 10364–10379.
9. Kakimoto, M., Grunzinger, S. J. and Hayakawa, T. (2010). Hyperbranchedpoly(ether sulfone)s: preparation and application to ion-exchange membranes, *Polym. J.*, **42**, pp. 697–705.
10. Konwarh, R.,Shail, M., Medhi, T., Mandal, M. and Karak, N. (2014). Sonication assisted assemblage of exotic polymer supported

nanostructured bio-hybrid system and prospective application, *Ultrason. Sonochem.*, **21**, pp. 634–642.

11. Konwarh, R., Karak, N. and Misra, M. (2013). Electrospun cellulose acetate nanofibers: the present status and gamut of biotechnological applications, *Biotechnol. Adv.*, **31**, pp. 421–437.
12. Konwarh, R., Pramanik, S., Devi, K. S. P., Saikia, N., Boruah, R., Maiti, T. K., Deka, R. C. and Karak, N. (2012). Lycopene coupled 'trifoliate' polyanilinenanofibers as multi-functional biomaterial, *J. Mater. Chem.*, **22**, pp. 15062–15070.
13. Konwarh, R., Saikia, J. P., Karak, N. and Konwar, B. K. (2010). 'Poly(ethylene glycol)-magnetic nanoparticles-curcumin' trio: directed morphogenesis and synergistic free radical scavenging, *Colloid Surf. B*, **81**, pp. 578–586.
14. Konwarh, R., Karak, N., Rai, S. K. and Mukherjee, A. K. (2009). Polymer-assisted iron oxide magnetic nanoparticle immobilized keratinase, *Nanotechnology*, **20**, p. 225107.
15. Pramanik, S., Konwarh, R., Barua, N., Buragohain, A. K. and Karak, N. (2014). Bio-based hyperbranched poly(ester amide)/MWCNT nanocomposites: multimodalities at biointerface, *Biomater. Sci.*, **2**, pp. 192–202.
16. Karak, N., Konwarh, R. and Voit, B. (2010). Catalytically active vegetable-oil based thermoplastic hyperbranched polyurethane/silver nanocomposite, *Macromol. Mater. Eng.*, **295**, pp. 159–169.
17. Christakiran M, J., Reardon, P. J. T., Konwarh, R., Knowles, J. C. and Mandal, B. B. (2017). Mimicking hierarchical complexity of the osteochondral interface using electrospun silk-bioactive glass composites, *ACS Appl. Mater. Interfaces*, **9**, pp. 8000–8013.
18. Pramanik, S., Konwarh, R., Deka, R. C., Aidew, L., Barua, N., Buragohain, A. K., Mohanta, D. and Karak, N. (2013). Microwave-assisted poly(glycidyl methacrylate)-functionalized multiwall carbon nanotubes with a 'tendrillar' nanofibrous polyaniline wrapping and their interaction at bio-interface, *Carbon*, **55**, pp. 34–43.
19. Kariduraganavar, M. Y., Nagarale, R. K., Kittur, A. A. and Kulkarni, S. S. (2006). Ion-exchange membranes: preparative methods for electrodialysis and fuel cell applications, *Desalination*, **197**, pp. 225–246.
20. Valero, F., Barceló, A. and Arbós, R. (2011). Electrodialysis technology: theory and applications, in *Desalination, Trends and Technologies*, Schorr, M., ed. (InTech), doi:10.5772/14297.

21. Geise, G. M., Cassady, H. J., Paul, D. R., Logan, B. E. and Hickner, M. A. (2004). Specific ion effects on membrane potential and the permselectivity of ion exchange membranes, *Phys. Chem. Chem. Phys.*, **16**, pp. 21673–21681.

22. Geise, G. M., Lee, H., Miller, D. J., Freeman, B. D., Mcgrath, J. E. and Paul, D. R. (2010). Water purification by membranes: the role of polymer science, *J. Polym. Sci., Part B: Polym. Phys.*, **48**, pp. 1685–1718.

23. Ferraz, N., Carlsson, D. O., Hong, J., Larsson, R., Fellstrom, B., Nyholm, L., Strømme, M. and Mihranyan, A. (2012). Haemocompatibility and ion exchange capability of nanocellulose polypyrrole membranes intended for blood purification, *J. R. Soc. Interface*, **9**, pp. 1943–1955.

24. Liu, C., Zhang, G., Zhao, C., Li, X., Li, M. and Na, H. (2014). MOFs synthesized by the ionothermal method addressing the leaching problem of IL-polymer composite membranes, *Chem. Commun.*, **50**, pp. 14121–14124.

25. Choi, D., Zhu, C., Fu, S., Du, D., Engelhard, M. H. and Lin, Y. (2017). Electrochemically controlled ion-exchange property of carbon nanotubes/polypyrrole nanocomposite in various electrolyte solutions, *Electroanalysis*, **29**, pp. 929–936.

26. Zhan, C., Yu, G., Lu, Y., Wang, L., Wujcik, E. and Wei, S. (2017). Conductive polymer nanocomposites: a critical review of modern advanced devices, *J. Mater. Chem. C*, **5**, pp. 1569–1585.

27. Tong, X., Zhang, B., Fan, Y. and Chen, Y. (20170. Mechanism exploration of ion transport in nanocomposite cation exchange membranes, *ACS Appl. Mater. Interfaces*, **9**, pp. 13491–13499.

28. Karak, N. (2010). *Fundamentals of Polymers* (PHI Learning Pvt. Ltd., New Delhi).

29. Barua, S., Dutta, N., Karmakar, S., Chattopadhyay, P., Aidew, L., Buragohain, A. K. and Karak, N. (2014). Biocompatible high performance hyperbranched epoxy/clay nanocomposite as an implantable material, *Biomed. Mater.*, **9**(2), p. 025006.

30. Xu, T. (2005). Ion exchange membrane: state of their development and perspective, *J. Membr. Sci.*, **263**, pp. 1–29.

31. Hosseini, S. M., Madaeni, S. S., Khodabakhshi, A. R. and Zendehnam, A. (2010). Preparation and surface modification of PVC/SBR heterogeneous cation exchange membrane with silver nanoparticles by plasma treatment, *J. Membr. Sci.*, **365**, pp. 438–446.

32. Zendehnam, A., Robatmili, N., Hosseini, S. M., Arabzadegan, M. and Madaeni, S. S. (2013). Fabrication of novel (acrylonitrile butadiene

styrene/activated carbon/silver nanoparticles) heterogeneous anion exchange membrane: physico-chemical and antibacterial characteristics, *J. Taiwan Inst. Chem. Eng.*, **44**, pp. 670–677.

33. Zarrinkhameh, M., Zendehnam, A. and Hosseini, S. M. (2014). Preparation and characterization of nanocomposite heterogeneous cation exchange membranes modified by silver nanoparticles, *Korean J. Chem. Eng.*, **31**, pp. 1187–1193.

34. Zendehnam, A., Mokhtari, S., Hosseini, S. M. and Rabieyan, M. (2014). Fabrication of novel heterogeneous cation exchange membrane by use of synthesized carbon nanotubes-co-copper nanolayer composite nanoparticles: characterization, performance in desalination, *Desalination*, **347**, pp. 86–93.

35. Fernandez-Gonzalez, C., Dominguez-Ramos, A., Ibañez, R., Chen, Y. and Irabien, A. (2017). Valorization of desalination brines by electrodialysis with bipolar membranes using nanocomposite anion exchange membranes, *Desalination*, **406**, pp. 16–24.

36. Daraei, P., Madaeni, S. S., Ghaemi, N., Salehi, E., Khadivi, M. A. Moradian, R. and Bandar, A. (2012). Novel polyethersulfone nanocomposites membrane prepared by PANI/Fe_3O_4 nanoparticles with enhanced performance for Cu(II) removal from water, *J. Membr. Sci.*, **415–416**, pp. 250–259.

37. Hosseini, S. M., Madaeni, S. S., Heidari, A. R. and Amirimehr, A. (2012). Preparation and characterization of ion-selective polyvinyl chloride based heterogeneous cation exchange membrane modified by magnetic iron-nickel oxide nanoparticles, *Desalination*, **284**, pp. 191–199.

38. Hong, J. G. and Chen, Y. (2014). Nanocomposite reverse electrodialysis (RED) ion-exchange membranes for salinity gradient power generation, *J. Membr. Sci.*, **460**, pp. 139–147.

39. Li, H., Gao, Y., Pan, L., Zhang, Y., Chen, Y. and Sun, Z. (2008). Electrosorptive desalination by carbon nanotubes and nanofibres electrodes and ion-exchange membranes, *Water Res.*, **42**, pp. 4923–4928.

40. Arai, S., Osaki, T., Hirota, T. and Uejima, M. (2016). Fabrication of copper/single-walled carbon nanotube composite film with homogeneously dispersed nanotubes by electroless deposition, *Mater. Today Commun.*, **7**, pp. 101–107.

41. Hosseini, S. M., Gholami, A., Madaeni, S. S., Moghadassi, A. R. and Hamidi, A. R. (2012). Fabrication of (polyvinyl chloride/cellulose acetate) electrodialysis heterogeneous cation exchange membrane:

characterization and performance in desalination process, *Desalination*, **306**, pp. 51–59.

42. Ferraz, N., Carlsson, D. O., Hong, J., Larsson, R., Fellstrom, B., Nyholm, L., Strømme, M. and Mihranyan, A. (2012). Haemocompatibility and ion exchange capability of nanocellulosepolypyrrole membranes intended for blood purification, *J. R. Soc. Interface*, **9**(73), pp. 1943–1955.
43. Ferraz, N., Leschinskaya, A., Toomadj, F., Fellstrom, B., Strømme, M. and Mihranyan, A. (2013). Membrane characterization and solute diffusion in porous composite nanocellulose membranes for hemodialysis, *Cellulose*, **20**, pp. 2959–2970.
44. Suman, Kardam, A., Gera, M. and Jain, V. K. (2015). A novel reusable nanocomposite for complete removal of dyes, heavy metals and microbial load from water based on nanocellulose and silver nano-embedded pebbles, *Environ. Technol.*, **36**, pp. 706–714.
45. Kohay, H., Sarisozen, C., Sawant, R., Jhaveri, A., Torchilin, V. P. and Mishael, Y. G. (2017). PEG-PE/clay composite carriers for doxorubicin: effect of composite structure on release, cell interaction and cytotoxicity, *Acta Biomater.*, **55**, pp. 443–454.
46. Shi, L., Zhang, X. and Chen, Z. (2011). Removal of chromium (VI) from wastewater using bentonite-supported nanoscale zero-valent iron, *Water Res.*, **45**, pp. 886–892.
47. Konatham, D., Yu, J., Ho, T. A. and Striolo, A. (2013). Simulation insights for graphene-based water desalination membranes, *Langmuir*, **29**, pp. 11884–11897.
48. Gahlot, S., Sharma, P. P., Gupta, H., Kulshrestha, V. and Jha, P. K. (2014). Preparation of graphene oxide nano-composite ion-exchange membranes for desalination application, *RSC Adv.*, **4**, pp. 24662–24670.
49. Thakur, S. and Karak, N. (2012). Green reduction of graphene oxide by aqueous phytoextracts, *Carbon*, **50**, pp. 5331–5339.
50. Wang, Y., Shi, Z. and Yin, J. (2011). Facile synthesis of soluble graphene via a green reduction of graphene oxide in tea solution and its biocomposites, *Appl. Mater. Interface*, **3**, pp. 1127–1133.
51. Akhavan, O., Kalaee, M., Alavi, Z. S., Ghiasi, S. M. A. and Esfandiar, A. (2012). Increasing the antioxidant activity of green tea polyphenols in the presence of iron for the reduction of graphene oxide, *Carbon*, **50**, pp. 3015–3025.
52. Kuila, T., Bose, S., Khanra, P., Mishra, A. K., Kim, N. H. and Lee, J. H. (2010). A green approach for the reduction of graphene oxide by wild carrot root, *Carbon*, **50**, pp. 914–921.

53. Esfandiar, A., Akhavan, O. and Irajizad, A. (2011). Melatonin as a powerful bio-antioxidant for reduction of graphene oxide, *J. Mater. Chem.*, **21**, pp. 10907–10914.

54. Barua, S., Thakur, S., Aidew, L., Buragohain, A. K., Chattopadhyay, P. and Karak, N. (2014). One step preparation of a biocompatible, antimicrobial reduced graphene oxide-silver nanohybrid as a topical antimicrobial agent, *RSC Adv.*, **4**, pp. 9777–9783.

55. Thakur, S. and Karak, N. (2015). Alternative methods and nature-based reagents for the reduction of graphene oxide: a review, *Carbon*, **94**, pp. 224–242.

56. Safarpour, M., Khataee, A. and Vatanpour, V. (2015). Thin film nanocomposite reverse osmosis membrane modified by reduced graphene oxide/TiO_2 with improved desalination performance, *J. Membr. Sci.*, **489**, pp. 43–54.

57. Hegab, H. M. and Zou, L. (2015). Graphene oxide-assisted membranes: fabrication and potential applications in desalination and water purification, *J. Membr. Sci.*, **484**, pp. 95–106.

58. Amiri, M. C. and Samiei, M. (2007). Enhancing permeate flux in a RO plant by controlling membrane fouling, *Desalination*, **207**, pp. 361–369.

59. Misdan, N., Lau, W. J. and Ismail, A. F. (2012). Seawater reverse osmosis (SWRO) desalination by thin-film composite membrane-current development, challenges and future prospects, *Desalination*, **287**, pp. 228–237.

60. Li, N., Yang, G., Sun, Y., Song, H., Cui, H., Yang, G. and Wang, C. (2015). Free-standing and transparent graphene membrane of polyhedron box-shaped basic building units directly grown using a NaCl template for flexible transparent and stretchable solid-state supercapacitors, *Nano Lett.*, **15**(5), pp. 3195–3203.

61. Hosseini, S. M., Nemati, M., Jeddi, F., Salehi, E., Khodabakhshi, A. R. and Madaeni, S. S. (2015). Fabrication of mixed matrix heterogeneous cation exchange membrane modified by titanium dioxide nanoparticles: mono/bivalent ionic transport property in desalination, *Desalination*, **359**, pp. 167–175.

62. Hosseini, S. M, Rafiei, S., Hamidi, A. R., Moghadassi, A. R. and Madaeni, S. S. (2014). Preparation and electrochemical characterization of mixed matrix heterogeneous cation exchange membranes filled with zeolite nanoparticles: ionic transport property in desalination, *Desalination*, **351**, pp. 138–144.

63. Son, M., Choi, H., Liu, L., Celik, E., Park, H. and Choi, H. (2015). Efficacy of carbon nanotube positioning in the polyethersulfone support layer on the performance of thin-film composite membrane for desalination, *Chem. Eng. J.*, **266**, pp. 376–384.

64. Madaeni, S. S., Zinadini, S. and Vatanpour, V. (2013). Preparation of superhydrophobic nanofiltration membrane by embedding multiwalled carbon nanotube and polydimethylsiloxane in pores of microfiltration membrane, *Sep. Purif. Technol.*, **111**, pp. 98–107.

65. Gu, J. C., Xiao, P., Chen, J., Liu, F., Huang, Y., Li, G., Zhang, J. and Chen, T. (2014). Robust preparation of superhydrophobic polymer/carbon nanotube hybrid membranes for highly effective removal of oils and separation of water-in-oil emulsions, *J. Mater. Chem. A*, **2**, pp. 15268–15272.

66. Gethard, K., Sae-Khow, O. and Mitra, S. (2011). Water desalination using carbon-nanotube-enhanced membrane distillation, *ACS Appl. Mater. Interfaces*, **3**, pp. 110–114.

67. Bae, T. H., Kim, I. C. and Tak, T. M. (2006). Preparation and characterization of fouling-resistant TiO_2 self-assembled nanocomposite membranes, *J. Membr. Sci.*, **275**, pp. 1–5.

68. Li, H. S., Shao, X. S., Zhou, Q., Li, M. Z. and Zhang, Q. Q. (2013). The double effects of silver nanoparticles on the PVDF membrane: surface hydrophilicity and antifouling performance, *Appl. Surf. Sci.*, **265**, pp. 663–670.

69. Homayoonfal, M., Mehrnia, M. R., Shariaty-Niassar, M., Akbari, A., Ismail, A. F. and Matsuura, T. (2014). A comparison between blending and surface deposition methods for the preparation of iron oxide/polysulfone nanocomposite membranes, *Desalination*, **354**, pp. 125–142.

70. Wara, N. M., Francis, L. F. and Velamakanni, B. V. (1995). Addition of alumina to cellulose acetate membranes, *J. Membr. Sci.*, **104**, pp. 43–49.

71. Genne, I., Kuypers, S. and Leysen, R. (1996). Effect of the addition of ZrO_2 to polysulfone based UF membranes, *J. Membr. Sci.*, **113**, pp. 343–350.

72. Arthanareeswaran, G., Devi, T. K. S. and Raajenthiren, M. (2008). Effect of silica particles on cellulose acetate blend ultrafiltration membranes: part I, *Sep. Purif. Technol.*, **64**, pp. 38–47.

73. Arthanareeswaran, G. and Thanikaivelan, P. (2010). Fabrication of cellulose acetate–zirconia hybrid membranes for ultrafiltration applications: performance, structure and fouling analysis, *Sep. Purif. Technol.*, **74**, pp. 230–235.

74. Chitpong, N. and Husson, S. M. (2017). High-capacity, nanofiber-based ion-exchange membranes for the selective recovery of heavy metals from impaired waters, *Sep. Purif. Technol.*, **179**, pp. 94–103.

75. Kaleekkal, N. J., Thanigaivelan, A., Durga, M., Girish, R., Rana, D., Soundararajan, P. and Mohan, D. (2015). Graphene oxide nano-composite incorporated polyetherimide mixed matrix membranes for *in-vitro* evaluation of its efficacy in blood purification applications, *Ind. Eng. Chem. Res.*, **54**, pp. 7899–7913.

76. Jiang, S. and Ladewig, B. P. (2017). High ion-exchange capacity semihomogeneous cation exchange membranes prepared via a novel polymerization and sulfonation approach in porous polypropylene, *ACS Appl. Mater. Interfaces*, **9**, pp. 38612–38620.

77. Park, H. B., Kamcev, J., Robeson, L. M., Elimelech, M. and Freeman, B. D. (2017). Maximizing the right stuff: the trade-off between membrane permeability and selectivity, *Science*, **356**, p. eaab0530.

78. Azamat, J. and Khataee, A. (2017). Improving the performance of heavy metal separation from water using MoS_2 membrane: molecular dynamics simulation, *Comput. Mater. Sci.*, **137**, pp. 201–207.

Chapter 6

Cellulose and Nanocellulose Derivatives from Lignocellulosic Biomass in Nanocomposite Applications

Nurhidayatullaili Binti Muhd Julkapli, You Wei Chen, and Hwei Voon Lee

Nanotechnology & Catalysis Research Centre (NANOCAT), Institute of Postgraduate Studies, University of Malaya, 50603 Kuala Lumpur, Malaysia

leehweivoon@um.edu.my

Lignocellulosic biomass is a renewable resource that consists mainly of three types of biopolymers: cellulose, hemicellulose, and lignin. It is a fossil alternative that is highly available in the form of agricultural residues, herbaceous crops, woody crops, forestry residues, waste paper, and other wastes from municipal and industrial uses. Cellulose is a major constituent with unique characteristics, especially depolymerization into its nanodimension (nanocellulose). These types of biomaterials are equipped with superior properties such as lightweight structure, high aspect ratio with durability, large surface area, tunable reactive surface, exceptional mechanical properties, high crystallinity, sustainability,

Hybrid Nanocomposites: Fundamentals, Synthesis, and Applications
Edited by Kaushik Pal

ISBN 978-981-4800-34-1 (Hardcover), 978-0-429-00096-6 (eBook)
www.panstanford.com

and biodegradability. The excellent features have made both cellulose and nanocellulose have a wide range of applications in the construction, packaging, automobile, transportation, and biomedical fields. However, there are also challenges in mimicking this biomaterial as a conventional petrol-based material in order to well suit and be compatible with existing matrices. Thus, tailoring the surface functionality of cellulose and nanocellulose via different types of functionalization (e.g., chemical functionalization, organic compound functionalization, macromolecular functionalization, and inorganic compound functionalization) is an important route to improving the performance of pure biomaterials. This chapter highlights the types of biomass resources for cellulose and nanocellulose; chemical functionalization to different types of derivatives; physicochemical properties of cellulose, nanocellulose, and its derivatives; and application of the functionalized materials for nanocomposites.

6.1 Introduction

With the recent developments in nanotechnology, a new scientific field of materials chemistry that lays emphasis on the rational synthesis and study of nanoscale materials has emerged. Nanotechnology is defined as a technology that allows the use of materials at dimensions between 1 and 100 nm, where unique phenomena (chemical, physical, or biological properties) are fundamentally different from the bulk material, which enables novel applications. Nanotechnology has rapidly become an interdisciplinary field, and one of its hot research topics is the production of cellulose nanomaterial from bioresources. The flexibility of cellulose has been evaluated as a valuable structural and functional green material with environmental benefits. In addition, the value of cellulose is also recognized as a versatile starting material for various chemical transformation processes in the production of a variety of cellulose derivatives for their application across different applications. It is known that functional nanomaterials are of particular importance as they allow the formation of novel green materials with enhanced or new properties by combining multiple ingredients as well as exploiting synergistic effects, such as adsorption, bioactivity, catalytic, electronic, magnetic, mechanical, optical, physicochemical, or selective permeation [1]. Due to these reasons, the production of nanocellulose materials and their surface modifications have become one of the most intensely investigated areas of nanocellulose research on nanocomposites, as the nanoscale cellulose material offers several unique and excellent physicochemical properties, even at low concentration loading, to the polymer matrix, is environmentally friendly, and is able to surpass other inorganic reinforcements such as boron nanowhiskers, carbon fibers, carbon nanotubes, and Kevlar [2]. Cellulose and its derivatives as promising materials are of greatest significance in various application fields and have been most extensively studied in industrial and academic research.

6.1.1 Current Trends in and Demand for Cellulose and Nanocellulose

In recent years, there is a growing interest in renewable biomaterials, particularly in cellulose due to its wide availability, biodegradability, renewability, and low-cost material sources. Cellulose is the most abundant and ubiquitous renewable natural polymer resource that originated in the biosphere with an annual production of 75 billion tons [3] used in the cellulose market for industrial conversion. This almost inexhaustible and fascinating sustainable biopolymer possesses remarkable physical and chemical traits, which allow it to be utilized in various forms for a wide range of products and materials. However, only around 2 billion tons of cellulose is being fully utilized per year, including all possible uses [4]. In this sense, great efforts are being concentrated on the use of cellulose materials due to their abundance, low energy cost, and economy, and the ease of surface modifications. Cellulose is known to have a main functional group, that is, a hydroxyl group, on cellulose surfaces, which is attributed to its properties. This hydroxyl group can be modified via chemical reactions in order to enhance its properties.

On the other hand, the research and development (R&D) activity on nanocellulose fields has been significantly increased due to technological advances and is mainly conducted by universities, research centers, and industries. It is worth mentioning that the technological milestone is mainly related to the efforts made by governments and public institutions to pursue a global strategy based on sustainable growth, driven by socioeconomic interests, and arising from climate change problems [5]. The interest in sustainability is due to the environmental and energy crises resulting from industrialization since past centuries, severe pollution, and increased consumption of nonrenewable depleting resources [6]. Currently, the disciplines of nanotechnology and nanoscience have been highlighted for being exceptionally focused by various funding agencies and governments. The worldwide market for products incorporating nanotechnology totaled about USD 254 billion in 2009, and this number is projected to double every 3 years until 2020, when this value could be USD 3 trillion [1]. Generally, the nanocellulose market size is forecasted to be approximately 35 million metric tons/year [7], driven by expanding its applications and the growing focus on green alternatives to petroleum in the manufacture of useful chemicals and synthetic polymers. In fact, the conservative estimate value for nanocellulose production in 2017 was approximately 1667 tons/year and the optimistic estimate value was twice the amount, 2548 tons/year [8].

6.1.2 Lignocellulosic Biomass and Its Resources

Lignocellulosic biomass is mainly composed of three biopolymers: lignin (15%–35%), hemicellulose (19%–45%), and cellulose (30%–50%). Depending on the type of biomass, these polymers are organized into complex and nonuniform structures in which cellulose fibers are embedded by a lignin and hemicellulose sheath [9]. These three major components of plants consist of around 90% of dry biomass. In addition, a small portion of extractives, pectin, and protein can also be found in plants. These materials, including nonstructural materials such as waxes, chlorophyll, nitrogenous materials, and nonstructural sugars, can be removed by using Soxhlet treatment with different solvents such as benzene, ethanol, and toluene. Generally, cellulose can be derived from a variety of renewable sources, such as woods, animal, annual plants, algal biomass, and bacteria. On the other hand, nanocellulose derived from cellulose is, therefore, of a biodegradable nature and allows recycling of nanocellulose-based products. It is known that one of the main parameters that impact the properties of resultant cellulose nanomaterials is their original source. Therefore, the following subsections provide a brief discussion of the different cellulosic sources that have been widely used for isolating cellulose materials.

6.1.2.1 Wood

The main source of cellulose is apparently contributed by the wood. Due to its abundance, wood is the most significant industrial source for isolating cellulose. Generally, the extraction of cellulose from wood sources begins with a biomass purification process, in which lignin, hemicellulose, and impurities are eliminated. However, the competition between various industrial sectors such as paper, pulp, furniture, and building materials, along with the combustion of wood for energy supply, limits the wood resource for all related sectors with the required quantities at a reasonable price. The increase of wood consumption directly results in overdeforestation, which further leads to imbalance in the ecology. Besides that, wood is not available in several regions. Henceforth, the interest in other cellulosic materials such as agriculture bioresidues, grass, herbaceous plants, agricultural crops, and aquatic plants has extensively stimulated significant interest. Most recently, nanocellulose was obtained from different lignocellulosic biomass sources by the latest technology.

6.1.2.2 Agriculture and bioresidues

Agricultural residues are considered an alternative cellulosic source for replacing the native wood material with acceptable properties. Agricultural residues have great potential for use as a cost-effective and appropriate source for producing nanocellulose. Compared to wood, these nonwood agricultural crops render positive characteristics such as low contents of hemicellulose and lignin, and thus less bleaching is required for removing noncellulosic components. Additionally, the bioresidue from industry can also be another by-product that has great potential for use as a low-cost and appropriate source for nanocellulose production. It is known that wastes generated by industrial or agriculture activities have an excellent advantage compared to other feedstocks because of low costs. Thus, the development of new value-added products such as nanocellulose is a potential route for improving the value of these wastes and can contribute to solving disposal problems for industries. In fact, it should be noted that cellulose microfibrils in agricultural fibers are less tightly bound to the primary cell wall compared to the secondary cell wall in wood; therefore less energy is demanded to produce nanocellulose [10]. In recent years, a wide variety of agricultural residues and nonwood sources has been investigated for the isolation of nanocellulose. There have been a number of review articles [1, 11] summarizing the potential sources of nanocellulose, which were previously reported by other researchers, including *Agave angustifolia* fibers, algae, banana rachis, bamboo, cotton, coconut husks, corncob, coir, citrus waste, *Calotropis procera* fibers, empty fruit bunches, flax fibers, groundnut shells, grass, jute, kapok, kenaf bast, *Miscanthus giganteus*, mango seed, the macrophyte *Typha domingensis*, onion waste, oil palm, olive stones, potato peel, pea hull fiber, pineapple leaf, rice husk, rice straw, sugarcane bagasse, sugar beet pulp, straw, swede root, sisal, soy hulls, sesame husk, spruce bark, tomato peels, and wheat straw. The nanocellulose products obtained from different types of cellulose sources of miscellaneous provenance commonly differ in their physicochemical properties such as crystallinity, the degree of polymerization, geometrical dimensions, morphology, mechanical properties, porosity, surface charge, surface area, and thermal stability.

6.1.2.3 Bacterial cellulose

The bacterial-derived cellulose possesses several outstanding properties such as good chemical stability, high chemical purity, high mechanical properties, large water-holding capacity, renewability, highly crystalline network structure, biodegradability, and nontoxicity, which does not require any further purification for removing hemicellulose and lignin [12]. Due to its special properties, cellulose produced by bacteria has grown in popularity since its discovery in 1886. One of the most studied types of microorganisms that is used to produce cellulose is *Acetobacter*. These microorganisms can procure glycerol, sugar, glucose, or other organic substrates and effectively convert them into pure cellulose. So far, it has continued to be the highest producer of bacterial cellulose (BC). It is strictly aerobic and nonphotosynthetic and is able to transform glucose and other organic substrates into cellulose in a few days [13]. BC is produced by a cultivation process in a culture medium. Even though the detailed mechanism of how bacteria generate cellulose is still unclear, it has been proposed that it is necessary for their survival, including acting as a barrier to yeasts, fungi, and other organisms or as a protector against ultraviolet light [14]. Normally, BC displays as a ribbon-shaped fibril structure. However, the detailed morphology of the cellulose microfibril and its network configuration within the pellicle are significantly affected by the variation in culture conditions [15].

6.1.2.4 Sea animals

Tunicates, which originated from the oceans, are the only animal source known to produce almost pure nanofibers via cellulose biosynthesis. In the natural phase, cellulose synthesis is performed by the cellulose-synthesizing enzyme complexes that are present in the plasma membrane of epidermal cells. Tunicates have a tunic comprising cellulose microfibrils embedded in the protein matrix. Basically, the tunic has been reported to consists of 27% nitrogen-containing components and 60% cellulose in the fresh condition [15]. It is worth noting that the cellulose present in the tunic is chemically identical to plant-derived cellulose. Tunicates are abundantly available; over 2300 species have been reported in the literature. Naturally, tunicate cellulose is composed of pure cellulose Iβ only. The cellulose microfibrils in the tunic are aggregated and deposited in a multilayered texture with a bundled structure parallel to the epidermis. To separate the cellulose mantles from the tunicate, the tunicate is treated with potassium hydroxide (KOH) solution at an elevated temperature [16].

6.1.2.5 Algae

In recent years, it has become known that nanocellulose is also produced by algae. In nature, algae (yellow-green, red, green, gray) vary from small unicellular microalgae (cyanobacteria and diatoms) to large, multicellular macroalgae (giant kelp) [17] and are mostly found in shallow seas and coastal rocks. Many researchers have confirmed that red algae such as *Gelidium*, mainly composed of agar and cellulose, are a sustainable resource for numerous applications due to a high carbohydrate content [18]. Among the available red algae species, *G. corneum*, *G. amansii* [19], and *G. elegans* [20] were utilized to produce bleached cellulose pulp and cellulose nanomaterials. Compared to terrestrial biomass, the cellulose present in the marine algae is in a smaller and more variable proportion. Consequently, *Gelidium* appear to be a new promising candidate for nanocellulose extraction. Moreover, green algae are reported to be suitable for cellulose production. Basically, cellulose-producing algae primarily belong generally to the orders Siphonocladales and Cladophorales [21]. The cellulose obtained from green algae normally presents exceptionally high crystallinity (up to 95%). The cellulose content in both brown and red algae is relatively low, whereas most of the green algae have a high cellulose content with a cellulosic cell wall, ranging up to 70% of the dry weight [15]. To some extent, the cellulose content of green algae is of the same order of magnitude as compared to wood sources. However, the real reason for the presence of highly crystallinity in cellulose in marine green algae is still ambiguous. So far, it has been suggested that a strong microfibril structure in the algae is necessary in order to maintain turgor, such as osmotic pressure inside the cells in the marine environment that features fluctuating salinity [15].

6.1.3 Cellulose and Nanocellulose

With the development and emergence of nanotechnology, cellulose, the most important and ancient natural biopolymer on earth, has been transformed into a more attractive new form of nanocellulose, which can be used as a novel and advanced green material. Cellulose is a linear chain of ringed glucose molecules with a flat ribbon-like conformation. The repeat unit of typical cellulose comprises two anhydroglucose rings with the formula $(C_6H_{10}O_5)_n$ (where n can vary from 10,000 to 15,000, depended on its original source) [22]. The glucose units are linked together by oxygen covalently bonded to C_1 of one glucose ring and C_4 of the adjoining ring via β-(1,4)-glycosidic linkage. It is a linear natural polymer consisting of 1,4-anhydro-D-glucopyranose units, as shown in Fig. 6.1.

During natural synthesis, the cellulose polymer does not exist as an individual unit but several chains of cellulose molecules tend to be packed together during extended chain conformation. The hydrogen bonding between oxygen atoms and hydroxyl groups of the adjoining ring molecules stabilizes the linkage, resulting in the linear configuration of the cellulose chain. During biosynthesis, intermolecular hydrogen bonds and van der Waals forces between oxygen atoms and hydroxyl groups of adjacent molecules promote parallel assembling of multiple cellulose chains (30–100 chains) to form the basic unit of cellulose fibers, which refers to elementary fibrils (protofibrils) at the nanoscale [23]. These protofibrils are further aggregated via inter- and intermolecular hydrogen bonding into hierarchical microstructures, which are usually known as microfibrils or nanosized fibrils (diameter: 5–50 nm; length: several microns) [24]. In fact, the relatively stable cellulose polymer is contributed to the inter- and intrachain hydrogen-bonding network and provides the cellulose fibrils with high axial stiffness.

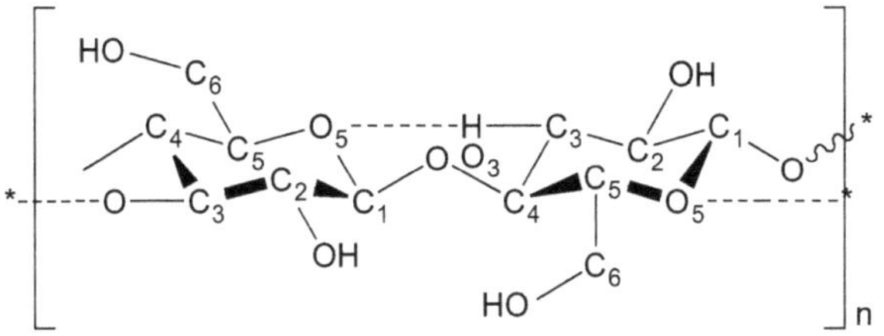

Figure 6.1 Schematic representation of a typical single cellulose chain repeating unit, presenting the 1→4 linkage and intrachain hydrogen bonding (dotted line).

It also known that cellulose is a linear semicrystalline polymer with both crystalline and amorphous regions. The strong and complex network of hydrogen bonding between the hydroxyl groups (–OH) of cellulose chains can stabilize and arrange the cellulose molecules into a well-organized structure (highly ordered structure) through crystalline packing [25]. On the other hand, another zone of cellulose molecules cannot be stabilized horizontally through hydrogen bonding; thus it would form less organized and disordered regions (amorphous-like) that are connected with cellulose crystals. These amorphous zones are normally characterized by lower density in comparison to the cellulose crystalline domains and are easily available for bonding with other molecules such as water (H_2O). On the application of an appropriate hydrolysis process, including biological (enzymatic), chemical, and mechanical treatments to the microfibrils to remove the amorphous regions, the highly crystalline regions of cellulose remain unaltered and thus can be easily isolated or extracted, resulting in the formation of tailor-made nanocellulose [26].

The development of cellulose nanomaterials has attracted great interest from the industrial and academic world. Its has wide acceptability because nanocellulose has many considerable features and unique characteristics, such as abundant hydroxyl groups, high tensile strength, light weight, high Young's modulus, high crystallinity, large specific surface area–to-volume ratio, nontoxicity, greater stiffness than Kevlar, high thermal stability, good mechanical properties, and low coefficient of thermal expansion, making it a good choice as a reinforcing agent [27, 28]. As a low-toxicity, renewable, and biodegradable material, nanocellulose has high potential to displace the petroleum-based packaging, metallic components, and other nonrenewable materials, which represent an important niche for more sustainable product development and design. In general, there are three main families of nanocellulose available [5]:

- Cellulose nanocrystals (CNCs), also referred to as cellulose nanowhiskers, nanocrystalline cellulose, rod-like cellulose microcrystals, are mainly obtained via acid hydrolysis and have a highly stiff-rod morphology.
- Cellulose nanofibrils (CNFs), with other names such as nanofibrillated cellulose fibers (NFCs), cellulose nanofibers, and microfibrillated cellulose (MFC), possess a spaghetti-like high-entangled network structure and are normally prepared by mechanical disintegration.
- BC, also called microbial cellulose, is produced by *Gluconacetobacter xylinus* bacteria, and its morphology can be engineered over length scales ranging from nano to macro, depending on the biosynthesis pathway.

6.1.4 Chemical Functionalization of Cellulose and Nanocellulose for Nanocomposites

6.1.4.1 Nanocellulose: Chemical functionalization

The functionalization of nanocellulose has obtained much attention because of the profound enhancement of properties, including increased mechanical and reduced gas barrier features [29, 30].

Besides, stable nanocellulose dispersion in a polar or low-polarity solvent is gained via physically coating the surface by using a surfactant or chemically grafting apolar moieties onto the surface (Fig. 6.2) [31]. The functionalized nanocellulose can be dispersed in a low-polarity organic liquid and mixed with a matrix solution. It allows an improvement in the dispersion of the functionalized nanocellulose in a continuous apolar medium, which is favorable for optimizing the final application of nanocellulose. However, it limits the interactions between nanocellulose particles through hydrogen bonding, which is the reason for the excellent dispersion features of nanocellulose [32, 33]. The nanocellulose surface functionalization includes plenty of surface –OH groups resulting from their nanoscale dimensions and ensuring a high specific surface area. Most of the surface functionalization experimental process is done in mild conditions. This is to preserve the integrity of the nanoparticles and the peeling effect of a surface-grafted chain that induces the dissolution in the reaction media [34, 35]. The surface functionalization of nanocellulose is categorized into two main groups, namely organic and inorganic compounds [36].

The key concern with nanocellulose functionalization would be to carry out the procedure to the extent that modifies the nanocellulose surface, while conserving the initial morphology to prevent any kind of polymorphic conversion and conserve the crystal integrity [37, 38].

6.1.4.2 Organic compound functionalization

The purpose of nanocellulose surface functionalization with organic compounds is typically to improve compatibility with the polar matrix, consequently to increase the dispersibility and functionality of nanocellulose particularly when utilized in combination with hydrophobic or nonpolar matrices [39]. This can be achieved by introducing highly stable positive or negative electrostatic charges on the nanocellulose surface [40].

6.1.4.2.1 *Molecular functionalization*

In general, molecular functionalization involves the substitution of –OH groups with small molecules: grafting-onto strategy with different coupling agents and grafting-from approach with radical polymerization involving ring-opening polymerization, atom transfer radical polymerization, and single-electron transfer living radical polymerization [41].

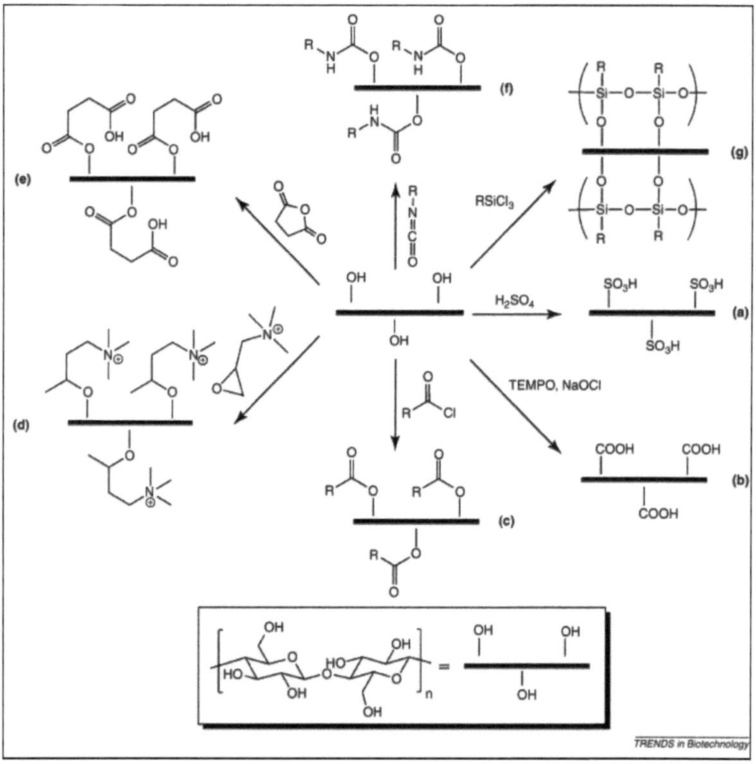

Figure 6.2 Route map on surface functionalization of nanocellulose: (a) sulfonation, (b) oxidation by 2,2,6,6-tetramethylpiperidine-1-oxyl radical (TEMPO), (c) ester linkages via acid chlorides, (d) cationization via epoxides, (e) ester linkages via acid anhydrides, (f) urethane linkages via isocyanates, and (g) silylation. Reprinted from Ref. [29], Copyright (2012), with permission from Elsevier.

6.1.4.2.1.1 *Nanocellulose-sulfoxide functionalization*

Water has been extensively used as a promising medium to stabilize the suspension of nanocellulose with negatively charged surface groups, normally produced by hydrolysis of the cellulose with sulfuric acid, obtained in various polar liquid media [42]. This has motivated some research work to produce stable nanocellulose with functionalized of dimethyl sulfoxide in *N,N*-dimethyl sulfoxide and *N*-methyl pyrrolidine, *m*-cresol, and formic acid [43].

6.1.4.2.1.2 *Nanocellulose-amide functionalization*

The hydrophobic nanocellulose surface is acquired by grafting of hexamethylene diisocyanate, following with a reaction with amines. Furthermore, the amine-functionalized nanocellulose with a positively charge is reported to be antimicrobially active in biomedical applications [44, 45]. The surface-functionalized nanocellulose presents antibacterial activity against both gram-positive and gram-negative bacteria, even at very low concentrations of the antimicrobial agent on the surface, killing more than 99% of *E. coli* and *S. aureus* at concentration of nitrogen up to 0.14% [46, 47].

6.1.4.2.1.3 *Nanocellulose-silane functionalization*

Silane functionalization usually improves the degree of crosslinking in the interface of nanocellulose as well as offering excellent bonding [48]. Silane surface functionalization undergoes hydrolysis, condensation, and bond formation stages of polysiloxane structures. In the presence of moisture, the hydroxylzable alkoxy group of silane derivatives results in silanol formation. Moreover, the silanol reacts with the nanocellulose –OH groups, forming stable covalent bonds to the cell wall that are chemisorbed onto the nanocellulose surface [3].

Utilization of silane derivatives, including isopropyl dimethylchlorosilane, 3-aminopropyltriethoxysilane, and 3-glycidoxypropyltrimethoxysilane for surface silylation of nanocellulose, results in the homogenization of its parenchymal cell walls [49]. Besides, nanocellulose is partially silylated by a series of alkyldimethylchlorosilanes, with the carbon backbone of the alkyl moieties ranging from a short carbon length of isopropyl to a longer length represented by n-butyl, n-octyl, and n-dodecyl. Infrared analysis proved that chemical links formed with all surface-functionalized compounds with aminopropyltriethoxysilane claimed the strongest link [50]. The morphological images of silane-nanocellulose have shown that microfibril morphology is retained under mild silylation conditions and could be dispersed in a nonflocculating manner into organic solvents [51]. It is demonstrated as well that with a degree of substitution between 0.6 and 1.0, the silylated nanocellulose becomes readily dispersible in solvents of low polarity, resulting in stable suspensions with birefringent behavior. However, at high silyation (degree of substitution >1), the chains in the core of the crystals became silylated, resulting in the disintegration of the crystal and subsequently the loss of original morphology [52]. The hydrocarbon chains provided by the application of silane restrain the swelling of nanocellulose by creating a crosslinked network. Therefore, the surface functionalization changes the character of nanocellulose from hydrophilic to hydrophobic, while the crystalline structure of nanocellulose remains intact. Indeed, the silylation process by using chlorodimethyl isopropylsilane is commonly employed to modify the surface utilization as a hydrophobic feature [53]. The hydrophobicity of the silylated nanocellulose performs with the reduction in its surface energy and increase in surface roughness. Owing to the nature of nanocellulose, it is commonly known that the –OH group was facile to adsorption water and it consequently decreased the performance of nanocellulose if it was fabricated for any application [49]. Therefore, hydrophobized nanocellulose via partial surface silylation utilizing the same silylation agent resulted in partial solubilization of nanocellulose and loss of nanostructure [54].

6.1.4.2.1.4 *Nanocellulose-acetyl functionalization*

The acetylation of nanocellulose improves the transparency and reduces hydroscopicity, which in turn reduces its moisture absorption [55, 56]. Acetylation is also reported to improve optical properties, thermal degradation resistance, dimension stability, and environmental degradation of cellulosic fibers. The pretreatment of nanocellulose with acetic anhydride substitutes the polymer –OH groups of the cell wall with acetyl groups (CH_3CO–R), which consequently modify the features of nanocellulose to become more hydrophobic [56]. The reaction is known to precede full esterification of all the three –OH of anhydro-D-glucose when carried out in the homogeneous phase. The –OH groups that react are those of the minor constituents of the nanocellulose and those of amorphous nanocellulose [57]. This is due to –OH groups in the crystalline region with close packing and strong interchain bonding. Homogeneous and heterogeneous acetylation of bacterial nanocellulose is possible by utilizing acetic anhydride in acetic acid [58]. For homogeneous acetylation, the partially acetylated molecules immediately partitioned into the acetylating medium once it is adequately soluble. Meanwhile, in heterogeneous conditions, the nanocellulose acetate remains insoluble and surrounds the crystalline core of unreacted nanocellulose chains [56]. This consequently induces an occurrence of nanocellulose hydrolysis and acetylation of –OH groups. Fischer esterification of –OH groups concurrently with the hydrolysis of amorphous nanocellulose domains has been introduced as a viable one-pot reaction methodology that allows isolation of acetylation nanocellulose in a one-step process [59].

The acetyl substitution degree has a critical effect on the final acetylated nanocellulose. However, beyond the optimum degree of substitution, excessive acetylation decreases the original features of nanocellulose [60]. Mostly, nanocellulose is partly acetylated to modify its physical properties, while preserving the microfibrillar morphology.

6.1.4.2.1.5 *Nanocellulose-carboxylic functionalization*

Nanocellulose-carboxylic functionalization represents a broadly utilized water-soluble nanocellulose derivative, applied where thickening, binding, suspending, stabilizing, and film-forming features are important [61]. Hydroxylmethyl groups of nanocellulose present on its structure can convert to the carboxylic form by using 2,2,6,6-tetramethylpiperidine-1-oxyl as an oxidation agent [61]. This oxidation reaction, which is extremely discriminative of primary –OH, is also simple and green to implement. It includes the application of 2,2,6,6-tetramethylpiperidine-1-oxyl as a stable nitroxyl radical in the presence of NaOCl and NaBr [62]. This carboxylic functionalization of nanocellulose includes a topologically confined reaction sequence, and because of the twofold screw axis of the nanocellulose chain, only half of the hydroxymethyl accessible groups are available to react, while the other half are buried within the crystalline particles (Fig. 6.3). This results in a repulsive force between individual nanocellulose and prevents agglomeration. The resulting carboxylated nanocellulose maintains its primary morphological integrity and forms a homogeneous suspension once dispersed in water. It is observed that the effect of different nanocellulose loadings had a significant effect on the mechanical, thermal, sorption, and barrier properties of functionalized nanocellulose [63]. The basis of these latter observations was the existence of the newly connected carboxyl groups that instructed negative charges at the nanocellulose surface and consequently prompted electrostatic stabilization. Meanwhile, there are some reports on the effect of pretreatments by using NaOH solution and dimethyl sulfoxide solvent on morphology, porous structure, and macro-/microstructures of carboxylated nanocellulose [64]. It was found that the pretreatment gave a uniform size of carboxylated nanocellulose (5–20 nm).

However, some reports on nanocellulose-carboxylic functionalization revealed that though this medium presents a number of peculiarities that are necessary for the high excess of reagents and a long reaction time, it is possible to prepare functionalized nanocellulose in the presence of solid NaOH particles [66]. Regarding the mole fractions of the different repeating units, the functionalized sample, which is prepared by using aqueous NaOH, possesses a static content. Nanocellulose exhibits an unconventional distribution of ether groups and unconventional features, which means nanocellulose displays a preferred substitution at position O6 and a block-like distribution of carboxymethyl groups along the nanocellulose backbone [67]. These molecular and supermolecular properties lead to some new macroscopic features with different rheological and colloidal behavior.

6.1.4.2.1.6 *Nanocellulose-aldehyde functionalization*

One of promising surface functionalizations of nanocellulose is to introduce reactive aldehyde functionalities with aqueous periodate oxidation [68]. The aldehyde groups of functionalized nanocellulose easily and selectively convert further into various functional groups, including carboxylic acids, sulfonates, and imines [69]. Indeed, acetic anhydride is added to a nanocellulose suspension in toluene after the solvent exchange process for obtaining hydrophobic features [70]. This functionalized nanocellulose showed good flocculation performance for wastewater treatment applications. Therefore, some studies used this type of modified nanocellulose to remove heavy metals from aqueous solutions with promising results [71].

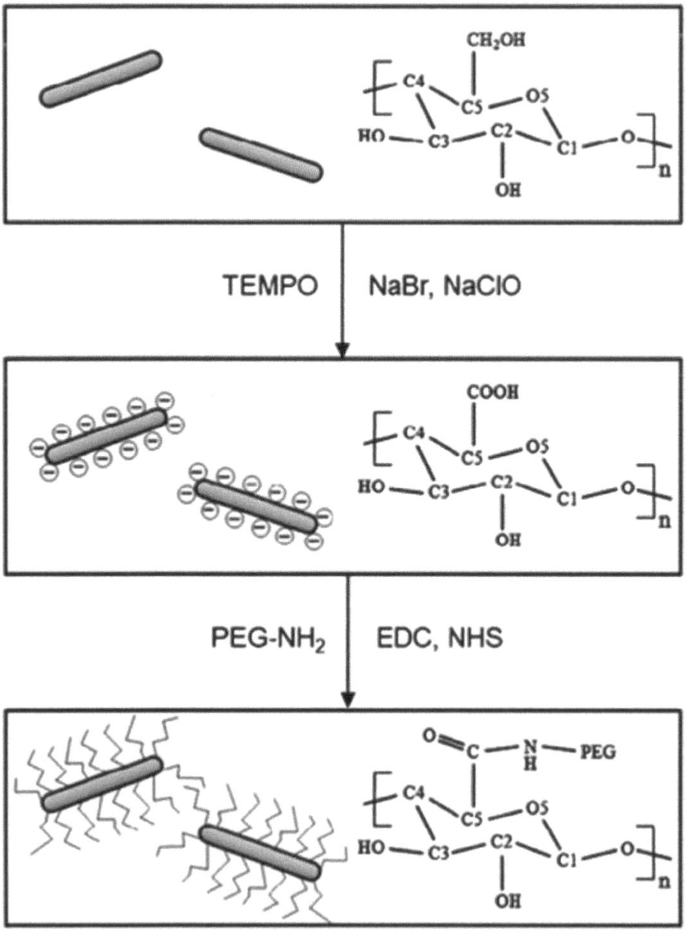

Figure 6.3 Functionalized colloidal nanocellulose with TEMPO and subsequent PEG grafting. EDC, 1-ethyl-3(3-dimethylaminopropyl)carbodiimide hydrochloride; NHS, *N*-hydroxysuccinimide. Reproduced from Ref. [65] with permission of The Royal Society of Chemistry.

6.1.4.2.1.7 *Nanocellulose-hyroxyapatite functionalization*

The adsorption ability of nanocellulose toward metal ions, including Ni, Cd, PO_4^{3-}, and NO_3^-, increased via its functionalization with carbonated hydroxyapatite [72]. Carbonated hydroxyapatite has a composition and structure analogous to bone apatite and displays greater bioactivity than pure hydroxyapatite [73]. Due to a high specific surface area and small size, carbonated hydroxyapatite nanostructures can efficiently interact with nanocellulose structures, leading to improvements [74].

6.1.4.3 Macromolecular functionalization

The funtionalization of nanocellulose with macromolecules has been currently investigated as a new way to produce good barrier materials and a possible solution to retain the advantages of nanocellulose and its surrounding medium [75]. The macromolecules used are normally defined as a material that could significantly decrease the surface tension of water when utilized in very low concentrations. The noncovalent surface functionalization of nanocellulose is typically made via adsorption of the macromolecules [76]. The obtained macromolecule-functionalized nanocellulose dispersed very well in a nonpolar solvent [77].

6.1.4.3.1 *Nanocellulose–cellulose derivative functionalization*

Cellulose derivatives have been used to functionalize the surface properties of nanocellulose because of their natural affinity toward nanocellulose [10,78,79]. Different approaches utilizing carboxymetyl cellulose for the surface functionalization of nanocellulose have been reported, but the negative charge of carboxymethyl cellulose is disruptive for high adsorption of nanocellulose. By contrast, unmodified hemicellulose derivatives, including xyloglucans, arabinoxylans, and O-acetyl galactoglucomannan, can be functionalized on the surface of nanocellulose in a inconsiderable amount and henceforth became promising starting materials for its functionalization [80]. To use hemicellulose derivatives as functionalizing agents for surface modification of nanocellulose, the main chain of hemicellulose derivatives should preserve its native structure in respect to molar mass, composition, and degree of substitution [81]. This is necessary to reveal high affinity of hemicellulose derivatives toward nanocellulose.

6.1.4.3.2 *Nanocellulose-polymer functionalization*

Mostly, physical properties of nanocellulose are changed by derivation, which involves chemical functionalization of the nanocellulose structure [74]. A good balance of features is obtained if the crystallinity of nanocellulose in the polymer network is reduced and/or the compatibility with a base polymer is improved [75, 76]. Besides, the main objectives of polymer-functionalized nanocellulose are to explore such polymer systems to give additional functionality to nanocellulose for better dispersion and solubility [77]. Lately, specific interest has grown in researching the soluble level of functionalized nanocellulose; there have been many efforts to fully understand and control the solution mechanism.

6.1.4.3.2.1 *Nanocellulose-polysulfone functionalization*

Polysulfone is a type of high-performance polymer with outstanding thermal and chemical stability, flexibility, and strength, as well as good film-forming properties and high glass transition temperature. In spite of a substantial improvement in its applications, polysulfone has some restrictions such as stress cracking, intrinsic hydrophilicity, and weathering features [82]. Therefore, the contribution of hydrophilicity functionalization to improve the hydrophilicity and antifouling properties of polysulfone membrane material is essentially required [83]. Therefore, some research works have brought functional nanocellulose into polysulfone networks not only to overcome these restrictions but also, more importantly, widen the potential application areas of polysulfone materials [84]. It is believed that the hydrophobization chain segment of amphilic nanocellulose provides compatibility between its polymer chains and polysulfone, while hydrophilic and antifouling protection are then created from the surface –OH of amphilic nanocellulose [83]. The flux of blend membranes revealed that the surface enrichment of amphilic nanocellulose expressively improves the hydrophilicity of the surface and polysulfone antipollution ability.

6.1.4.3.2.2 *Nanocellulose-polypropylene functionalization*

The grafting-onto the approach to graft maleated polypropylene onto the surface of tunicate-extracted nanocellulose has resulted in grafted nanocellulose that displays very good compatibility and high adhesion when dispersed in atactic polypropylene [85, 86].

6.1.4.3.2.3 *Nanocellulose–polylactic acid functionalization*

The surface functionalization of nanocellulose with polylactic acid is done via a ring-opening polymerization approach. Polylactic-functionalized nanocellulose displayed stable colloidal behavior in organic solvents in comparison to native nanocellulose that formed aggregates and sediments over time. In addition, as shown by a polarized light microscope, the dispersion of polylactic functionalized nanocellulose was more homogeneous prior to solvent evaporation [87]. The thermal measurement suggested better interaction between functionalized nanocellulose and the nonpolar matrix, whereby the functionalized nanocellulose functions as a nucleating agent, which in turn could increase its crystallinity [88]. Recent studies on polylactic acid–functionalized nanocellulose prove also the positive impact of nanocellulose on water vapor barrier properties. However, the polylatic acid–functionalized nanocellulose did not display a transparent appearance, which might be a result of pore formation [89]. It reported that an increase in the number of pores is related to the increase in a number of nanocellulose concentrations [87].

6.1.4.3.2.4 *Nanocellulose-polyurethane functionalization*

Polyurethane, which is prepared from isocyanate and polyol, is broadly utilized in many applications. In a commercial sense, polyol is utilized for developing polyurethane predominantly derived from petroleum-based resources [90]. With the rising problem of fossil energy resource depletion and also environmental footprints, there is a robust worldwide interest to explore renewable bioresources as alternative feedstock for making polyurethane. Nanocellulose is prepared with phosphoric acid and entirely utilized to modify polyurethane [90]. Nanocellulose is a reinforcement material and oligosaccharides from the hydrolyzed cellulose partly replace polyol [91].

The functionalization process starts with the fabrication of nanocellulose in an anhydrous phosphoric acid system with medical absorbent cotton as its raw material. After ammonia neutralization, the whole system with produced phosphates and hydrolyzed saccharides is utilized as a modifier for preparing polyurethane foam [92]. Adding the modifier meaningfully enhances mechanical properties and flame retardancy of nanocellulose-functionalized polyurethane without inferior thermal conductivity. X-ray and micrograph analysis confirmed that the nanocellulose reacts well with polyurethane with a diameter of 10 nm and had more uniform cells and a regular skeleton structure as compared to neat polyurethane [93].

6.1.4.3.2.5 *Nanocellulose-chitosan functionalization*

Chitosan is traditionally used in water purification; it is most effective toward negatively charged acidic dyes due to the functional group present (NH_2^+). However, the water permeability and water stability of chitosan in different pH conditions, especially after crosslinking, will be of advantage in fabricating water-cleaning membranes [94]. The biggest advantage with the process was the fabrication of a loose and nonaggregated network, which is expected to provide easy availability of surface groups on nanocellulose as adsorption sites for contaminates [95]. High concentration of nanocellulose as a functional entity is used with an aim to have high process efficiency [96].

6.1.4.4 Nanocellulose: Inorganic compound functionalization

Functionalization of inorganic compounds toward the nanocellulose structure is strongly considered by the grafting of metal/metal oxide particles at its –OH positions. This functionalization process strongly induces the surface functionality of nanocellulose if it is fabricated as a composite structure. In recent years, researchers have strongly attempted to functionalize metal/metal oxide at the –OH position of nanocellulose for dielectric and piezoelectric responses, which is considered to result in the electromechanical characteristic of nanocellulose. Structural characterization of inorganic functionalized nanocellulose is mainly carried out by its solids by well-developed solid-state techniques such as Fourier transform infrared (FTIR), X-ray photoelectron spectroscopy (XPS), energy-dispersive X-ray spectroscopy (EDX), and carbon nuclear magnetic resonance (C NMR).

6.1.4.4.1 *Nanocellulose–titanium oxide functionalization*

Nanocellulose-functionalized TiO_2 strongly enhances the photocatalytic antimicrobial effect of TiO_2. It has been proved that it is better to use functionalized nanocellulose either alone or for functionalization with TiO_2 if antibacterial properties are desired [97]. The chemical surface functionalization applied on nanocellulose did not negatively influence this valuable property of nanocellulose but helped in or monitoring this property, which could be very useful for paper, packaging, and composites [98].

6.1.4.4.2 *Nanocellulose-fluorine functionalization*

In general, hydrophobicity of nanocellulose is attained by lowering the surface free energy [99]. For this purpose, surface functionalization of nanocellulose with fluorine is the most effective approach to lower the surface free energy because of its small atomic radius and the biggest electronegativity among atoms [100]. Once fluorine is replaced by other elements of nanocellulose, including C and H, the surface free energy reduces in the order of $CH_2 > CH_3 > CF > CF_2H > CF_3$, in which the CF_3 groups on the surface give the lowest surface free energy of the functionalized nanocellulose [99].

6.1.4.4.3 *Nanocellulose-gold functionalization*

Nanocellulose-functionalized gold (Au) nanoparticles assist as an outstanding support for enzyme immobilization, including cyclodextrin glycosyltransferase and alcohol oxidase, which are immobilized on nanocellulose with high enzyme-loading capacity [101]. The improvement in enzyme loading is because of the greater exposed specific surface area provided by nanocellulose-Au nanoparticles [102]. It was reported that Au nanoparticles with a size around 2 to 7 nm are able to be deposited on nanocellulose by the reduction of $AuCl_3.3H_2O$ with $NaBH_4$, which resulted in covalent binding of thiotic acid to the nanocellulose-functionalized Au [103].

6.1.4.4.4 *Nanocellulose-silver functionalization*

Nanocellulose with functionalization of silver (Ag) is used in wound-dressing applications to mitigate bacterial growth in areas of high humidity [104]. The synthesis of nanocellulose-functionalized Ag started from the reduction of $AgNO_3$ with $NaBH_4$ to CNFs [105]. The nanocellulose fibrils excreted by bacteria, including *Gluconacterobacter xylinum*, are 200 times finer than cotton fiber. This resulted in the presence of extraordinarily high surface area due to their high aspect ratio (length:diameter ratio). The nanocellulose-functionalized Ag also demonstrated antimicrobial performance of more than 99.99% against *E. coli* and *S. aureus* [106].

6.1.4.4.5 *Nanocellulose-Pd functionalization*

Functionalization of nanocellulose with palladium (Pd) nanoparticles with an average size of 3.6 ± 0.8 nm is done by reduction of $PdCl_2$ with H in the presence of nanocellulose [107]. Nanocellulose serves as a support matrix for the formation of stable Pd nanoparticles and provides the necessary sites for the substrate to absorb and participate in further chemical reaction [108]. The fast rate of reaction in comparison to other Pd-functionalized materials could be attributed to both smaller Pd nanoparticles and the positive charge on the surface of nanocellulose [109].

6.1.4.4.6 *Nanocellulose-CdS functionalization*

As a semiconductor material, cadmium sulfide (CdS) has found application in solar cells and optoelectronic and electronic devices [110]. Furthermore, functionalization of nanocellulose with CdS using electroless deposition has become a universal platform for producing nanoscale functional material with advantages over protein or DNA templating in terms of cost, versatility, and simplicity [111]. The morphology-controlled CdS nanocrystals with nanocellulose, which have been prepared by a hydrothermal method, act as high-efficiency photocatalysts [110].

6.2 Applications of Nanocomposites from Cellulose and Nanocellulose Derivatives

On the basis of its unique properties, functionalized nanocellulose is used in numerous applications ranging from bulk applications, including as a rheological modifier, composite reinforcement, or paper additive, to high-end applications such as tissue engineering, drug delivery, and functional materials.

6.2.1 Wastewater Treatment

The wastewater produced from different kinds of industries normally contains very fine suspended solids, dissolved solids, inorganic and organic particles, metals, and other impurities. Due to the very small size of the particles and the presence of surface charges, the task to bring these particles closer to the heavier mass for settling and filtration becomes challenging [112]. Functionalized nanocellulose has been employed as a nanocomposite filter for the removal of organic/inorganic pollutants from industrial effluents via chemical precipitation, membrane separation, ion exchange, flocculation, electrolysis, and evaporation. Native nanocellulose has packed aggregates and high fractal dimensions, whereas functionalized nanocellulose has lower fractal dimensions due to large, highly branched, and loosely bound structures [112]. Besides, few functional groups in functionalized nanocellulose are able to capture metal ions through some derivatization. Some of these techniques are based on utilizing amine and carboxylate groups as chelating agents and/or catalytic and selective oxidation of primary –OH groups of nanocellulose [113]. The succinylation reaction has also been exposed to be an alternative to cellulose functionalization. Therefore, functionalized nanocellulose has recently been utilized in the coagulation-flocculation treatment of wastewater. The combined coagulation-flocculation treatment of municipal wastewater led to lower residual turbidity and chemical oxygen demand (COD) in a settled suspension, with significantly decreased total chemical consumption [114].

For example, the dicarboxylic acid–nanocellulose showed a reduction in turbidity and COD removal performance of wastewater than a commercial reference polymer in low dosage, with considerably decreased chemical consumption relative to coagulation [114, 115]. The results showed that the dicarboxylic acid–nanocellulose is able to flocculate wastewater very proficiently. The wastewater flocs produced with functionalized nanocellulose are smaller and rounder than those produced with the commercial reference polymer, with the flocs produced with anionic nanocellulose being more stable under shear force than the flocs produced with the reference polymer [116]. This in turn makes dicarboxylic acid–nanocellulose have good performance within the chosen pH range and high stability in aqueous suspensions over a long period of time.

6.2.2 Biomedical Applications

Nanocellulose-functionalized Ag with antimicrobial properties has been found as a biobased nanocomposite to inhibit the growth of both *E. coli* and *S. aureus*. The greater effectiveness of the nanocellulose-functionalized Ag solution suggests a favorable interaction between nanocellulose and bacterial growth inhibition [104]. The smaller nanocellulose particle sizes are predisposed to Ag nanoparticle suspension use in antiseptic solutions or in wound-healing gels at greater nanocellulose concentrations. Isolating a solid material by freeze-drying allows it to be utilized for the manufacture of biodegradable wound dressings [105].

Nunctionalized nanocellulose has been applied also as an agent for enzyme or protein immobilization because of its large surface area and porous structure [117]. For example, nanocellulose-functionalized peroxidase through activation with cyanogen bromide has been used for the removal of chlorinated phenolic compounds in an aqueous medium. The immobilized peroxidase demonstrates improved removal of chlorinated phenolic compounds compared to its soluble counterpart [117]. This probably is because of protective effects of the immobilization toward enzyme deactivation, as well as product precipitation induced by the conjugate amino groups.

6.2.3 Biosensor and Bioimaging

The functional groups on the surface of nanocellulose could be conjugated with different biological moieties or serve as binding sites for inorganic nanoparticles, which enable its utilization in biosensing or bioimaging. One class of biomolecules conjugated to functionalized nanocellulose is nucleic acids using 2,2,6,6-tetramethylpiperidine-1-oxyl radical (TEMPO)-mediated oxidation and an amino modifier. This allows hybridizing reversibly using the molecular recognition ability of the nucleic acid to form a duplex that decouples at high temperature (Fig. 6.4) [68].

Another efficient method of attaching nanocellulose to nucleic acids is through the functionalization of bifunctional oxirane 1,4-butane-diol diglycidyl ether. This functionalization product is used to purify complementary nucleic acid compounds by affinity chromatography. This method could probably as well be adapted for use with functionalized nanocellulose to developed chromatographic materials with a high surface area for a variety of applications [118]. Meanwhile, nanocellulose-functionalized chitosan with competitive binding assays by using triclosan and dodecylsulfate anions demonstrates great sensitivity and potential utilization in surfactant detection.

Figure 6.4 Model of peptide-conjugated cellulose nanocrystal. Reprinted by permission from Springer Nature Customer Service Centre: Springer Nature, *Cellulose*, Ref. [65], copyright (2013).

Furthermore, inorganic materials functionalization with nanocellulose can be used as labels for electrical detection of nucleic acid hybridization. For example, Au-carboxylated nanocellulose utilized labeled nucleic acid probes to identify the complementary target of the nucleic acid sequence [119]. The carboxyl and hydroxyl groups of carboxylated nanocellulose trigger a coordination effect to adsorb metallic cations and alloy nanoparticles, preventing the agglomeration of nanoparticles. Meanwhile, nanocellulose-functionalized TiO_2 with promising conducting pathways for an electron in a relatively open nanocellulose structure is suitable for methemoglobin immobilization.

6.2.4 Catalysis

The uses of functionalized nanocellulose-based nanocomposites as a support matrix for new heterogeneous catalysis are growing. The advantage of highly dispersed inorganic nanoparticles is efficient contact among substrates and the inorganic material surface for reactions to occur. The catalytic properties of nanocellulose-functionalized Pd have been exploited for the hydrogenation of phenol to cyclohexanone and the Heck coupling reaction of styrene with iodobenzene. It is recorded that up to 90% conversion of phenol to cyclohexanone is achieved after 24 h at room temperature using H with a 7:1 substrate-to-catalyst ratio [107].

6.3 Conclusions

In summary, this chapter was divided into three parts: The first part briefly discussed lignocellulosic biomass–derived cellulose and nanocellulose, followed by a part that reviewed the progress of functionalized cellulose/nanocellulose. The last part focused on the applications of functionalized nanocellulose for nanocomposte applications (e.g., water treatment, biomedicine, biosensors, and catalysis). The functionalized cellulose/nanocellulose products with excellent characteristics (optical, mechanical, and thermal properties), which intergrate with their ecofriendliness and biodegradability, make them potential biomaterials of choice in the area of bionanotechnology, opening up major commercial markets in line with the green chemistry trend.

Acknowledgments

The authors are grateful for research support from the SATU Joint Research Scheme (ST015-2017) and Ajinomoto Co., Inc. (IF010-2017).

References

1. Trache, D., et al. (2017). Recent progress in cellulose nanocrystals: sources and production, *Nanoscale*, **9**(5), pp. 1763–1786.
2. Moon, R. J., et al. (2011). Cellulose nanomaterials review: structure, properties and nanocomposites, *Chem. Soc. Rev.*, **40**(7), pp. 3941–3994.
3. Habibi, Y. (2014). Key advances in the chemical modification of nanocelluloses, *Chem. Soc. Rev.*, **43**(5), pp. 1519–1542.
4. Li, F., Mascheroni, E. and Piergiovanni, L. (2015). The potential of nanocellulose in the packaging field: a review, *Packag. Technol. Sci.*, **28**(6), pp. 475–508.
5. García, A., et al. (2016). Industrial and crop wastes: a new source for nanocellulose biorefinery, *Ind. Crops Prod.*, **93**, pp. 26–38.
6. Pacheco-Torgal, F. and Labrincha, J. (2014). Biotechnologies and bioinspired materials for the construction industry: an overview, *Int. J. Sustainable Eng.*, **7**(3), pp. 235–244.
7. Faradilla, R. F., et al. (2016). Nanocellulose characteristics from the inner and outer layer of banana pseudo-stem prepared by TEMPO-mediated oxidation, *Cellulose*, **23**(5), pp. 3023–3037.
8. Bajpai, P. (2017). Introduction, in *Pulp and Paper Industry* (Elsevier), pp. 1–13.
9. Mood, S. H., et al. (2013). Lignocellulosic biomass to bioethanol, a comprehensive review with a focus on pretreatment, *Renewable Sustainable Energy Rev.*, **27**, pp. 77–93.
10. Siró, I. and Plackett, D. (2010). Microfibrillated cellulose and new nanocomposite materials: a review, *Cellulose*, **17**(3), pp. 459–494.
11. Jonoobi, M., et al. (2015). Different preparation methods and properties of nanostructured cellulose from various natural resources and residues: a review, *Cellulose*, **22**, pp. 935–969.
12. Charreau, H., Foresti, M. L. and Vázquez, A. (2013). Nanocellulose patents trends: a comprehensive review on patents on cellulose nanocrystals, microfibrillated and bacterial cellulose, *Recent Pat. Nanotechnol.*, **7**(1), pp. 56–80.
13. Keshk, S. (2014). Bacterial cellulose production and its industrial applications, *J Bioprocess. Biotech.*, **4**(2), p. 150.
14. Kim, H. C., et al. (2016). Renewable smart materials, *Smart Mater. Struct.*, **25**(7), p. 073001.

15. dos Santos, F. A., Iulianelli, G. C. and Tavares, M. I. B. (2016). The use of cellulose nanofillers in obtaining polymer nanocomposites: properties, processing, and applications, *Mater. Sci. Appl.*, **7**(5), p. 257.
16. Zhang, D., et al. (2013). A nanocellulose polypyrrole composite based on tunicate cellulose, *Int. J. Polym. Sci.*, **2013**, p. 6.
17. Kouzuma, A. and Watanabe, K. (2015). Exploring the potential of algae/bacteria interactions, *Curr. Opin. Biotechnol.*, **33**, pp. 125–129.
18. Kim, H. M., et al. (2015). Efficient approach for bioethanol production from red seaweed Gelidium amansii, *Bioresour. Technol.*, **175**, pp. 128–134.
19. Seo, Y.-B., et al. (2010). Red algae and their use in papermaking, *Bioresour. Technol.*, **101**(7), pp. 2549–2553.
20. Chen, Y. W., et al. (2016). Production of new cellulose nanomaterial from red algae marine biomass Gelidium elegans, *Carbohydr. Polym.*, **151**, pp. 1210–1219.
21. Mihranyan, A. (2011). Cellulose from cladophorales green algae: from environmental problem to high-tech composite materials, *J. Appl. Polym. Sci.*, **119**(4), pp. 2449–2460.
22. Thakur, V. K., Thakur, M. K. and Kessler, M. R. (2017). *Handbook of Composites from Renewable Materials, Biodegradable Materials*, Vol. 5 (John Wiley & Sons).
23. Ng, H.-M., et al. (2015). Extraction of cellulose nanocrystals from plant sources for application as reinforcing agent in polymers, *Composites Part B*, **75**, pp. 176–200.
24. Dufresne, A. (2013). *Nanocellulose: From Nature to High Performance Tailored Materials* (Walter de Gruyter).
25. César, N. R., et al. (2015). Cellulose nanocrystals from natural fiber of the macrophyte Typha domingensis: extraction and characterization, *Cellulose*, **22**(1), pp. 449–460.
26. Domingues, R. M., Gomes, M. E. and Reis, R. L. (2014). The potential of cellulose nanocrystals in tissue engineering strategies, *Biomacromolecules*, **15**(7), pp. 2327–2346.
27. Silvério, H. A., et al. (2013). Extraction and characterization of cellulose nanocrystals from corncob for application as reinforcing agent in nanocomposites, *Ind. Crops Prod.*, **44**, pp. 427–436.
28. Zhang, W., et al. (2011). Mechanochemical activation of cellulose and its thermoplastic polyvinyl alcohol ecocomposites with enhanced physicochemical properties, *Carbohydr. Polym.*, **83**(1), pp. 257–263.

29. Lam, E., et al. (2012). Applications of functionalized and nanoparticle-modified nanocrystalline cellulose, *Trends Biotechnol.*, **30**(5), pp. 283–290.

30. Pei, A., Zhou, Q. and Berglund, L. A. (2010). Functionalized cellulose nanocrystals as biobased nucleation agents in poly (l-lactide)(PLLA)–Crystallization and mechanical property effects, *Compos. Sci. Technol.*, **70**(5), pp. 815–821.

31. Rebouillat, S. and Pla, F. (2013). State of the art manufacturing and engineering of nanocellulose: a review of available data and industrial applications, *J. Biomater. Nanobiotechnol.*, **4**(2), p. 165.

32. Filpponen, I. and Argyropoulos, D. S. (2010). Regular linking of cellulose nanocrystals via click chemistry: synthesis and formation of cellulose nanoplatelet gels, *Biomacromolecules*, **11**(4), pp. 1060–1066.

33. Siqueira, G., Bras, J. and Dufresne, A. (2008). Cellulose whiskers versus microfibrils: influence of the nature of the nanoparticle and its surface functionalization on the thermal and mechanical properties of nanocomposites, *Biomacromolecules*, **10**(2), pp. 425–432.

34. Kloser, E. and Gray, D. G. (2010). Surface grafting of cellulose nanocrystals with poly (ethylene oxide) in aqueous media, *Langmuir*, **26**(16), pp. 13450–13456.

35. Braun, B. and Dorgan, J. R. (2008). Single-step method for the isolation and surface functionalization of cellulosic nanowhiskers, *Biomacromolecules*, **10**(2), pp. 334–341.

36. Tingaut, P., Zimmermann, T. and Sèbe, G. (2012). Cellulose nanocrystals and microfibrillated cellulose as building blocks for the design of hierarchical functional materials, *J. Mater. Chem.*, **22**(38), pp. 20105–20111.

37. Pahimanolis, N., et al. (2011). Surface functionalization of nanofibrillated cellulose using click-chemistry approach in aqueous media, *Cellulose*, **18**(5), p. 1201.

38. Niederberger, M., et al. (2004). Tailoring the surface and solubility properties of nanocrystalline titania by a nonaqueous in situ functionalization process, *Chem. Mater.*, **16**(7), pp. 1202–1208.

39. Zaman, M., et al. (2012). Synthesis and characterization of cationically modified nanocrystalline cellulose, *Carbohydr. Polym.*, **89**(1), pp. 163–170.

40. Yang, W., et al. (2005). Electrically addressable biomolecular functionalization of conductive nanocrystalline diamond thin films, *Chem. Mater.*, **17**(5), pp. 938–940.

41. Biyani, M. V., Foster, E. J. and Weder, C. (2013). Light-healable supramolecular nanocomposites based on modified cellulose nanocrystals, *ACS Macro Lett.*, **2**(3), pp. 236–240.

42. Jiang, F., et al. (2014). 1D and 2D NMR of nanocellulose in aqueous colloidal suspensions, *Carbohydr. Polym.*, **110**, pp. 360–366.

43. Hua, K., et al. (2014). Translational study between structure and biological response of nanocellulose from wood and green algae, *RSC Adv.*, **4**(6), pp. 2892–2903.

44. Anirudhan, T. S. and Rejeena, S. R. (2014). Poly(acrylic acid-co-acrylamide-co-2-acrylamido-2-methyl-1-propanesulfonic acid)-grafted nanocellulose/poly (vinyl alcohol) composite for the in vitro gastrointestinal release of amoxicillin, *J. Appl. Polym. Sci.*, **131**(17).

45. Jebali, A., et al. (2013). Antimicrobial activity of nanocellulose conjugated with allicin and lysozyme, *Cellulose*, **20**(6), pp. 2897–2907.

46. Jafary, R., et al. (2015). Antibacterial property of cellulose fabric finished by allicin-conjugated nanocellulose, *J. Text. Inst.*, **106**(7), pp. 683–689.

47. Moritz, S., et al. (2014). Active wound dressings based on bacterial nanocellulose as drug delivery system for octenidine, *Int. J. Pharm.*, **471**(1), pp. 45–55.

48. Mabrouk, A. B., et al. (2014). Cellulose-based nanocomposites prepared via mini-emulsion polymerization: understanding the chemistry of the nanocellulose/matrix interface, *Colloids Surf., A*, **448**, pp. 1–8.

49. Jin, H., et al. (2011). Superhydrophobic and superoleophobic nanocellulose aerogel membranes as bioinspired cargo carriers on water and oil, *Langmuir*, **27**(5), pp. 1930–1934.

50. Lee, J.-A., et al. (2014). Preparation and characterization of cellulose nanofibers (CNFs) from microcrystalline cellulose (MCC) and CNF/polyamide 6 composites, *Macromol. Res.*, **22**(7), pp. 738–745.

51. Surip, S. N., et al. (2012). Microscopy observation on nanocellulose from kenaf fibre, *Adv. Mater. Res.*, **488–489**, pp. 72–75.

52. Khan, R. A., et al. (2012). Improvement of the mechanical and barrier properties of methylcellulose-based films by treatment with HEMA and silane monomers under gamma radiation, *Radiat. Phys. Chem.*, **81**(8), pp. 927–931.

53. Ben Mabrouk, A., et al. (2011). Preparation of nanocomposite dispersions based on cellulose whiskers and acrylic copolymer by miniemulsion polymerization: effect of the silane content, *Polym. Eng. Sci.*, **51**(1), pp. 62–70.

54. Dai, D., Fan, M. and Collins, P. (2013). Fabrication of nanocelluloses from hemp fibers and their application for the reinforcement of hemp fibers, *Ind. Crops Prod.*, **44**, pp. 192–199.
55. Cunha, A. G., et al. (2014). Topochemical acetylation of cellulose nanopaper structures for biocomposites: mechanisms for reduced water vapour sorption, *Cellulose*, **21**(4), pp. 2773–2787.
56. Khalil, H. A., et al. (2014). Production and modification of nanofibrillated cellulose using various mechanical processes: a review, *Carbohydr. Polym.*, **99**, pp. 649–665.
57. Isogai, A. (2013). Wood nanocelluloses: fundamentals and applications as new bio-based nanomaterials, *J. Wood Sci.*, **59**(6), pp. 449–459.
58. Cherian, B. M., et al. (2010). Isolation of nanocellulose from pineapple leaf fibres by steam explosion, *Carbohydr. Polym.*, **81**(3), pp. 720–725.
59. Heßler, N. and Klemm, D. (2009). Alteration of bacterial nanocellulose structure by in situ modification using polyethylene glycol and carbohydrate additives, *Cellulose*, **16**(5), pp. 899–910.
60. Rehman, N., et al. (2014). Cellulose and nanocellulose from maize straw: an insight on the crystal properties, *J. Polym. Environ.*, **22**(2), pp. 252–259.
61. Anirudhan, T. and Rejeena, S. (2012). Adsorption and hydrolytic activity of trypsin on a carboxylate-functionalized cation exchanger prepared from nanocellulose, *J. Colloid Interface Sci.*, **381**(1), pp. 125–136.
62. Wang, M., et al. (2011). Colloidal ionic assembly between anionic native cellulose nanofibrils and cationic block copolymer micelles into biomimetic nanocomposites, *Biomacromolecules*, **12**(6), pp. 2074–2081.
63. Mishra, S. P., et al. (2011). Production of nanocellulose from native cellulose–various options utilizing ultrasound, *BioResources*, **7**(1), pp. 0422–0436.
64. Fujisawa, S., et al. (2011). Preparation and characterization of TEMPO-oxidized cellulose nanofibril films with free carboxyl groups, *Carbohydr. Polym.*, **84**(1), pp. 579–583.
65. Holt, B. L., et al. (2010). Novel anisotropic materials from functionalised colloidal cellulose and cellulose derivatives, *J. Mater. Chem.*, **20**(45), pp. 10058–10070.
66. Benkaddour, A., et al. (2013). Grafting of polycaprolactone on oxidized nanocelluloses by click chemistry, *Nanomaterials*, **3**(1), pp. 141–157.

67. Barazzouk, S. and Daneault, C. (2012). Tryptophan-based peptides grafted onto oxidized nanocellulose, *Cellulose*, **19**(2), pp. 481–493.

68. Edwards, J. V., Prevost, N., Sethumadhavan, K., Ullah, A. and Condon, B. (2013). Peptide conjugated cellulose nanocrystals with sensitive human neutrophil elastase sensor activity, *Cellulose*, **20**(3), pp. 1223–1235.

69. Carlsson, D. O., et al. (2014). Cooxidant-free TEMPO-mediated oxidation of highly crystalline nanocellulose in water, *RSC Adv.*, **4**(94), pp. 52289–52298.

70. Lu, T., et al. (2014). Composite aerogels based on dialdehyde nanocellulose and collagen for potential applications as wound dressing and tissue engineering scaffold, *Compos. Sci. Technol.*, **94**, pp. 132–138.

71. Sirviö, J. A., et al. (2014). Strong, self-standing oxygen barrier films from nanocelluloses modified with regioselective oxidative treatments, *ACS Appl. Mater. Interfaces*, **6**(16), pp. 14384–14390.

72. Zimmermann, K. A., et al. (2011). Biomimetic design of a bacterial cellulose/hydroxyapatite nanocomposite for bone healing applications, *Mater. Sci. Eng., C*, **31**(1), pp. 43–49.

73. Hokkanen, S., et al. (2014). Adsorption of Ni^{2+}, Cd^{2+}, PO_4^{3-} and NO^{3-} from aqueous solutions by nanostructured microfibrillated cellulose modified with carbonated hydroxyapatite, *Chem. Eng. J.*, **252**, pp. 64–74.

74. Taokaew, S., et al. (2013). Biosynthesis and characterization of nanocellulose-gelatin films, *Materials*, **6**(3), pp. 782–794.

75. Anirudhan, T. S. and Rejeena, S. R. (2013). Selective adsorption of hemoglobin using polymer-grafted-magnetite nanocellulose composite, *Carbohydr. Polym.*, **93**(2), pp. 518–527.

76. Lin, N. and Dufresne, A. (2014). Nanocellulose in biomedicine: current status and future prospect, *Eur. Polym. J.*, **59**, pp. 302–325.

77. Bodin, A., et al. (2007). Modification of nanocellulose with a xyloglucan–RGD conjugate enhances adhesion and proliferation of endothelial cells: implications for tissue engineering, *Biomacromolecules*, **8**(12), pp. 3697–3704.

78. Qamhia, I. I., Sabo, R. C. and Elhajjar, R. F. (2013). Static and dynamic characterization of cellulose nanofibril scaffold-based composites, *BioResources*, **9**(1), pp. 381–392.

79. Ansari, F., et al. (2014). Cellulose nanofiber network for moisture stable, strong and ductile biocomposites and increased epoxy curing rate, *Composites Part A*, **63**, pp. 35–44.

80. Liu, A. and Berglund, L. A. (2013). Fire-retardant and ductile clay nanopaper biocomposites based on montmorrilonite in matrix of cellulose nanofibers and carboxymethyl cellulose, *Eur. Polym. J.*, **49**(4), pp. 940–949.

81. Pahimanolis, N., et al. (2013). Nanofibrillated cellulose/carboxymethyl cellulose composite with improved wet strength, *Cellulose*, **20**(3), pp. 1459–1468.

82. Bai, H., Zhou, Y. and Zhang, L. (2015). Morphology and mechanical properties of a new nanocrystalline cellulose/polysulfone composite membrane, *Adv. Polym. Technol.*, **34**(1).

83. Gao, Y., et al. (2014). Effect of nano-amphiphilic cellulose as a modifier to PSf composite membranes, *Vacuum*, **107**, pp. 199–203.

84. Grygiel, K., et al. (2014). Omnidispersible poly (ionic liquid)-functionalized cellulose nanofibrils: surface grafting and polymer membrane reinforcement, *Chem. Commun.*, **50**(83), pp. 12486–12489.

85. Savadekar, N. and Mhaske, S. (2012). Synthesis of nano cellulose fibers and effect on thermoplastics starch based films, *Carbohydr. Polym.*, **89**(1), pp. 146–151.

86. Xie, K., Gao, X. and Zhao, W. (2010). Thermal degradation of nano-cellulose hybrid materials containing reactive polyhedral oligomeric silsesquioxane, *Carbohydr. Polym.*, **81**(2), pp. 300–304.

87. Aulin, C., et al. (2013). Transparent nanocellulosic multilayer thin films on polylactic acid with tunable gas barrier properties, *ACS Appl. Mater. Interfaces*, **5**(15), pp. 7352–7359.

88. Jonoobi, M., et al. (2012). A comparison of modified and unmodified cellulose nanofiber reinforced polylactic acid (PLA) prepared by twin screw extrusion, *J. Polym. Environ.*, **20**(4), pp. 991–997.

89. Baheti, V., et al. (2014). Influence of noncellulosic contents on nano scale refinement of waste jute fibers for reinforcement in polylactic acid films, *Fibers Polym.*, **15**(7), pp. 1500–1506.

90. Pei, A., et al. (2011). Strong nanocomposite reinforcement effects in polyurethane elastomer with low volume fraction of cellulose nanocrystals, *Macromolecules*, **44**(11), pp. 4422–4427.

91. Aranguren, M. I., et al. (2013). Effect of the nano-cellulose content on the properties of reinforced polyurethanes. A study using mechanical

tests and positron anihilation spectroscopy, *Polym. Test.*, **32**(1), pp. 115–122.

92. Juntaro, J., et al. (2012). Bacterial cellulose reinforced polyurethane-based resin nanocomposite: a study of how ethanol and processing pressure affect physical, mechanical and dielectric properties, *Carbohydr. Polym.*, **87**(4), pp. 2464–2469.
93. Liu, H., et al. (2012). Cellulose nanocrystal/silver nanoparticle composites as bifunctional nanofillers within waterborne polyurethane, *ACS Appl. Mater. Interfaces*, **4**(5), pp. 2413–2419.
94. Khan, A., et al. (2012). Mechanical and barrier properties of nanocrystalline cellulose reinforced chitosan based nanocomposite films, *Carbohydr. Polym.*, **90**(4), pp. 1601–1608.
95. Dehnad, D., et al. (2014). Optimization of physical and mechanical properties for chitosan–nanocellulose biocomposites, *Carbohydr. Polym.*, **105**, pp. 222–228.
96. Pereda, M., et al. (2014). Polyelectrolyte films based on chitosan/olive oil and reinforced with cellulose nanocrystals, *Carbohydr. Polym.*, **101**, pp. 1018–1026.
97. Bardet, R., Belgacem, M. N. and Bras, J. (2013). Different strategies for obtaining high opacity films of MFC with TiO_2 pigments, *Cellulose*, **20**(6), pp. 3025–3037.
98. Miettunen, K., et al. (2014). Nanocellulose aerogel membranes for optimal electrolyte filling in dye solar cells, *Nano Energy*, **8**, pp. 95–102.
99. Pandey, J. K., et al. (2014). Cellulose nanofiber assisted deposition of titanium dioxide on fluorine-doped tin oxide glass, *RSC Adv.*, **4**(2), pp. 987–991.
100. Korhonen, J. T., et al. (2011). Inorganic hollow nanotube aerogels by atomic layer deposition onto native nanocellulose templates, *ACS Nano*, **5**(3), pp. 1967–1974.
101. Shi, Z., Phillips, G. O. and Yang, G. (2013). Nanocellulose electroconductive composites, *Nanoscale*, **5**(8), pp. 3194–3201.
102. Lokanathan, A. R., et al. (2013). Cilia-Mimetic hairy surfaces based on end-immobilized nanocellulose colloidal rods, *Biomacromolecules*, **14**(8), pp. 2807–2813.
103. Stevanic, J. S., et al. (2011). Bacterial nanocellulose-reinforced arabinoxylan films, *J. Appl. Polym. Sci.*, **122**(2), pp. 1030–1039.

104. Berndt, S., et al. (2013). Antimicrobial porous hybrids consisting of bacterial nanocellulose and silver nanoparticles, *Cellulose*, **20**(2), pp. 771–783.

105. Dong, H., et al. (2013). Hydrogel, aerogel and film of cellulose nanofibrils functionalized with silver nanoparticles, *Carbohydr. Polym.*, **95**(2), pp. 760–767.

106. Suman, et al. (2015). A novel reusable nanocomposite for complete removal of dyes, heavy metals and microbial load from water based on nanocellulose and silver nano-embedded pebbles, *Environ. Technol.*, **36**(6), pp. 706–714.

107. Zhou, P., et al. (2012). Bacteria cellulose nanofibers supported palladium (0) nanocomposite and its catalysis evaluation in Heck reaction, *Ind. Eng. Chem. Res.*, **51**(16), pp. 5743–5748.

108. Zander, N. E., et al. (2014). Metal cation cross-linked nanocellulose hydrogels as tissue engineering substrates, *ACS Appl. Mater. Interfaces*, **6**(21), pp. 18502–18510.

109. Rezayat, M., et al. (2014). Green one-step synthesis of catalytically active palladium nanoparticles supported on cellulose nanocrystals, *ACS Sustainable Chem. Eng.*, **2**(5), pp. 1241–1250.

110. Rajawat, D. S., et al. (2013). Nanocellulosic fiber-modified carbon paste electrode for ultra trace determination of Cd (II) and Pb (II) in aqueous solution, *Environ. Sci. Pollut. Res.*, **20**(5), pp. 3068–3076.

111. Fazilova, S., Yugai, S. and Rashidova, S. S. (2011). Structural investigation of polysaccharides and nanocompositions based on them, *Russ. J. Bioorg. Chem.*, **37**(7), pp. 786–790.

112. Suopajärvi, T., et al. (2015). Lead adsorption with sulfonated wheat pulp nanocelluloses, *J. Water Process Eng.*, **5**, pp. 136–142.

113. Zhen, W. J. (2011). Study on nanocellulose/starch composites, *Adv. Mater. Res.*, **187**, pp. 544–547.

114. Suopajärvi, T., et al. (2013). Coagulation–flocculation treatment of municipal wastewater based on anionized nanocelluloses, *Chem. Eng. J.*, **231**, pp. 59–67.

115. Mishra, S. P., et al. (2010). Ultrasound-catalyzed TEMPO-mediated oxidation of native cellulose for the production of nanocellulose: effect of process variables, *BioResources*, **6**(1), pp. 121–143.

116. Lin, N., Bruzzese, C. C. and Dufresne, A. (2012). TEMPO-oxidized nanocellulose participating as crosslinking aid for alginate-based sponges, *ACS Appl. Mater. Interfaces*, **4**(9), pp. 4948–4959.

117. Barud, H. S., et al. (2011). Antimicrobial bacterial cellulose-silver nanoparticles composite membranes, *J. Nanomater.*, **2011**, p. 10.

118. Zhang, R., et al. (2014). Reversible cross-linking, microdomain structure, and heterogeneous dynamics in thermally reversible cross-linked polyurethane as revealed by solid-state NMR, *J. Phys. Chem. B*, **118**(4), pp. 1126–1137.

119. Edwards, J. V., et al. (2013). Nanocellulose-based biosensors: design, preparation, and activity of peptide-linked cotton cellulose nanocrystals having fluorimetric and colorimetric elastase detection sensitivity, *Engineering*, **5**(9), p. 20.

Chapter 7

Gold–Iron Oxide Nanohybrids: Characterization and Biomedical Applications

Yasir Javed,[a] **M. Irfan Hussain,**[a] **Muhammad Yaseen,**[b] **and Muhammad Asif**[a]

[a]*Magnetic Materials Laboratory, Department of Physics, University of Agriculture, University Main Rd. Faisalabad, Faisalabad 38000, Pakistan*

[b]*Department of Physics, University of Agriculture, University Main Rd. Faisalabad, Faisalabad 38000, Pakistan*

myasi60@hotmail.com

Multifunctional nanostructures containing two or more different materials have received great interest in recent years because of their high prospects for applications as advanced nanomaterials. These nanocomposites exhibited innovative physical and chemical characteristics due to a controlled structure and interface interactions. These nanohybrids can exist in different forms such as core–shell, dimers, dumbbell shape, and Janus morphology systems and presented improved optical, magnetic, and catalytic properties. Gold–iron oxide nanostructures in hybrid form emerged as a

Hybrid Nanocomposites: Fundamentals, Synthesis, and Applications
Edited by Kaushik Pal

ISBN 978-981-4800-34-1 (Hardcover), 978-0-429-00096-6 (eBook)
www.panstanford.com

promising class for enhanced applications, especially in the medical field where a single nanosystem can be used for multitherapy, for example, magnetic hyperthermia and plasmon thermotherapy. This type of nanostructure is usually obtained by physical deposition methods, but in recent years solution-phase chemical synthesis has also been applied for the formation of these bifunctional nanoparticles. In this chapter, we will review different properties of nanohybrids based on gold–iron oxide nanostructures and their applications in biological and medical fields.

7.1 Introduction

Nanomaterials are considered the backbone of nanoscience and nanotechnology. Nanoscience mostly deals with synthesis, characterization, and applications of nanostructure-based materials [1, 2]. The materials developed during all these years have shown excellent magnetic, optical, and electrical properties and are used in bioengineering, information technology, and energy-related applications [3–6]. In the past two decades, nanohybrid structures have received a lot of attention in the scientific community due to conjugation of different properties in a single nanosystem and as a result have become more efficient with respect to technologically related applications [7–10]. At present, there is great interest in developing nanoparticles (NPs) with multiple functional or properties that cannot be obtained by individual particles [11]. This development of materials is very fascinating and facile because the resulting hybrid material not only maintains its individual electronic, magnetic, semiconducting, or plasmonic properties but also improves its optical, plasmonic, or catalytic properties compared to individual components [12, 13]. Nanoshells, for example, with a spherical dielectric core and a metallic shell are ideal nanostructures that provide new collective structural properties from the constituents that are employed for the fabrication of multifunctional probes [14, 15].

There are a number of different dielectric NPs used as a core for gold shells [16–19], but iron oxide is used a lot because of its numerous advantages. First, iron oxide is a magnetic material. Particularly, magnetite NPs have exhibited high susceptibility, low coercive field, low retentivity, and high saturation magnetization [20, 21]. Magnetic iron oxide nanoparticles (IONPs), particularly maghemite (γ-Fe_2O_3) and magnetite (Fe_3O_4) having a size less than 20 nm, show superparamagnetic behavior [22]. IONPs are playing a vital role in biological systems by forming complexes in the human body such as hemoglobin and other myoglobins [23]. The structural properties make these particles ideal for different biomedical applications such as magnetic resonance imaging (MRI), magnetically guided drug delivery, magnetic hyperthermia, biosensors, etc. [24, 25]. The high chemical stability of magnetite NPs against oxidation enables them to internalize in the body through blood circulation and can be directed to a particular target site by applying a magnetic field [26, 27].

For high-performance applications, the magnetic NPs should possess normal size distribution, smoother surface, spherical morphology, and potential to make colloidals with physiological fluids [28, 29]. But these particles tend to aggregate in liquid form, which decreases their efficiency [30]. This difficulty can be controlled by coating a layer of gold on the surface of magnetite NPs. A gold shell layer also yields an energetic plasmonic optical response to the NPs [31]. The resonant frequency is influenced by the architecture of the clusters, the dielectric core, and the surrounding environment of the NPs [32, 33].

Gold NPs have gained much attention due to their altogether different behavior at the nanoscale than their bulk counterpart. The reason is a large surface area, which induces unique chemical and physically properties and can be modified according to the requirement on the basis of shape, size, and composition [34]. Among other interesting properties in gold NPs, the most commonly known characteristic is the collective behavior of their electrons that induce plasmon resonance. This idea emerged in 1990s and is now a separated established field [35, 36]. In the meantime, biophysics developed as one of the major application fields for plasmonics. Gold NPs have applications as contrast agents in bioimaging for cancer diagnosis, biomolecule sensing, dark-field microscopy, enhanced fluorescence microscopy, etc. [37, 38].

The plasmonic properties of gold NPs can be adjusted for the spectrum of the near-infrared (NIR) region to which tissues are normally permeable [39]. This can be done by tuning the core and shell dimensions. This provides a wide range of information about diagnostic, sensing, optical imaging, and therapeutic-related applications. It is reported that NIR light penetrates deeply into soft tissues, nearly ~10 cm passing along the breast and 4 cm through brain tissue [40].

Plasmonic properties can be understood by using the plasmonic hybridization concept. According to this, dielectric-nature core/metal spherical shell-like NPs have changeable plasmon resonances that appear by the interaction from cavity plasmons of the inside shell surface having sphere plasmons [41, 42]. When the outer and inner surfaces of the shell structure plasmon hybridize and mix, a low-energy bright or bonding plasmon resonance is shown due to strong interaction with incident light, whereas high-energy dark or antibonding plasmons are shown due to weak coupling with incident light [43]. In a thinner layer, the hybridization becomes stronger due to interaction, which leads to a strongly red-phase-shifted resonance at a specific wavelength calculated for the thickness of the shell-like structure and cumulative particle radius. Consequently, it is very important to synthesize metallic core–shell particles of thin and uniform shape [44].

It is well identified that the interface between gold and iron oxide hybrid nanostructures influences the diffusion of free charge carriers and forms electrical junctions. Surface electronic states can also be affected by the coated polymer, solvent, and surrounding NPs [45, 46]. The behavior of the metal nanocrystal's electrons in the external environment is an active research area [47, 48]. One example of nanocontact between gold–iron oxides is the red shift in the plasmon peak and other is a slow rise in magnetization of the iron oxide moiety. In this chapter, we will summarize properties of iron oxide, gold, and hybrid structures; effects of different types of coatings; synthesis protocols; and characterization and applications in the medical field.

7.2 Properties of Iron Oxide NPs

Iron is most common element by mass on earth. It has strong tendency to oxidize in different oxidation states, but more common oxidation states are 2+ and 3+, representing ferrous and ferric ions, respectively [25]. There are 16 types of iron oxides and oxyhydroxides in nature. Iron oxides are being used extensively in the environment and many other applications in society [49]. Iron is also present in the body to perform different biological functions and extra iron stored in ferritin proteins in the form of ferrihydrite (Fh) [50]. IONPs emerged as a major type of NPs in the past decade, which showed applications in almost every walk of life, especially in the biomedical field. IONPs have shown very promising applications as a contrast agent in MRI, nanoheaters in magnetic hyperthermia, drug delivery, gene delivery, etc. [51]. Being part of the body constituents, IONPs show very low toxicity [29, 52]. All these applications emerged thanks to excellent magnetic and catalytic characteristics of IONPs [53]. In the following section, we will discuss different IONPs from different aspects such as type, structure, and magnetism.

7.2.1 Crystal Structure

Iron oxide exists in 16 different types which have different orders of crystallinity and atomic arrangement [54]. Magnetite, maghemite, and hematite are more important among these forms. In general, anions usually occupied regular positions in the crystal structures, whereas iron ions are distributed at the tetrahedral and octahedral interstitial sites. Iron oxide has hexagonal closed-packed (hcp) and cubic lattice-type structures [55]. Now we will discuss the crystal structure of these three major iron oxide types.

7.2.1.1 Magnetite (Fe_3O_4)

Magnetite has a face-centered cubic (fcc) structure with 32 O^{2-} ions arranged periodically along [111]. It has an inverse spinel structure. It is one of those forms of iron oxide that have Fe^{2+} ions along with – FeO and $Fe(OH)_2$. All the divalent iron ions are located at tetrahedral sites, whereas half of the trivalent ions are situated at tetrahedral sites and the remaining half ions are positioned at octahedral sites [20]. The number of trivalent cations at the tetrahedral sites provides information about the disorderliness of the inverse spinel structure. Ferrites are a modified form of magnetite, where Fe^{2+} ions are being replaced by new metal ions (Fig. 7.1a) [56]. These new metal ions are adapted in the structure by contracting or expanding the oxygen framework to make up the size difference produced from divalent iron ions [21].

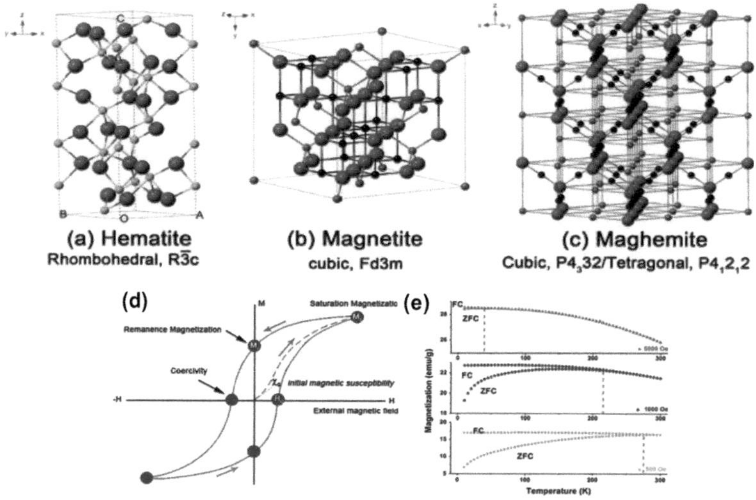

Figure 7.1 Properties of iron oxide NPs. (a–c) Crystal structure of different iron oxide types: hematite, magnetite, and maghemite. (d) Typical magnetization curve of iron oxide NPs. (e) Temperature-dependent magnetic properties: field-cooled and zero field–cooled measurements [64].

7.2.1.2 Maghemite (γ-Fe_2O_3)

Due to same structure as magnetite, it is impossible to identify structure of the NPs between the two iron oxide forms using conventional structural techniques such as by X-ray diffraction (XRD) or electron diffraction. On the other hand, electron energy loss spectroscopy (EELS) is one of the technique that can be employed to identify two oxidation states [57, 58]. Maghemite crystallizes in both cubic and hexagonal systems. In a unit cell, eight Fe^{3+} ions are positioned at the tetrahedral sites, whereas the left over are distributed at octahedral sites at random. The remaining sites are occupied by vacancies to compensate charge. The value of its lattice parameter is 0.834 nm. In its unit cell, there are 32 O^{2-} ions, 21 one-third parts are trivalent ions, and the remaining 2 one-third parts are vacancies (Fig. 7.1b) [54, 59].

7.2.1.3 Hematite (α-Fe_2O_3)

Its structure has similarity with corundum [60]. It has an hcp structure. The lattice parameters of a unit cell are a = 0.5034 nm and

c = 1.375 nm. So their c/a ratio is 2.731. It has six formula units per unit cell. The hcp structure is stacked with O ions in the direction of [001]. The anions are parallel to the (001) plane. The hcp structure is occupied by Fe^{3+} ions in two-third sites. So there is one vacancy and two occupied sites; hence it makes sixfold rings (Fig. 7.1c) [54]. All octahedrons share their edges with adjacent octahedral sites. Similarly one side of the face is shared with an octahedron to the attached plane. The distortions of cations are handled by face sharing [61].

7.2.2 Magnetic Properties

Iron is ferromagnetic in nature and has a strong response to the external applied field and nonlinear magnetic behavior. This strong magnetic behavior is due to electrostatic exchange interaction in the iron oxide material [62]. Both magnetite and maghemite show ferromagnetic behavior at room temperature. The Curie temperature of magnetite is 850 K. It is hard to determine the Curie temperature of maghemite because it becomes hematite at a temperature above 800 K. When the temperature is below room temperature, spins at the tetrahedral sites filled by Fe^{3+} and Fe^{2+} become antiparallel [25]. This makes them a ferromagnetic material. In maghemite, the magnetic structure has two sublattices. These sublattices become antiparallel and result in ferromagnetic behavior. Hematite is also weakly ferromagnetic at room temperature but changes into antiferromagnetic at 260 K (Fig. 7.1d,e). Its Curie temperature is above 956 K [63]. Another interesting phenomenon that emerged in the nanosized iron oxide is called superparamagnetic behavior. When the dimensions of IONPs cuts down to 20 nm or less, ferromagnetic material contains a single domain and its magnetic anisotropic energy reaches close to its thermal energy. This causes the magnetization to flip along easy axes, as observed in original paramagnetic materials [52].

7.2.3 Properties of Gold NPs

Physical properties such as color reflection, chemical stability, and higher redox potentials and nanoscale features such as electromagnetic confinement with optical waves and quantum effects change gold conduction behavior from metallic to semiconductor and make this material a strong candidate for biomedical applications in diagnosis and treatment [65]. In the periodic table, gold is present at the second position in the eleventh group among copper and silver but its behavior is totally unpredictable that substantiates its uniqueness. Gold chemistry shows that it's an inert or nonreactive metal. Its inertness is another important property that makes it suitable for in vivo applications. Moreover, its mechanical softness allows it to be alloyed with silver, copper, and some other metals. In alloy form, the color of gold can be modified by addition of aluminum, indium, and cobalt, where it shows purple, blue, and black coloration, respectively [34].

With respect to its structural properties, gold has a usual fcc lattice arrangement and space group Fm-3mc. It has a lattice parameter a = 0.407 nm and four gold atoms per unit cell [66, 67]. Figure 7.2 shows the crystal structure, diffraction pattern, and powder electron diffraction of gold. When its density is compared to other elements of the same group, it is lower in interatomic distance and consequently higher in density than silver and copper. Due to unique optoelectronics properties, it exhibits greater electrical resistivity than silver [68].

By decreasing the bandgap between the Fermi level and the center of the 5d band, gold shows good optical absorption in the visible spectrum. In the case of the interband threshold for gold, when the electron is moved from th e5d valence band to the 6p conduction band, it requires 1.84 eV energy [69, 70]. On the other hand, the interband transition in silver represents visible light after reflection from the surface and has no effect for the UV range. The electronegativity of gold is 2.4 in Pauling units, which is nearly equal to sulfur, selenium, and iodine (2.5 Pauling units) and mostly their properties match halogen [71].

There are considerable variations in physical and chemical traits when the particle size is reduced to 10 nm or less. These unique and astonishing properties of gold provide it plasmonic behavior due to the combined effect of conduction electrons. This effect is usually used in plasmon thermotherapy. The field of plasmons is now well established and researchers are exploring other applications in the medical field [72, 73]. In addition, these plasmons add different colors to gold NP solutions according to their sizes, from blue to crystal red (Fig. 7.2A) [74].

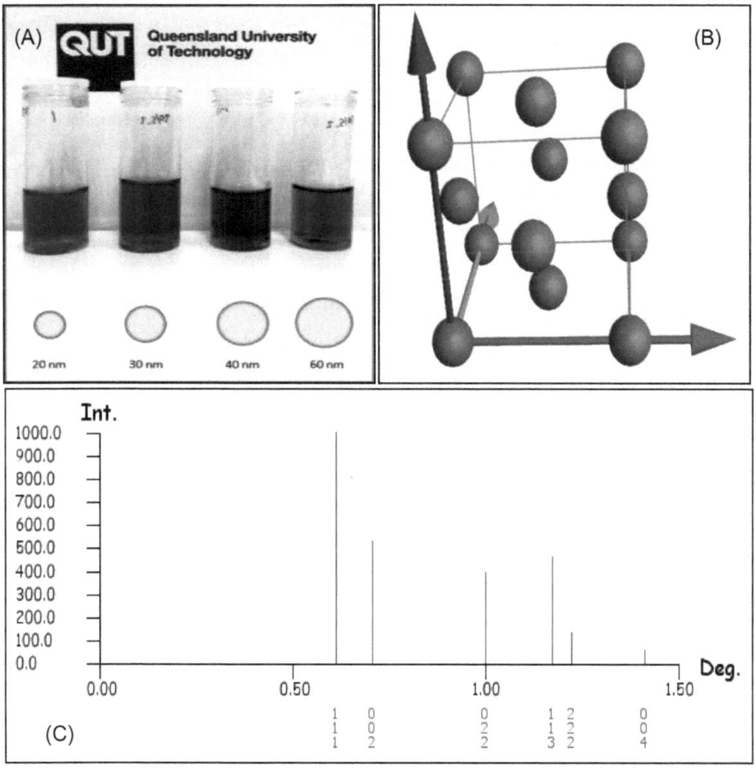

Figure 7.2 Properties of gold NPs. (A) Change in solution color of gold NPs with size. Reprinted from Ref. [74], Copyright (2014), with permission from Elsevier. (B) Typical fcc crystal structure of gold. (C) XRD peaks relevent to gold [75].

7.3 Characteristics of Two Moieties in Hybrid Form

7.3.1 Structural Analysis

The crystal structure of iron oxide–gold heterostructures is very interesting. As discussed in the previous sections, lattice parameters of magnetite (Fe_3O_4; 0.839 nm) or maghemite (Fe_2O_3; 0.835) are approximately double than those of gold (Au; 0.408 nm). This provides a good opportunity for the epitaxial growth of magnetite or maghemite on gold nanoseeds [76]. This is clearer in Table 7.1, which shows different planes of iron oxide (magnetite or maghemite) and gold with roughly identical lattice spacing. These heterostructures are produced by the in situ nucleation of gold seeds on which the iron oxide nanocrystals can grow selectively. High-resolution transmission electron microscopy (HRTEM) is the best way to elaborate the crystal structure [77]. The structural analysis of the images presents a cube-on-cube epitaxial growth of iron oxide on a gold moiety (Fig. 7.2B,C) [8]. This growth mechanism is energetically favorable due to a two times larger lattice parameter of inverse spinel iron oxide than fcc gold. This can be seen in Fig. 7.3a, which exhibits a gold–iron oxide dimer oriented along the [001] direction. The fast Fourier transformations (FFT) are calculated on different parts of nanostructures: 1, iron oxide part; 2, superposed gold/iron oxide. Both FFTs show the same reflections because (004) and (440) planes of magnetite/maghemite overlap with the (002) and (220) planes of gold. These superimposing planes indicate the cube-on-cube epitaxial growth [76].

Table 7.1 Energetically favorable lattice spacings and corresponding plans of iron oxide magnetite or maghemite and gold

Sr. No. \ Species	Fe_3O_4 or γ-Fe_2O_3 Lattice Spacing (Plans)	Au Lattice Spacing (Plans)
1	1.47 nm (044)	1.44 nm (022)
2	2.41 nm (222)	2.35 nm (111)
3	1.26 nm (226)	1.23 nm (113)
4	2.08 nm (004)	2.04 nm (002)

7.3.2 Magnetic Properties

An important factor that needs to be evaluated in the case of nanoscale contact between iron oxide and gold species is magnetic behavior of the iron oxide moiety. The interface link of gold influences the magnetization of the magnetite NPs, particularly when the size is less than 8 nm [78]. These changes have been attributed to thermal disturbance and surface spin canting induced in NPs. Examples of such effect are shown in Fig. 7.3b. There are two types of iron oxide–gold nanohetrostructures: first with 3 nm gold and 14 nm magnetite and second with 3 nm gold and 6 nm magnetite. The hysteresis loop of the first system is similar to 14 nm IONPs, whereas the 3–6 nm dimer system shows a slow loop rise in the moment with the applied magnetic field (Fig. 7.3b) [10].

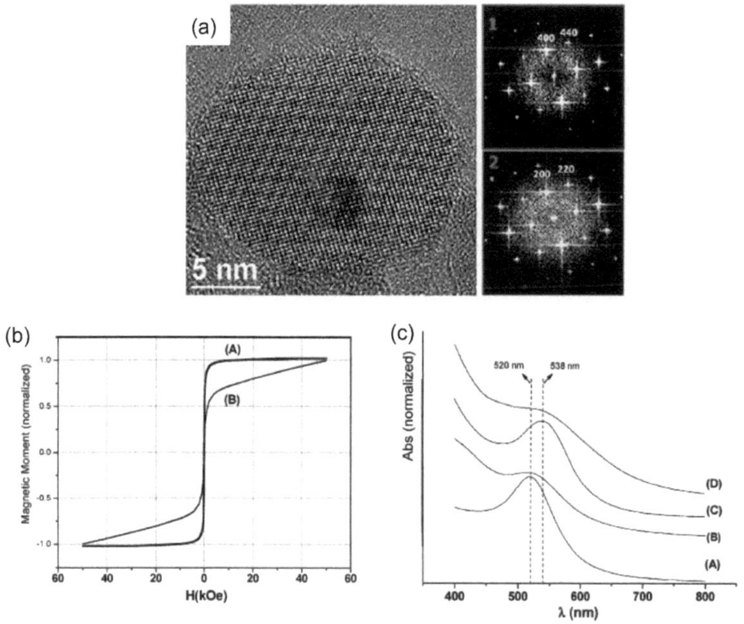

Figure 7.3 Properties of gold–iron oxide hybrids. (a) HRTEM image of nanohybrids. The gold part is dark in the image due to a higher atomic number, whereas iron oxide has a light contrast. (1 and 2) Fast Fourier transformations calculated at two different portions of the transmission electron microscopy (TEM) image. Reprinted with permission from Ref. [76]. Copyright (2015) American Chemical Society. (b) Magnetization curves of two different types of gold–iron oxide: (A) 3–14 nm gold–iron oxide moiety and (B) 3–6 nm gold–iron oxide moiety. (c) Optical spectra of (A) 8 nm gold, (B) 4 nm gold, (C) 7–14 nm gold–iron oxide, and (D) 3–14 nm gold–iron oxide. (b, c) Reprinted with permission from Ref. [10]. Copyright (2005) American Chemical Society.

7.3.3 Optical Properties

It is well known that gold NPs with a size range of 5–20 nm show plasmon resonances due to combined vibration frequency of the trapped electrons. This results in the form of an absorption peak in the optical spectrum. The actual absorption peak can vary with the particle's shape and surface functionalization [79]. Figure 7.3c shows the spectra of gold NPs of different sizes and gold–iron oxide NPs with different aspect ratios. Spectra A and B belong to gold NPs of size 8 nm and 4 nm, respectively. The peak position at 520 nm is invariable for the two sizes, but the peak width increases with size reduction and a similar peak is observed with gold NPs prepared by various protocols. On the other hand, when gold NPs are attached with iron oxide, the plasmon absorption is observed at 538 nm (spectra C and D). A red shift of approximately 18 nm is observed from pure gold NPs. Charge variation of gold NPs surrounded by iron oxide results in this peak shifting in the absorption spectrum [10]. There are studies that showed that excess electrons on gold NPs cause a blue shift of the absorption spectrum. Therefore, a red shift is due to deficiency of electron population on gold, which is caused by the interface link between gold and the iron oxide moiety [9].

7.4 Synthesis Protocol for Nanohybrids

The main focus of today's research is to generate multifunctional nanomaterials that exhibit novel nanomagnetism, plasmonics, and thiol-based conjugation chemistry [80]. For biomedical applications, gold–iron-based hybrid NPs are gained special attention. The basic structures of gold–iron-based nanomaterials are classified into two categories, monodispersed NPs and aggregate NPs [81]. By using different synthesis methods, size-controlled gold–iron-based hybrid NPs can be synthesized [82]. The gold moiety in the dimers can be utilized for two-photon imaging and plasmon thermotherapy; on the other hand, iron oxide can be utilized as a contrast agent in MRI or as a heat mediator in hyperthermia. At the nanoscale, the size of the nanostructure is very important, considering technologically relevant properties [83] and the same can be observed in the case of gold–iron oxide nanohybrids, as discussed in Section 7.3.3. The basic structure of hybrid NPs is in the form of a core–shell or binary nanostructure [84]. The stability and compatibility of hybrid NPs are increased with surface modification via charges, reactive groups, or functional moieties [85]. The diversity of their properties highly depends upon their nanostructure, composition, stability, and disparity of particles under different conditions [86]. In this section, we will discuss different synthesis protocols that are used to synthesize gold–iron oxide heterostructures.

7.4.1 Chemical Synthesis

Multifunctional nanostructures based on gold–iron oxide have been first reported by Yu et al. in 2005 [10]. They prepared these nanostructures by mixing gold NPs with $Fe(CO)_5$ in 1-octadecane as a solvent with oleic acid and oleylamine. This mixture was heated at 300°C and then oxidized in air. This led to the formation of gold–iron oxide heterostructures in nanocontact with each other. Gold NPs were prepared separately and injected in the solution. They prepared two types of structures, one with 3–14 nm gold–iron oxide and other 8–14 nm gold–iron oxide. They also observed multi-iron oxide moiety attachment with single gold NPs, indicating multinucleation on the different faces of gold seeds. This is one of the most commonly used methods to synthesize gold–iron oxide nanohybrids till date. Yin et al. [8] also used the same method and by adjusting the molar ration of Au/Fe obtained nanohybrids with size 2.5–3.5 nm and 15–16 nm of gold and iron oxide, respectively. These nanostructures were used for CO oxidation. Wang et al. [87] proposed a similar method with slight modifications for the synthesis of noble metal–metal oxide NPs. They introduced absolute ethanol after heating the mixture at 300°C for 20 min.

Recently, Guardia et al. [88] presented two approaches for large-scale production of nanodimers named as one-pot and two-pot synthesis (Fig. 7.4A–E). In two-pot synthesis, gold NPs were prepared separately and injected in the mixture of 1-octadecane, oleic acid, and oleylamine being used as a solvent. Iron salt was introduced at 150°C and heated till 300°C. They proposed that a chlorine-bearing compound could be used to adjust the dimension of th eiron oxide moiety in the dimers. In one-pot synthesis, they further simplified the reaction by using a gold precursor solution in the solvent mixture. Gold salt, $HAuCl_4$, served the purpose of the gold precursor and also provided Cl^- ions. Gold salt can be added till the temperature reaches 120°C. By adding iron salt at 150°C, the reaction can be stopped at 200°C to 300°C to obtain narrow size distribution dimers.

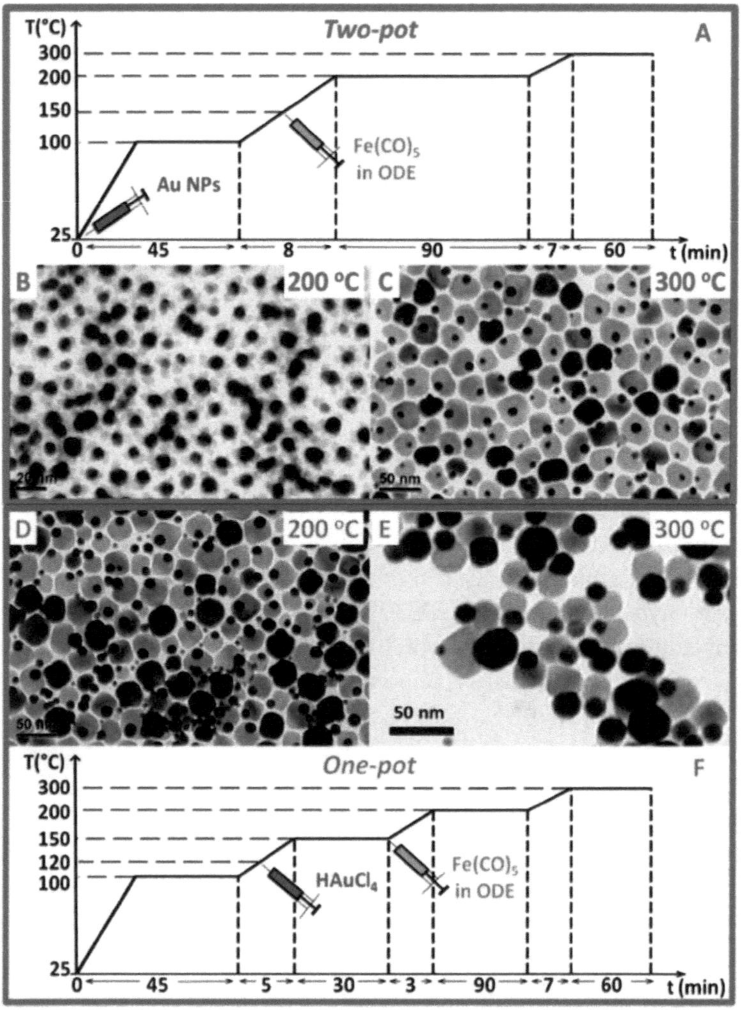

Figure 7.4 Temperature profiles of one-pot (A) and two-pot (F) synthesis of gold–iron oxide heterostructures. In the second protocol, gold NPs were added in the beginning of the reaction and iron salt was added after 150°C. (B, C) TEM images of dimers obtained after 200°C and 300°C. In one-pot synthesis, gold salt is introduced at 120°C instead of gold nanoparticles. (D, E) TEM images of gold–iron oxide dimers obtained at 200°C and 300°C [88].

7.4.2 Physical Method

Combined chemical and physical nanofabrication methods have emerged as another potential way to synthesize NPs [89–91]. Material ablation by lasers in a liquid environment provides certain benefits such as confinement of vapors and plasma, production of debris-free surfaces, and lowering of heat load [92, 93]. In a typical experiment, metal precursors are mixed in aqueous solution and then irradiated with laser. Formation of NPs depends on different laser parameters, for example, laser fluence, pulsed duration to continuous-wave laser, and solution concentration. For focusing the incident laser beam, a convex lens can be used between the sample quartz tube and the laser beam. The sample is usually positioned following the back focal plane of a convex lens. To control the fluence on the sample, the distance *D* between the focusing lens and the sample can be adjusted. If *D* is large, fluence will be lower on the quartz cell and vice versa. The number of shots can be increased periodically [94]. The schematic of flash laser annealing is shown in Fig. 7.5.

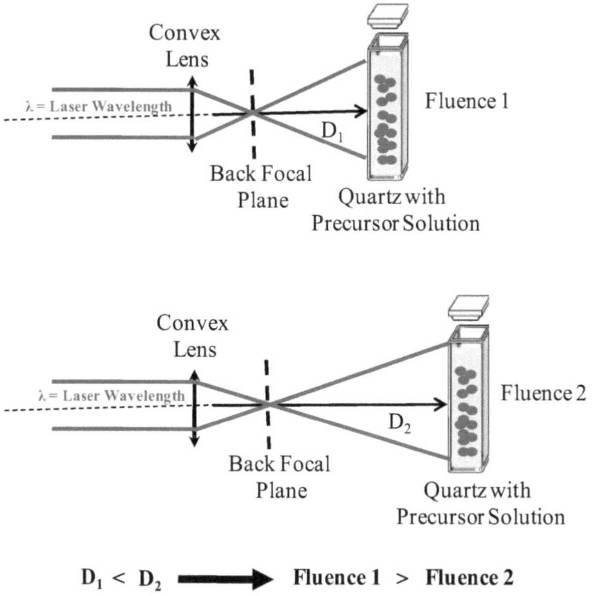

Figure 7.5 Schematic representation of laser-assisted synthesis. Laser energy or fluence is affected by the distance between the quartz cell and the convex lens [94].

7.5 Surface Modification by Functionalization

Surface functionalization is an important precondition for every potential application of inorganic NPs. This surface modification governs the interaction between NPs and the exposed environment and as a result influences the colloidal stability [95]. This is also important in certain cases to control the morphology of NPs or targeted delivery of NPs by attaching appropriate biomolecules. Due to surface charge, a repulsive force can be induced between the NPs, which stabilizes them from aggregation [96]. In this case, ligand molecules are attached first by some attractive force such as chemisorption, electrostaticity, or hydrophobicity, and then an opposite charge of the molecule, on the other hand, creates repulsion between NPs due to the same charge [97]. There are different types of polymers available for functionalization [98, 99]. In this section, we will discuss different methods of surface functionalization of gold and iron oxide NPs.

7.5.1 Poly(Ethylene Glycol)

Poly(ethylene glycol) (PEG) consists of replicated units of $-CH_2-CH_2-O-$, with different molecular weights. It is also named poly(ethylene oxide) or polyoxyethylene [100]. It is well soluble in many organic solvents and also in water, where it forms random spirals of diameter relatively larger than proteins of the same molecular weight [101]. It is one of the basic biocompatible polymer due to chemical stability and a simple structure [102]. The inertness and nontoxicity of PEG make it applicable in medicines. PEG is also being used as a presertive in cosmetics, pharmaceuticals, and food [103]. When PEG binds to the surface of NPs, it repels nearby molecules by steric effects, that is, neighboring molecules are not influenced by electrostatic forces and as a result cannot infiltrate the hydrated PEG layer [104]. This effect is also useful for increasing the half-life of the drug or NPs in the bloodstream by reducing the attachment of antibodies from blood plasma [105].

7.5.2 Biomolecules

There also exists a variety of organic molecules with diverse composition, size, and complexity. These molecules offer structures, specify biological processes, and organisms [106, 107]. Examples of such biomolecules are lipids, vitamins, sugars, proteins, enzymes, DNA, and RNA [108, 109]. The functionalization with biomolecules enables the NPs to interact with specific biological systems [110, 111]. Targeted drug delivery is an example of such a system in which specific biomolecules are usually attached to transport the drug at a precise location in the body [112, 113]. The association of biomolecules to NPs can be done by four different ways [114, 115]:

- By chemisorption of a thiol group (R-S-H) on the surface of the NPs using ligand-like binding
- By electrostatic attraction of positively charged biomoleclues on the negatively charged surface on NPs
- By covalent bonding of functional groups present on NPs and biomolecules
- By affinity-based receptor–ligand systems

7.6 Different Types of Coatings Used in Nanohybrids

Now we will discuss different types of functionalization of gold–iron oxide NPs. Chowdhury et al. [116] recently reported separation of specific DNA from a DNA–protein mixture. NPs were functionlized with thiol-linked single-stranded DNA (ssDNA), and then a magnetic field was applied to separate the targeted DNA from the mixture. In another study, Reguera et al. [117] used a thiol-linked PEG coating on gold–iron oxide NPs for studying their potential utilization in multimodel imaging. NPs were also functionlized by mercaptobenzioc acid, which is being used as a tag for surface-enhanced Raman spectroscopy (SERS). The coating of the dimer system with silica also enhances stability in the aqueous phase and makes these NPs potentially viable for biomedical applications [15]. PEG coatings have also been used on gold–iron oxide NPs that were utilized to boost the heating efficiency of magnetic hyperthermia [118]. Heterostructures have also been coated with cetyltrimethylammonium bromide (CTAB) for studying their photocatalytic performance. Table 7.2 detailing different coatings used for functionalization of gold–iron oxide nanoheterostrcutres is provided.

Table 7.2 Coatings used for functionalization of various gold–iron oxide nanostructures and use of nanostructures in different applications

Coating [Reference]	Structure	Size (nm) Gold–iron oxide	Application
PEI [119]	Core–shell	4–108 nm	SERS
PEI [120]	Heterostructures	6–65 nm	–
Citric acid [121]	Heterostructures	5–12 nm	–
Oleylamine [122]	Dumbbell	3.1–18.1 nm 5.4–18.4 nm 10.5–23.6 nm	CO oxidation catalytic activity
PEG [123]	Micelles	1.9–15 nm	MRI
PEG + polystyrene [124]	Janus	6.5–6.0 nm	SERS, MRI
PAA [125]	Flower-like	35 nm	Protein detection
PAA [126]	Janus	127 nm	Dual-model imaging, photothermal therapy
PMAL [127]	Dumbbell	8–25 nm	Targeted cancer therapy
Epidermal growth factor receptor antibody-PEG [128]	Dumbbell	8–20 nm	MRI
Silica [8]	Dumbbell	3.5–16 nm	Nanocatalyst
PEG3000-CONH-herceptin [129]	Dumbbell	8–25 nm	Drug delivery
Poly-L-histidine [130]	Core–shell	3–40 nm	Photothermal therapy

Coating [Reference]	Structure	Size (nm) Gold–iron oxide	Application
Antidigoxin monoclonal antibody [14]	Core–shell	20 nm	Detection of immunological interaction
PEG [131]	Dimers	5–10 nm	MRI
$^{99m}T_c$, radiolabeling [132]	Dumbbell	10–28 nm	Photon emission CT
Myoglobin [133]	Core–shell	85 nm	Immunosensor
Citric acid [108]	Hybrid hollow spheres	100 nm	SERS
CTAB [134]	Hybrid	3–120 nm	SERS
PEI [135]	Composites	3–9.7 nm	MRI, CT
PEG [136]	Dumbbell	10 nm	MRI

PAA, poly(acrylic acid); PEI, polyethylenimine; PMAL, dimethylaminopropylamine ; CT, computed tomography.

7.7 Examples of Different Nanohybrids

In this section, we will present different morphologies of gold nanohybrids. Figure 7.6 shows transmission electron microscopy (TEM) images of gold–iron oxide nanohybrids prepared by two different iron sources, that is, iron pentacarbonyl and iron acetate. Figure 7.7 presents gold–iron oxide core–shell NPs and where different diffraction spots for gold and iron oxide can be identified. Guardia et al. studied the effect of different HCl concentrations on the formation of gold–iron oxide hybrid nanostructures (Fig. 7.8). Figure 7.9 shows another type of gold–iron oxide hybrids, that is, dumbbell-shape NPs. Figure 7.10 depicts strawberry-like nanostructures.

Figure 7.11 presents attachement of small gold NPs on the interface of iron oxide NPs and characterized by different microscopic tools.

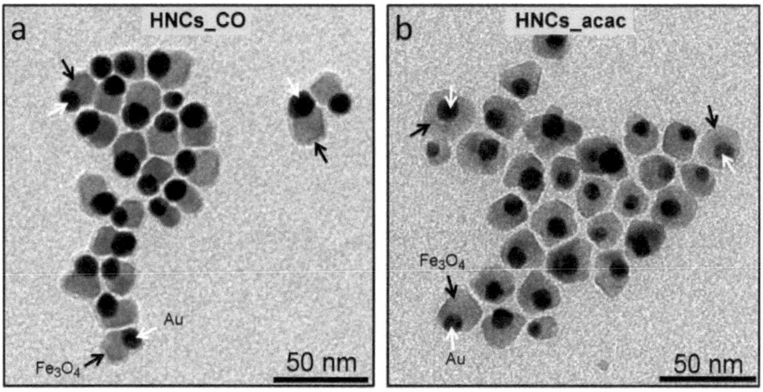

Figure 7.6 TEM images of gold–iron oxide nanohybrids prepared by iron pentacarbonyle (a) and iron acetate (b). Reprinted with permission from Ref. [15]. Copyright (2017) American Chemical Society.

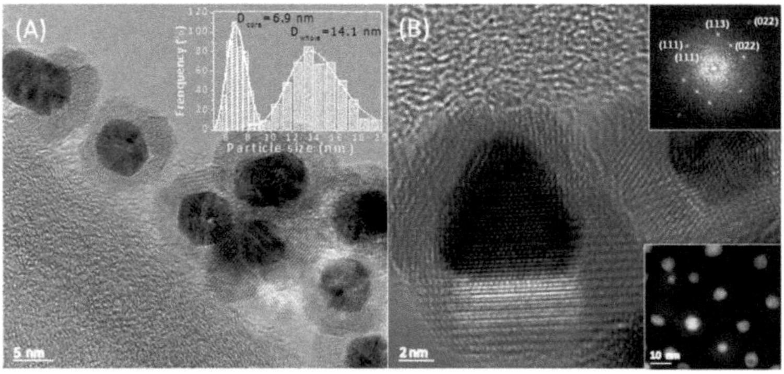

Figure 7.7 (A) TEM iamges of gold–iron oxide core–shell NPs. Inset: Histogram of size distribution of core and whole particles. (B) HRTEM image of core–shell nanoparticles. Inset (upper): Calculated FFT of the NP; (lower): scanning transmission electron microscopy (STEM) dark-field image of the nanostructures [137].

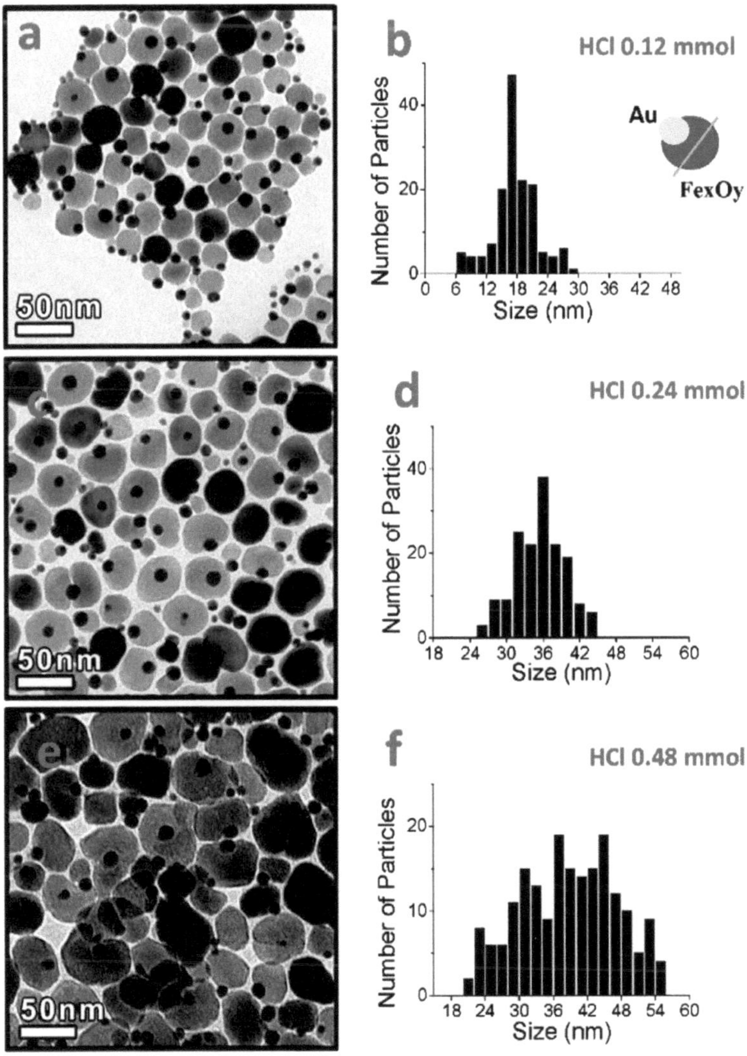

Figure 7.8 Gold–iron oxide dimers produced by different concentrations of HCl. (a, b) 0.12 mmol, (c, d) 0.24 mmol, and (e. f) 0.48 mmol. From Ref. [118]. Published by The Royal Society of Chemistry.

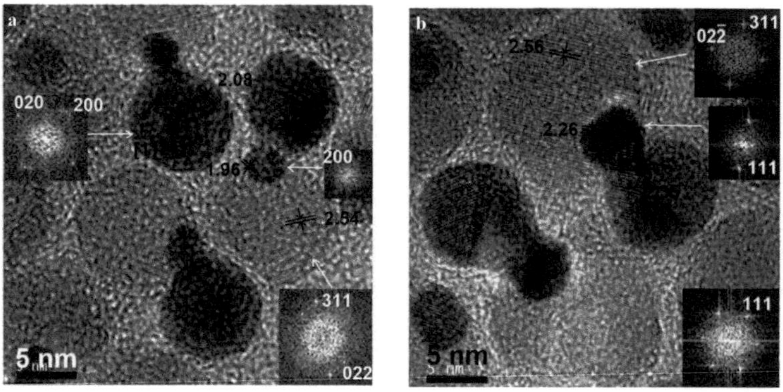

Figure 7.9 HRTEM image of dumbbell hybrid nanostructures. (Insets) FFTs of different moieties of the structures. Repinted with permission from Ref. [131]. Copyright (2014) American Chemical Society <https://pubs.acs.org/doi/10.1021/nn500188y>. Further permissions related to the material excerpted should be directed to the American Chemical Society.

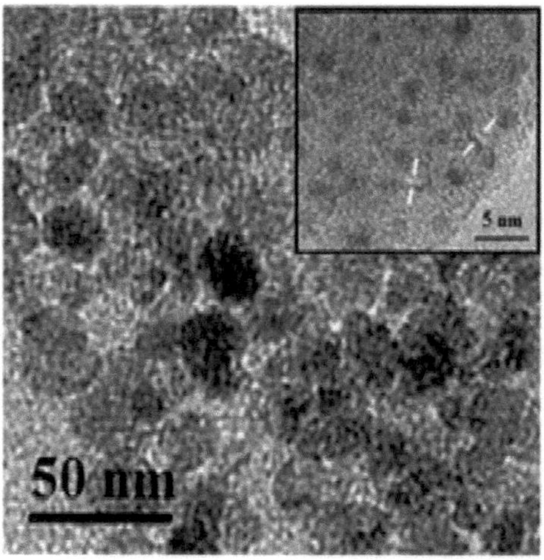

Figure 7.10 Strawberry-like iron oxide–gold nanoparticles. Inset: High-resolution image with d spacing. Reprinted from Ref. [138], Copyright (2015), with permission from Elsevier.

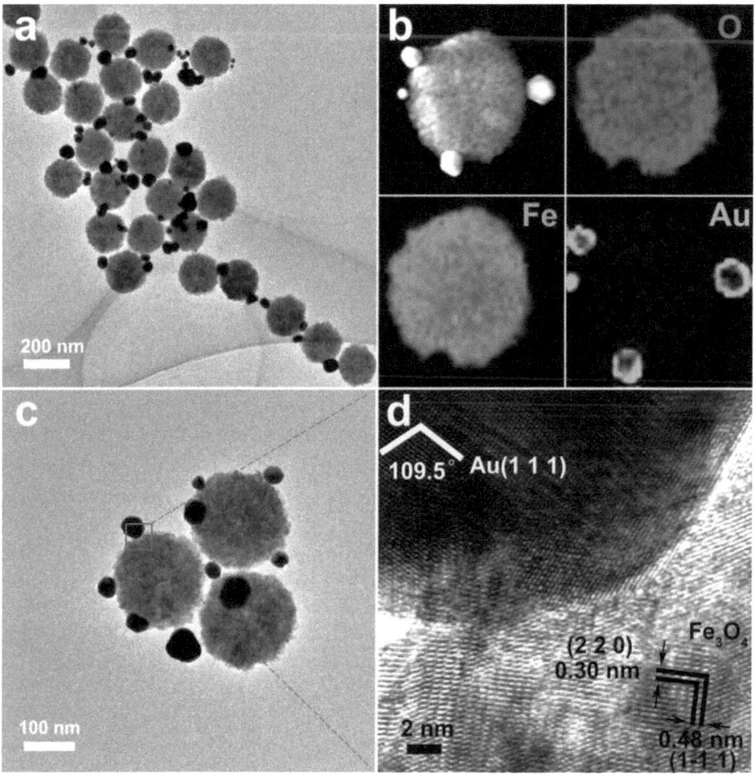

Figure 7.11 (a) TEM image of iron oxide–gold nanoparticles; (b) STEM dark-field images and chemical mapping of oxygen, iron, and gold; and (c-d) TEM and HRTEM images, respectively. Reprinted by permission from Springer Nature Customer Service Centre GmbH: Springer Nature, *Scientific Reports*, Ref. [134], Copyright (2014).

7.8 Applications of Nanohybrids in the Medical Field

Gold NPs show efficient optical properties, while the magnetic behavior of magnetite or maghemite is well known; therefore, gold–iron oxide hybrid NPs behave as a multifunctional material ensemble [139]. The combined vibrations of free electrons in gold NPs produce an electron resonance that is known as plasmons. The plasmons get excited when light of a suitable wavelength falls on them and cause absorption on that wavelength [140, 141]. That's why gold–iron oxide NPs are being used for optical detection to visualize specific regions and act as a novel contrast agent in MRI [10]. These NPs are also used in photothermal therapy and DNA/protein sensing [133]. The magnetic properties of gold–iron oxide NPs are structure dependent, and the nanointerface of gold–iron oxide hybrid NPs may alter their magnetic properties (as discussed in the previous section). Kim et al. [142] worked on thermosensitive gold–iron oxide NP–loaded micelles to attain a cumulative chemotherapy and magnetic hyperthermia synergic anticancer effect. The core micelles provide heat for magnetic hyperthermia and an optical imaging source for diagnosis. The micelle's structure of gold–iron oxide NPs enhances the bioavailability of hydrophobic drugs and also improves their stability in an aqueous environment because of their hydrophilic shells. The gold-coated IONPs intercept the core oxidation and enzymatic degradation. These hybrid NPs provide real-time imaging from surgery to localized tissue heating, which opens up a wide range of biomedical applications such as tumor ablation [143], drug delivery [144], multimodel imaging, photothermal therapy, and SERS. In this section, we will discuss applications of these nanostructures in magnetic hyperthermia, multimodel imaging, and SERS.

7.8.1 Magnetic Hyperthermia

Hyperthermia has undergone four decades of experimental and theoretical research for application in cancer treatment [145]. Usually, hyperthermia means having a higher temperature rather than the normal body temperature. The procedure followed in magnetic hyperthermia is associated with energy loss when a ferromagnetic material is exposed to an alternating current (AC) magnetic field because of the hypersensibility of tumor cells to heating [146]. In this process, body tissues are exposed to high temperature to kill the malignant cells, leaving the healthy cells intact. The range of temperature is fixed between 42°C and 48°C, which is enough for cancer cells, whereas healthy cells can bear temperatures up to 50°C. Electromagnetic radiation exhibits strong interactions with tissues; owing to this, it permits potential applications in thermal therapies. This interaction has some limitations in deep-seated tissues and can be improved by a frequency range up to 10 MHz. Human tissues have diamagnetic behavior, so they has negligible magnetic influence. For cancer treatment by temperature, there are two broadly followed ways: one is hyperthermia for which the temperature range is 41°C–47°C, while the second is thermoablation with a range of 47°C–56°C [147].

There are three categories of hyperthermia treatment: whole-body, regional, and local hyperthermia. In the whole-body treatment, the complete body is subjected to an AC magnetic field, while the other two treatments are localized. Localized hyperthermia can be further classified into external, interstitial, and endocavity hyperthermia; additionally, heat generation sequences can also be adjusted according to the target region [148].

Magnetic fluids can be supplied efficiently to specific regions inside an organism. This delivery can be carried out with the help of different routes [149]. The noninvasive mechanisms for delivery are most efficient in hyperthermia. These delivery mechanisms are highly tissue specific and, as a result, have the capability to generate high-intensity heat to localized regions in deep-seated tissues [150]. Magnetic NPs have also an aptness to be used inside certain types of cancer cells [151]. Superparamagnetic NPs are used in hyperthermia due to their high absorbance power, acceptable frequency, and magnetic field [152].

Magnetic NPs exhibit heating effects owing to heating losses when subjected to an AC magnetic field. There are three mechanisms involved in these losses: eddy current due to frictional heating [153], magnetic heating owing to hysteresis losses [154], and Néel and Brownian relaxation [155]. The last mechanism is relatively more important for superparamagnetic NPs. The heat produced due to these mechanisms can be calculated by the product of the specific absorption rate (SAR) and concentration of NPs. The heat generation and relaxation mechanism are size dependent [154], and the peak site of the SAR based on the anisotropy of NPs; nevertheless other minor factors also have an influence, which include frequency, viscosity, and temperature [147].

To generate high heating power for each unit mass of particle, mediators are developed for remarkable heat loss under an AC magnetic field. Therefore, several types and shapes of nanomaterials [156, 157] are proposed for heat production, in which mostly are iron, manganese, cobalt, nickel, zinc, gadolinium, magnesium, and their oxides [158]. Especially, iron oxide nanomaterials have been evaluated widely for their promising use in hyperthermia [159]. Different ferrites are also involved in hyperthermia [160]. In addition, iron and gold nanohybrids have also been studied [161]. However, IONPs have much attention because of their low toxicity and biocompatibility. There are many factors responsible for heating efficiency, such as anisotropy of the NPs, dispersion media, biological environment, synthesis protocols, multifunctionality, biocompatibility, opsonization process, surface charge, protein corona, and toxicity [147].

Here we review some results of researchers, where they characterized the gold–iron oxide heterostructures for magnetic hyperthermia. Kim et al. [142] probed the magnetic, optical, and thermal sensitivity of gold–iron oxide micelles. The synthesized micelles were helpful for combined hyperthermia therapy and chemotherapy. The hyperthermia temperature was observed in the range from 42°C to 45°C by optimizing the ratio of different polymers. In another study, Chung and coworkers [162] reported that denser concentrations presented better heating efficacy. They calculated that each coated particle produces 1.33×10^{-16} J by 20-min exposure of a high-frequency induction wave. Guardia et al. [118] synthesized gold–iron oxide dimers by two different protocols and determined the SAR for various iron oxide–gold domains with a size range of 17–26 nm and 11–15 nm, respectively. The highest SAR value, 1330 ± 20 w/g, was measured for gold–iron oxide nanostructures with an iron moiety of 23 nm at 300 kHz. This SAR value was higher than that observed in the case of iron oxide nanocubes with the same frequency conditions [88]. Mohammad et al. [161] revealed that the amount of heat increases four- to fivefold when a gold nanoshell around superparamagnetic IONPs was used with low-frequency ranges. It was also observed that water was a more efficient solvent than ethanol and toluene (Fig. 7.12C). The SAR value of 697.5 w/g was observed at 44 Hz with water as a solvent in comparison to 67.8 w/g with toluene at the same frequency. The highest SAR value, 1199.7 w/g, was observed in water at 430 Hz.

Hoskins et al. [163], conversely, probed plasmonic hyperthermia by irradiating gold–iron oxide nanohybrids by continuous laser light of 532 nm wavelength. A temperature rise up to 65 °C was observed after 90 s irradiation for gold–iron oxide NPs concentration of 50 μgmL^{-1} (Figure 12A-B). In another study, Chung and Shih [164] evaluated the temperature elevation of gold–iron core-shell NPs under high frequency induction wave for different concentrations. The temperature was raised up to 53°C in 10 min for 30 mg/mL concentration and presented good application in magnetic hyperthermia. Khosroshahi and Tajabadi [165] investigated a plasmonic–magneto dendrimer (superparamagnetic IONPs functionalized with polyamide-o-amine dendrimer) conjugated with gold NPs and used it in laser-induced hyperthermia. Bell et al. [166] exhibited substantial influence on the magnetoheating properties of the IONPs by a gold moiety. Iron oxide–gold nanocomposites showed threefold enhanced heating efficiency (SAR = 88.3 w/g).

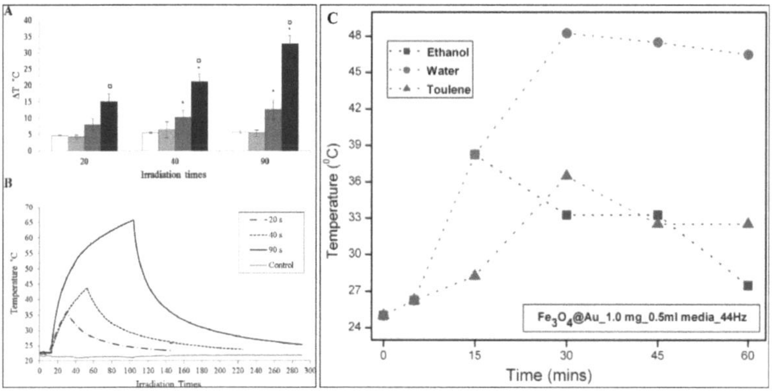

Figure 7.12 Hyperthermia results presented by two groups. (A) Temperature variation of gold–iron oxide nanoparticles for different concentrations under the effect of continuous laser light of 532 nm [163]. (B) Characteristic real-time 50 μg/mL concentration temperature curves for different times: 20, 40, and 90 s [163]. (C) Comparison of heat release due to different solvent conditions of gold–iron oxide nanoparticles at magnetic field frequency 44 Hz. (A, B) http://creativecommons.org/licenses/by/2.0. (C) Reprinted with permission from Ref. [161]. Copyright (2010) American Chemical Society.

7.8.2 Multimodel Imaging

Owing to the early detection and screening of diverse pathologies and therapeutic treatments, the noninvasive imaging methodology is the current big challenge for biomedicine [167, 168]. Several types of imaging modalities have been employed for cancer diagnosis in vivo, for instance, optical imaging, MRI, computed tomography (CT), and positron emission tomography (PET) are imaging techniques that have certain advantages such as spatial resolution, imaging time, sensitivity, ease of utilization, and penetration [169]. A major limitation with these techniques is selective imaging; therefore, a single technique cannot provide sufficient information required for diagnosis [170]. Consequently, a combinational technique that is a multimodal imaging technique was devised based on hybrid nanostructures. Gold–iron oxide hybrid nanostructures provide this multifunctional property in a single entity. This multifunctionality is successfully employed in hyperthermia (magnetic and plasmonic hyperthermia), as discussed in the previous section.

The basic principle in MRI is the alignment of the water proton and precession by the application of a magnetic field. Two types of magnetization, longitudinal and transverse, are generated. These magnetizations are responsible for contrast enhancement as T1 (positive contrast) or T2 (negative contrast), depending upon the material being used as a contrast agent. This categorization is based on the relaxation, that is, longitudinal relaxation and transverse relaxation, respectively. Factors that can control the contrast properties of NPs are magnetic properties, compositional properties, surface properties, size, and architecture [147].

Superparamagnetic gadolinium is used usually as a positive contrast agent, but a major limitation is low retention time, as it can be excreted very rapidly from the body by the renal system. IONPs are useful in MRI as contrast agents due to their biochemical behavior, low toxicity, and biodegradation [171]. Longer times in the body can be maintained by using certain organic coatings [172]. IONPs exhibited both type of relaxation mechanisms, that is, dual contrast, but their negative contrast has been studied extensively [173]. Mostly nanostructure-based multimodal imaging analysis is a combination of MRI with other optical imaging techniques. In this way, MRI provides high spatial resolution, and optical imaging allows rapid visualization. Similar to dual-imaging modalities by mixing magnetic resonance and optical imaging probes, triple-model imaging is also used by introducing an additional imaging modality. PET is commonly used as the third imaging modality [174].

Colloidal gold NPs revealed distinctive surface plasmon resonance properties by the correlation of electromagnetic waves and electrons in the conduction band. These properties can be used for noninvasive photothermal therapy to treat localized tumors [175]. This kind of therapy takes an enormous absorption cross section of the nanomaterial in the near-infrared (NIR) region; the region has enough ability to penetrate the skin without disturbing normal tissues and treat localized cells [176]. Gold nanostructures with shells [177] and rods [178] are more promising for ignition with NIR light and can convert light to heat to destroy malignant cells. West et al. [179] used gold nanoshells to enhance the ability of optical coherence tomography in vivo and utilized nanoshells to absorb NIR light for photothermal ablation. These nanoshells are used as a contrast agent to improve optical CT. Here we reviewed the use of gold–iron oxide NPs for multimodel imaging. Reguera et al. [180] investigated Janus magnetoplasmonic nanostructures for multimodel imaging. Higher relaxivity values between 180 and 300 m M^{-1} s^{-1} were observed for a smaller iron oxide moiety. In actual, relaxivity was varied more with concentration for smaller size and remained constant for larger NPs. Nanostructures produced higher attenuation in CT imaging than currently used standards. They also increased the measurement time window because of slow diffusion. Ju et al. [181] studied Au–Fe_2C nanostructures in triple-model imaging, that is, MRI/multispectral optical tomography (MSOT)/CT. For photothermal therapy, the temperature of the solution was elevated from 4.99°C to 49.95°C after 5 min of 808 nm laser irradiation. The relaxivity value measured ba y 3T clinical MRI scanner was 210.6 m M^{-1} s^{-1}. Nanostructures have shown promising application for photoacoustic imaging, which was evaluated by an MSOT imaging system. The accumulation and penetration of these nanostructures was then probed in the tumor site (Fig. 7.13).

Song et al. [124] investigated vesicles made up of Janus plasmonic-magnetic NPs for photoacoustic imaging and MRI. These vesicles presented higher r_2 values (405±12.4 m M^{-1} s^{-1}), and this value was about seven times more than single Janus NPs. The photoacoustic signal was five times enhanced by vesicles when irradiated with 750 nm laser. Strawberry-like Fe_3O_4-Au hybrid nanostructures were tested for MRI and X-ray attenuation property in a fatty liver animal model. The in vivo imaging showed almost 35 times contrast enhancement in T2-weighted MRI images, whereas 174 Hounsfield units (HU) were observed for CT at 30 min postinjection [182]. In another study, Au/Fe_3O_4@C with Janus morphology was used for MRI and photothermal therapy. The negative contrast value was 100.71 m M^{-1} s^{-1}, whereas a temperature rise from 23°C to 63°C was observed for 0.8 mg/mL concentration by NIR laser radiation. The change in color of the photothermal images was also noticed by an IR thermal camera [126].

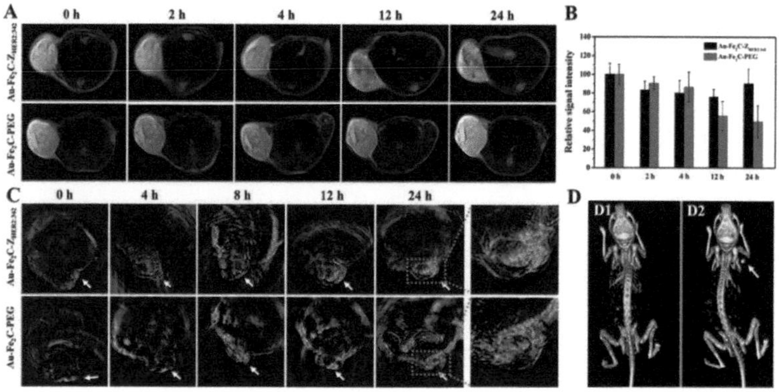

Figure 7.13 In vivo results of MRI/MSOT/CT. (A) T2-weighted images of antibody-targeted and nontargeted gold–iron oxide nanostructures at different time intervals. (B) Relative MRI signal intensity at specific time points in the tumor. (C) MSOT images of targeted and nontargeted gold–iron oxide nanostructures. (D) 3D construction of CT images (D1) pre- and (D2) postinjection. Reprinted with permission from Ref. [181]. Copyright (2017) American Chemical Society.

7.8.3 Surface-Enhanced Raman Spectroscopy

SERS has been developed over the past few decades since its discovery, because of single-molecule sensitivity, enhanced mechanism, and improvement in substrates for its applications. It is a highly surface-sensitive technique, which allows detection of molecules in low concentrations and delivers rich structural data via intensification of electromagnetic fields produced by excitation of localized surface plasmons. This excitation enhances Raman scattering due to adsorption of molecules at rough surfaces. It happens when directed molecules are brought a few nanometers close to an active substrate and have high spatial resolution. It is influenced by the nature of the metal and surface roughness. The SERS mechanism depends upon two factors, electromagnetic theory (i.e., excitation of localized surface plasmons) and chemical theory (i.e., effect of charge transfer). The choice of surface metal depends on the plasmon resonance frequency, which is why silver, gold, and copper are selected for SERS experiments due to their enhanced plasmon resonance frequencies and their wavelengths in the visible-to-IR range. SERS-active substrate fabrication has a very significant role in the amplification of the Raman effect and is highly dependent on the substrate structure, especially using a gold dimer. Gao et al. [108] reported Au–Fe_3O_4 hybrid hollow spheres have higher SERS sensitivity for rhodamine 6G exposure. The nanostructures with molar ratio (Au/Fe) 0.2 amplify the detection limit of the Au–Fe_3O_4 hybrid hollow spheres to R6G molecules up to 10^{-10} M. Hu et al. [134] studied liquid SERS substrates containing suspensions of Fe_3O_4/Au NPs, which offer high spot-to-spot uniformity, reproducibility, and reversibility. The enhancement factor of these substrates can be tuned by an external magnetic field. Reguera et al. [180] detected a crystal violet analyte concentration down to 15 nM (Fig. 7.14). In this way, these nanohybrids are emerging as new analytes for the detection of low concentration of molecules.

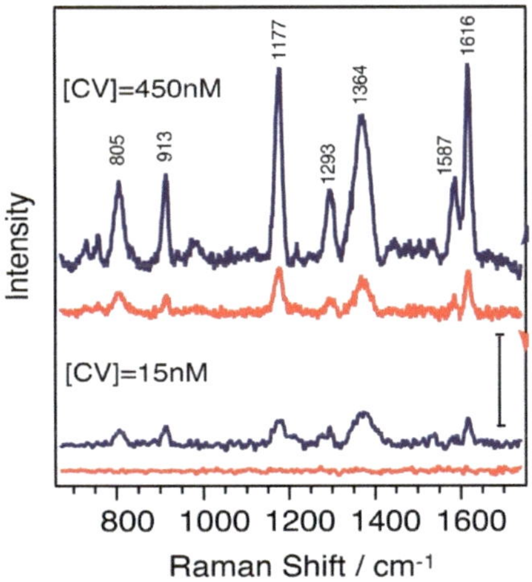

Figure 7.14 SERS spectra of crystal violet analyte present in a Janus gold–iron oxide nanoparticle solution (red) and after magnetic concentration (blue). Graphs are with two different dye concentrations: upper, 450 nm; lower, 15 nm. From Ref. [180]. Published by The Royal Society of Chemistry.

Conclusion

Multifunctional NPs are an emerging field due to their applications in different fields of life and more specifically in the medical field, where one-time injection of NPs in the biological environment can be utilized for different types of therapies. The properties of these gold and iron oxide hybrid nanostructures provide a new class of nanomaterials that can be used for different biomedical applications. There are different preparation methods, both chemical and physical, involved in the formation of hybrid structures. Although hybrid nanostructures with nanocontact are easy to produce, a core–shell nanostructure based on an iron oxide core and a gold shell is still challenging. More controlled synthesis protocols are required to make a continuous shell of gold on an iron oxide core. These hybrid nanostructures have been employed for magnetic hyperthermia and photothermal therapy applications for cancer treatment. These nanostructures have shown promising results for multimodel imaging, that is, in MRI and optical imaging. On a final note, the field of hybrid nanostructures is progressing rapidly, but certain points are still needed to be addressed, such as relevant advances in synthesis protocols, self-assembling mechanisms, and physical properties.

References

1. Landsiedel, R., Ma-Hock, L., Kroll, A., Hahn, D., Schnekenburger, J., Wiench, K. and Wohlleben, W. (2010). Testing metal-oxide nanomaterials for human safety, *Adv. Mater.*, **22**(24), pp. 2601–2627.
2. Nalwa, H. S. (2014). A special issue on reviews in biomedical applications of nanomaterials, tissue engineering, stem cells, bioimaging, and toxicity, *J. Biomed. Nanotechnol.*, **10**(10), pp. 2421–2423.
3. Bhattacharyya, D., Singh, S., Satnalika, N., Khandelwal, A. and Jeon, S.-H. (2009). Nanotechnology, big things from a tiny world: a review, *Nanotechnology*, **2**(3), pp. 29–38.
4. Byrappa, K., Ohara, S. and Adschiri, T. (2008). Nanoparticles synthesis using supercritical fluid technology - towards biomedical applications, *Adv. Drug Delivery Rev.*, **60**(3), pp. 299–327.

5. Vaddiraju, S., Tomazos, I., Burgess, D. J., Jain, F. C. and Papadimitrakopoulos, F. (2010). Emerging synergy between nanotechnology and implantable biosensors: a review, *Biosens. Bioelectron.*, **25**(7), pp. 1553–1565.
6. Xu, P., Zeng, G. M., Huang, D. L., Feng, C. L., Hu, S., Zhao, M. H., Lai, C., Wei, Z., Huang, C. and Xie, G. X. (2012). Use of iron oxide nanomaterials in wastewater treatment: a review, *Sci. Total Environ.*, **424**, pp. 1–10.
7. Colfen, H. and Mann, S. (2003). Higher-order organization by mesoscale self-assembly and transformation of hybrid nanostructures, *Angew. Chem. Int. Ed.* **42**(21), pp. 2350–2365.
8. Yin, H., Wang, C., Zhu, H., Overbury, S. H., Sun, S. and Dai, S. (2008). Colloidal deposition synthesis of supported gold nanocatalysts based on Au-Fe_3O_4 dumbbell nanoparticles, *Chem. Commun.*, (36), pp. 4357–4359.
9. Yong, H., Jing-Chao, L., Ming-Wu, S. and Xiang-Yang, S. (2014). Formation of multifunctional Fe_3O_4/Au composite nanoparticles for dual-mode MR/CT imaging applications, *Chin. Phys. B*, **23**(7), p. 078704.
10. Yu, H., Chen, M., Rice, P. M., Wang, S. X., White, R. L. and Sun, S. (2005). Dumbbell-like bifunctional Au-Fe_3O_4 nanoparticles, *Nano Lett.*, **5**(2), pp. 379–382.
11. Costi, R., Saunders, A. E. and Banin, U. (2010). Colloidal hybrid nanostructures: a new type of functional materials, *Angew. Chem. Int. Ed.*, **49**(29), pp. 4878–4897.
12. Pariti, A., Desai, P., Maddirala, S. K. Y., Ercal, N., Katti, K. V., Liang, X. and Nath, M. (2014). Superparamagnetic Au-Fe_3O_4 nanoparticles: one-pot synthesis, biofunctionalization and toxicity evaluation, *Mater. Res. Express*, **1**(3), p. 035023.
13. Pineider, F., de Julián Fernández, C., Videtta, V., Carlino, E., al Hourani, A., Wilhelm, F., Rogalev, A., Cozzoli, P. D., Ghigna, P. and Sangregorio, C. (2013). Spin-polarization transfer in colloidal magnetic-plasmonic Au/iron oxide hetero-nanocrystals, *ACS Nano*, **7**(1), pp. 857–866.
14. Ahmadi, A., Shirazi, H., Pourbagher, N. and Omidfar, K. (2014). Synthesis and characterization of core-shell Au Fe oxide nanocomposites and their application for detecting immunological interaction, *Monoclon. Antib. Immunodiagn. Immunother.*, **33**(2), pp. 74–79.
15. Fantechi, E., Roca, A. G., Sepúlveda, B., Torruella, P., Estradé, S., Peiró, F., Coy, L. E., Jurga, S., Bastús, N. G., Nogués, J. and Puntes, V. (2017). Seeded growth synthesis of Au-Fe_3O_4 heterostructured nanocrystals: rational design and mechanistic insights, *Chem. Mater.*, **29**(9), pp. 4022–4035.

16. Averitt, R. D., Sarkar, D. and Halas, N. J. (1997). Plasmon resonance shifts of Au-coated Au2S nanoshells: insight into multicomponent nanoparticle growth, *Phys. Rev. Lett.*, **78**(22), p. 4217.

17. Bao, Y., Calderon, H. and Krishnan, K. M. (2007). Synthesis and characterization of magnetic-optical Co–Au core–shell nanoparticles, *J. Phys. Chem. C*, **111**(5), pp. 1941–1944.

18. Yang, Y., Shi, J., Kawamura, G. and Nogami, M. (2008). Preparation of Au–Ag, Ag–Au core–shell bimetallic nanoparticles for surface-enhanced Raman scattering, *Scr. Mater.*, **58**(10), pp. 862–865.

19. Yu, H., Jiao, Y., Li, N., Pang, J., Li, W., Zhang, X., Li, X. and Li, C. (2017). Au-CeO_2 Janus-like nanoparticles fabricated by block copolymer templates and their catalytic activity in the degradation of methyl orange, *Appl. Surf. Sci.*, **427**(Part A), pp. 771–778.

20. Majewski, P. and Thierry, B. (2007). Functionalized magnetite nanoparticles–synthesis, properties, and bio-applications, *Crit. Rev. Solid State Mater. Sci.*, **32**(3–4), pp. 203–215.

21. Sun, S. and Zeng, H. (2002). Size-controlled synthesis of magnetite nanoparticles, *J. Am. Chem. Soc.*, **124**(28), pp. 8204–8205.

22. Reddy, L. H., Arias, J. L., Nicolas, J. and Couvreur, P. (2012). Magnetic nanoparticles: design and characterization, toxicity and biocompatibility, pharmaceutical and biomedical applications, *Chem. Rev.*, **112**(11), pp. 5818–5878.

23. Beard, J. L. (2001). Iron biology in immune function, muscle metabolism and neuronal functioning, *J. Nutr.*, **131**(2), pp. 568S–580S.

24. Gupta, A. K. and Gupta, M. (2005). Synthesis and surface engineering of iron oxide nanoparticles for biomedical applications, *Biomaterials*, **26**(18), pp. 3995–4021.

25. Laurent, S., Forge, D., Port, M., Roch, A., Robic, C., Vander Elst, L. and Muller, R. N. (2008). Magnetic iron oxide nanoparticles: synthesis, stabilization, vectorization, physicochemical characterizations, and biological applications, *Chem. Rev.*, **108**(6), pp. 2064–2110.

26. Gupta, A. K. and Curtis, A. S. G. (2004). Lactoferrin and ceruloplasmin derivatized superparamagnetic iron oxide nanoparticles for targeting cell surface receptors, *Biomaterials*, **25**(15), pp. 3029–3040.

27. Wahajuddin, S. A. (2012). Superparamagnetic iron oxide nanoparticles: magnetic nanoplatforms as drug carriers, *Int. J. Nanomed.*, **7**, p. 3445.

28. Cengelli, F., Maysinger, D., Tschudi-Monnet, F., Montet, X., Corot, C., Petri-Fink, A., Hofmann, H. and Juillerat-Jeanneret, L. (2006). Interaction of

functionalized superparamagnetic iron oxide nanoparticles with brain structures, *J. Pharmacol. Exp. Ther.*, **318**(1), pp. 108–116.

29. Singh, N., Jenkins, G. J. S., Asadi, R. and Doak, S. H. (2010). Potential toxicity of superparamagnetic iron oxide nanoparticles (SPION), *Nano Rev.*, **1**(1), p. 5358.
30. Meng, X., Li, B., Ren, X., Tan, L., Huang, Z. and Tang, F. (2013). One-pot gradient solvothermal synthesis of Au-Fe_3O_4 hybrid nanoparticles for magnetically recyclable catalytic applications, *J. Mater. Chem. A*, **1**(35), pp. 10513–10517.
31. Sood, A., Arora, V., Shah, J., Kotnala, R. K. and Jain, T. K. (2017). Multifunctional gold coated iron oxide core-shell nanoparticles stabilized using thiolated sodium alginate for biomedical applications, *Mater. Sci. Eng., C*, **80**, pp. 274–281.
32. Hemery, G., Keyes Jr, A. C., Garaio, E., Rodrigo, I., Garcia, J. A., Plazaola, F., Garanger, E. and Sandre, O. (2017). Tuning sizes, morphologies, and magnetic properties of monocore versus multicore iron oxide nanoparticles through the controlled addition of water in the polyol synthesis, *Inorg. Chem.*, **56**(14), pp. 8232–8243.
33. Javed, Y., Ali, K. and Jamil, Y. (2017). Magnetic nanoparticle-based hyperthermia for cancer treatment: factors affecting heat generation efficiency, in *Complex Magnetic Nanostructures* (Springer), pp. 393–424.
34. Louis, C. and Pluchery, O. (2017). *Gold Nanoparticles for Physics, Chemistry and Biology* (World Scientific).
35. Lamprecht, B., Schider, G., Lechner, R. T., Ditlbacher, H., Krenn, J. R., Leitner, A. and Aussenegg, F. R. (2000). Metal nanoparticle gratings: influence of dipolar particle interaction on the plasmon resonance, *Phys. Rev. Lett.*, **84**(20), p. 4721.
36. Rechberger, W., Hohenau, A., Leitner, A., Krenn, J. R., Lamprecht, B. and Aussenegg, F. R. (2003). Optical properties of two interacting gold nanoparticles, *Opt. Commun.*, **220**(1), pp. 137–141.
37. Ghosh, S. K. and Pal, T. (2007). Interparticle coupling effect on the surface plasmon resonance of gold nanoparticles: from theory to applications, *Chem. Rev.*, **107**(11), pp. 4797–4862.
38. Jain, P. K., Huang, X., El-Sayed, I. H. and El-Sayed, M. A. (2007). Review of some interesting surface plasmon resonance-enhanced properties of noble metal nanoparticles and their applications to biosystems, *Plasmonics*, **2**(3), pp. 107–118.

39. Huang, X., Jain, P. K., El-Sayed, I. H. and El-Sayed, M. A. (2007). Gold nanoparticles: interesting optical properties and recent applications in cancer diagnostics and therapy, *Nanomedicine (Lond)*, **2**(5), pp. 681–693.

40. Giljohann, D. A., Seferos, D. S., Daniel, W. L., Massich, M. D., Patel, P. C. and Mirkin, C. A. (2010). Gold nanoparticles for biology and medicine, *Angew. Chem. Int. Ed.*, **49**(19), pp. 3280–3294.

41. Nordlander, P., Oubre, C., Prodan, E., Li, K. and Stockman, M. I. (2004). Plasmon hybridization in nanoparticle dimers, *Nano Lett.*, **4**(5), pp. 899–903.

42. Yang, S.-C., Kobori, H., He, C.-L., Lin, M.-H., Chen, H.-Y., Li, C., Kanehara, M., Teranishi, T. and Gwo, S. (2010). Plasmon hybridization in individual gold nanocrystal dimers: direct observation of bright and dark modes, *Nano Lett.*, **10**(2), pp. 632–637.

43. Prodan, E., Radloff, C., Halas, N. J. and Nordlander, P. (2003). A hybridization model for the plasmon response of complex nanostructures, *Science*, **302**(5644), pp. 419–422.

44. Nehl, C. L. and Hafner, J. H. (2008). Shape-dependent plasmon resonances of gold nanoparticles, *J. Mater. Chem.*, **18**(21), pp. 2415–2419.

45. Comin, A., Korobchevskaya, K., George, C., Diaspro, A. and Manna, L. (2012). Plasmon bleaching dynamics in colloidal gold–iron oxide nanocrystal heterodimers, *Nano Lett.*, **12**(2), pp. 921–926.

46. Jain, P. K., Xiao, Y., Walsworth, R. and Cohen, A. E. (2009). Surface plasmon resonance enhanced magneto-optics (SuPREMO): Faraday rotation enhancement in gold-coated iron oxide nanocrystals, *Nano Lett.*, **9**(4), pp. 1644–1650.

47. Brus, L. (2008). Noble metal nanocrystals: plasmon electron transfer photochemistry and single-molecule Raman spectroscopy, *Acc. Chem. Res.*, **41**(12), pp. 1742–1749.

48. Xie, Y., Chen, W., Bertoni, G., Kriegel, I., Xiong, M., Li, N., Prato, M., Riedinger, A., Sathya, A. and Manna, L. (2017). Tuning and locking the localized surface plasmon resonances of CuS (covellite) nanocrystals by an amorphous CuPd x S shell, *Chem. Mater.*, **29**(4), pp. 1716–1723.

49. Wu, W., He, Q. and Jiang, C. (2008). Magnetic iron oxide nanoparticles: synthesis and surface functionalization strategies, *Nanoscale Res. Lett.*, **3**(11), p. 397.

50. Guo, H. and Barnard, A. S. (2013). Naturally occurring iron oxide nanoparticles: morphology, surface chemistry and environmental stability, *J. Mater. Chem. A*, **1**(1), pp. 27–42.

51. Teja, A. S. and Koh, P.-Y. (2009). Synthesis, properties, and applications of magnetic iron oxide nanoparticles, *Prog. Cryst. Growth Charact. Mater.*, **55**(1), pp. 22–45.

52. Mahmoudi, M., Simchi, A., Milani, A. S. and Stroeve, P. (2009). Cell toxicity of superparamagnetic iron oxide nanoparticles, *J. Colloid Interface Sci.*, **336**(2), pp. 510–518.

53. Schwertmann, U. and Cornell, R. M. (2008). *Iron Oxides in the Laboratory: Preparation and Characterization* (John Wiley & Sons).

54. Cornell, R. M. and Schwertmann, U. (2003). *The Iron Oxides: Structure, Properties, Reactions, Occurrences and Uses* (John Wiley & Sons).

55. Rohrer, G. S. (2001). *Structure and Bonding in Crystalline Materials* (Cambridge University Press).

56. Daou, T. J., Pourroy, G., Begin-Colin, S., Greneche, J. M., Ulhaq-Bouillet, C., Legaré, P., Bernhardt, P., Leuvrey, C. and Rogez, G. (2006). Hydrothermal synthesis of monodisperse magnetite nanoparticles, *Chem. Mater.*, **18**(18), pp. 4399–4404.

57. Colliex, C., Manoubi, T. and Ortiz, C. (1991). Electron-energy-loss-spectroscopy near-edge fine structures in the iron-oxygen system, *Phys. Rev. B*, **44**(20), p. 11402.

58. Van Aken, P. A., Liebscher, B. and Styrsa, V. J. (1998). Quantitative determination of iron oxidation states in minerals using Fe L 2, 3-edge electron energy-loss near-edge structure spectroscopy, *Phys. Chem. Miner.*, **25**(5), pp. 323–327.

59. Hyeon, T., Lee, S. S., Park, J., Chung, Y. and Na, H. B. (2001). Synthesis of highly crystalline and monodisperse maghemite nanocrystallites without a size-selection process, *J. Am. Chem. Soc.*, **123**(51), pp. 12798–12801.

60. Pauling, L. and Hendricks, S. B. (1925). The crystal structures of hematite and corundum, *J. Am. Chem. Soc.*, **47**(3), pp. 781–790.

61. Bodker, F., Hansen, M. F., Koch, C. B., Lefmann, K. and Morup, S. (2000). Magnetic properties of hematite nanoparticles, *Phys. Rev. B*, **61**(10), p. 6826.

62. Woo, K., Hong, J., Choi, S., Lee, H.-W., Ahn, J.-P., Kim, C. S. and Lee, S. W. (2004). Easy synthesis and magnetic properties of iron oxide nanoparticles, *Chem. Mater.*, **16**(14), pp. 2814–2818.

63. Chatterjee, J., Haik, Y. and Chen, C.-J. (2003). Size dependent magnetic properties of iron oxide nanoparticles, *J. Magn. Magn. Mater.*, **257**(1), pp. 113–118.

64. Wei, W., Zhaohui, W., Taekyung, Y., Changzhong, J. and Woo-Sik, K. (2015). Recent progress on magnetic iron oxide nanoparticles: synthesis, surface functional strategies and biomedical applications, *Sci. Technol. Adv. Mater.*, **16**(2), p. 023501.

65. Chow, P. E. (2010). *Gold Nanoparticles: Properties, Characterization and Fabrication* (Nova Science).

66. Claus, P., Brueckner, A., Mohr, C. and Hofmeister, H. (2000). Supported gold nanoparticles from quantum dot to mesoscopic size scale: effect of electronic and structural properties on catalytic hydrogenation of conjugated functional groups, *J. Am. Chem. Soc.*, **122**(46), pp. 11430–11439.

67. Grzelczak, M., Perez-Juste, J., Mulvaney, P. and Liz-Marzán, L. M. (2008). Shape control in gold nanoparticle synthesis, *Chem. Soc. Rev.*, **37**(9), pp. 1783–1791.

68. Wessels, J. M., Nothofer, H.-G., Ford, W. E., von Wrochem, F., Scholz, F., Vossmeyer, T., Schroedter, A., Weller, H. and Yasuda, A. (2004). Optical and electrical properties of three-dimensional interlinked gold nanoparticle assemblies, *J. Am. Chem. Soc.*, **126**(10), pp. 3349–3356.

69. Link, S. and El-Sayed, M. A. (1999). Size and temperature dependence of the plasmon absorption of colloidal gold nanoparticles, *J. Phys. Chem. B*, **103**(21), pp. 4212–4217.

70. Link, S., Mohamed, M. B. and El-Sayed, M. A. (1999). Simulation of the optical absorption spectra of gold nanorods as a function of their aspect ratio and the effect of the medium dielectric constant, *J. Phys. Chem. B*, **103**(16), pp. 3073–3077.

71. Das, S. K., Das, A. R. and Guha, A. K. (2009). Gold nanoparticles: microbial synthesis and application in water hygiene management, *Langmuir*, **25**(14), pp. 8192–8199.

72. Homola, J., Yee, S. S. and Gauglitz, G. (1999). Surface plasmon resonance sensors, *Sens. Actuators, B*, **54**(1), pp. 3–15.

73. Willets, K. A. and Van Duyne, R. P. (2007). Localized surface plasmon resonance spectroscopy and sensing, *Annu. Rev. Phys. Chem.*, **58**, pp. 267–297.

74. Notarianni, M., Vernon, K., Chou, A., Aljada, M., Liu, J. and Motta, N. (2014). Plasmonic effect of gold nanoparticles in organic solar cells, *Solar Energy*, **106**(Suppl C), pp. 23–37.

75. Stadelmann, P. A. (1987). EMS - a software package for electron diffraction analysis and HREM image simulation in materials science, *Ultramicroscopy*, **21**(2), pp. 131–145.

76. Kolosnjaj-Tabi, J., Javed, Y., Lartigue, L., Volatron, J., Elgrabli, D., Marangon, I., Pugliese, G., Caron, B., Figuerola, A. and Luciani, N. (2015). The one year fate of iron oxide coated gold nanoparticles in mice, *ACS Nano*, **9**(8), pp. 7925–7939.

77. Wang, Z. L. (2000). Transmission electron microscopy of shape-controlled nanocrystals and their assemblies, *J. Phys. Chem. B*, **104**(6), pp. 1153–1175.

78. Yallapu, M. M., Othman, S. F., Curtis, E. T., Gupta, B. K., Jaggi, M. and Chauhan, S. C. (2011). Multi-functional magnetic nanoparticles for magnetic resonance imaging and cancer therapy, *Biomaterials*, **32**(7), pp. 1890–1905.

79. Zhang, Q., Ge, J., Goebl, J., Hu, Y., Sun, Y. and Yin, Y. (2010). Tailored synthesis of superparamagnetic gold nanoshells with tunable optical properties, *Adv. Mater.*, **22**, pp. 1905–1909.

80. Jiang, H., Zeng, X., Xi, Z., Liu, M., Li, C., Li, Z., Jin, L., Wang, Z., Deng, Y. and He, N. (2013). Improvement on controllable fabrication of streptavidin-modified three-layer core–shell Fe_3O_4@ SiO_2@ Au magnetic nanocomposites with low fluorescence background, *J. Biomed. Nanotechnol.*, **9**(4), pp. 674–684.

81. Kim, M. and Song, H. (2014). Precise adjustment of structural anisotropy and crystallinity on metal-Fe_3O_4 hybrid nanoparticles and its influence on magnetic and catalytic properties, *J. Mater. Chem. C*, **2**(25), pp. 4997–5004.

82. Song, H., Zhang, X., Zhang, J., Liu, D., Shu, T., Yang, X., Du, L. and Liao, S. (2015). Facile synthesis of high dispersion [gamma]-Fe_2O_3-Au nanoparticles within mesoporous silica spheres, *RSC Adv.*, **5**(62), pp. 49914–49919.

83. Ali, K., Javed, Y. and Jamil, Y. (2017). Size and shape control synthesis of iron oxide-based nanoparticles: current status and future possibility, in *Complex Magnetic Nanostructures* (Springer), pp. 39–81.

84. Feygenson, M., Bauer, J. C., Gai, Z., Marques, C., Aronson, M. C., Teng, X., Su, D., Stanic, V., Urban, V. S., Beyer, K. A. and Dai, S. (2015). Exchange bias effect in Au-Fe_3O_4 dumbbell nanoparticles induced by the charge transfer from gold, *Phys. Rev. B*, **92**(5), p. 054416.

85. Sebastian, V., Pilar Calatayud, M., Goya, G. F. and Santamaria, J. (2013). Magnetically-driven selective synthesis of Au clusters on Fe_3O_4 nanoparticles, *Chem. Commun.*, **49**(7), pp. 716–718.

86. Li, L., Mak, K. Y., Leung, C. W., Leung, C. H., Ruotolo, A., Chan, K. Y., Chan, W. K. and Pong, P. W. T. (2014). Synthesis and morphology control

of gold/iron oxide magnetic nanocomposites via a simple aqueous method, *IEEE Trans. Magn.*, **50**(1), pp. 1–5.

87. Wang, C., Yin, H., Dai, S. and Sun, S. (2010). A General approach to noble metal–metal oxide dumbbell nanoparticles and their catalytic application for CO oxidation, *Chem. Mater.*, **22**(10), pp. 3277–3282.

88. Guardia, P., Riedinger, A., Nitti, S., Pugliese, G., Marras, S., Genovese, A., Materia, M. E., Lefevre, C., Manna, L. and Pellegrino, T. (2014). One pot synthesis of monodisperse water soluble iron oxide nanocrystals with high values of the specific absorption rate, *J. Mater. Chem. B*, **2**(28), pp. 4426–4434.

89. Kuladeep, R., Jyothi, L., Alee, K. S., Deepak, K. L. N. and Rao, D. N. (2012). Laser-assisted synthesis of Au-Ag alloy nanoparticles with tunable surface plasmon resonance frequency, *Opt. Mater. Express*, **2**(2), pp. 161–172.

90. Peng, Z., Spliethoff, B., Tesche, B., Walther, T. and Kleinermanns, K. (2006). Laser-assisted synthesis of Au–Ag alloy nanoparticles in solution, *J. Phys. Chem. B*, **110**(6), pp. 2549–2554.

91. Zhang, J., Post, M., Veres, T., Jakubek, Z. J., Guan, J., Wang, D., Normandin, F., Deslandes, Y. and Simard, B. (2006). Laser-assisted synthesis of superparamagnetic Fe@ Au core–shell nanoparticles, *J. Phys. Chem. B*, **110**(14), pp. 7122–7128.

92. Duan, X. and Lieber, C. M. (2000). Laser-assisted catalytic growth of single crystal GaN nanowires, *J. Am. Chem. Soc.*, **122**(1), pp. 188–189.

93. Fantoni, R., Borsella, E., Piccirillo, S., Nannetti, C. A., Ceccato, R. and Enzo, S. (1989). Laser assisted synthesis of ultrafine silicon powder, *Appl. Surf. Sci.*, **43**(1), pp. 308–315.

94. Alloyeau, D., Ricolleau, C., Langlois, C., Le Bouar, Y. and Loiseau, A. (2010). Flash laser annealing for controlling size and shape of magnetic alloy nanoparticles, *Beilstein J. Nanotechnol.*, **1**, pp. 55–59.

95. Li, L., Du, Y. M., Mak, K. Y., Leung, C. W. and Pong, P. W. T. (2014). Novel hybrid magnetic octahedron-like nanoparticles with tunable size, *IEEE Trans. Magn.*, **50**(1), pp. 1–5.

96. Verma, A. and Stellacci, F. (2010). Effect of surface properties on nanoparticle–cell interactions, *Small*, **6**(1), pp. 12–21.

97. He, C., Hu, Y., Yin, L., Tang, C. and Yin, C. (2010). Effects of particle size and surface charge on cellular uptake and biodistribution of polymeric nanoparticles, *Biomaterials*, **31**(13), pp. 3657–3666.

98. Kobayashi, Y., Katakami, H., Mine, E., Nagao, D., Konno, M. and Liz-Marzán, L. M. (2005). Silica coating of silver nanoparticles using a modified Stöber method, *J. Colloid Interface Sci.*, **283**(2), pp. 392–396.

99. Sahoo, Y., Pizem, H., Fried, T., Golodnitsky, D., Burstein, L., Sukenik, C. N. and Markovich, G. (2001). Alkyl phosphonate/phosphate coating on magnetite nanoparticles: a comparison with fatty acids, *Langmuir*, **17**(25), pp. 7907–7911.

100. Sperling, R. A. and Parak, W. J. (2010). Surface modification, functionalization and bioconjugation of colloidal inorganic nanoparticles, *Philos. Trans. R. Soc. London, Ser. A*, **368**(1915), pp. 1333–1383.

101. Butterworth, M. D., Illum, L. and Davis, S. S. (2001). Preparation of ultrafine silica- and PEG-coated magnetite particles, *Colloids Surf., A*, **179**(1), pp. 93–102.

102. Veronese, F. M. (2001). Peptide and protein PEGylation: a review of problems and solutions, *Biomaterials*, **22**(5), pp. 405–417.

103. Daou, T. J., Li, L., Reiss, P., Josserand, V. and Texier, I. (2009). Effect of poly(ethylene glycol) length on the in vivo behavior of coated quantum dots, *Langmuir*, **25**(5), pp. 3040–3044.

104. Zareie, H. M., Boyer, C., Bulmus, V., Nateghi, E. and Davis, T. P. (2008). Temperature-responsive self-assembled monolayers of oligo(ethylene glycol): control of biomolecular recognition, *ACS Nano*, **2**(4), pp. 757–765.

105. van Vlerken, L. E., Vyas, T. K. and Amiji, M. M. (2007). Poly(ethylene glycol)-modified nanocarriers for tumor-targeted and intracellular delivery, *Pharm. Res.*, **24**(8), pp. 1405–1414.

106. Mout, R., Moyano, D. F., Rana, S. and Rotello, V. M. (2012). Surface functionalization of nanoparticles for nanomedicine, *Chem. Soc. Rev.*, **41**(7), pp. 2539–2544.

107. Rana, S., Bajaj, A., Mout, R. and Rotello, V. M. (2012). Monolayer coated gold nanoparticles for delivery applications, *Adv. Drug Delivery Rev.*, **64**(2), pp. 200–216.

108. Gao, Q., Zhao, A., Guo, H., Chen, X., Gan, Z., Tao, W., Zhang, M., Wu, R. and Li, Z. (2014). Controlled synthesis of Au-Fe_3O_4 hybrid hollow spheres with excellent SERS activity and catalytic properties, *Dalton Trans.*, **43**(21), pp. 7998–8006.

109. Ghosh Chaudhuri, R. and Paria, S. (2011). Core/shell nanoparticles: classes, properties, synthesis mechanisms, characterization, and applications, *Chem. Rev.*, **112**(4), pp. 2373–2433.

110. Carlson, C., Hussain, S. M., Schrand, A. M., Braydich-Stolle, L. K., Hess, K. L., Jones, R. L. and Schlager, J. J. (2008). Unique cellular interaction of silver nanoparticles: size-dependent generation of reactive oxygen species, *J. Phys. Chem. B*, **112**(43), pp. 13608–13619.

111. Nath, N. and Chilkoti, A. (2002). A colorimetric gold nanoparticle sensor to interrogate biomolecular interactions in real time on a surface, *Anal. Chem.*, **74**(3), pp. 504–509.

112. Bareford, L. M. and Swaan, P. W. (2007). Endocytic mechanisms for targeted drug delivery, *Adv. Drug Delivery Rev.*, **59**(8), pp. 748–758.

113. Veiseh, O., Gunn, J. W. and Zhang, M. (2010). Design and fabrication of magnetic nanoparticles for targeted drug delivery and imaging, *Adv. Drug Delivery Rev.*, **62**(3), pp. 284–304.

114. Katz, E. and Willner, I. (2004). Integrated nanoparticle–biomolecule hybrid systems: synthesis, properties, and applications, *Angew. Chem. Int. Ed.*, **43**(45), pp. 6042–6108.

115. Tansil, N. C. and Gao, Z. (2006). Nanoparticles in biomolecular detection, *Nano Today*, **1**(1), pp. 28–37.

116. Chowdhury, A., Agnihotri, N., Doong, R. A. and De, A. (2017). Label-free and nondestructive separation technique for isolation of targeted DNA from DNA-protein mixture using magnetic Au-Fe_3O_4 nanoprobes, *Anal. Chem.*, **89**(22), pp. 12244–12251.

117. Reguera, J., Jimenez de Aberasturi, D., Henriksen-Lacey, M., Langer, J., Espinosa, A., Szczupak, B., Wilhelm, C. and Liz-Marzán, L. M. (2017). Janus plasmonic-magnetic gold-iron oxide nanoparticles as contrast agents for multimodal imaging, *Nanoscale*, **9**(27), pp. 9467–9480.

118. Guardia, P., Nitti, S., Materia, M. E., Pugliese, G., Yaacoub, N., Greneche, J. M., Lefevre, C., Manna, L. and Pellegrino, T. (2017). Gold-iron oxide dimers for magnetic hyperthermia: the key role of chloride ions in the synthesis to boost the heating efficiency, *J. Mater. Chem. B*, **5**, pp. 4587–4594.

119. Lee, D. K., Song, Y., Kim, J., Park, E. Y. and Lee, J. (2017). Preparation of concave magnetoplasmonic core-shell supraparticles of gold-coated iron oxide via ion-reducible layer-by-layer method for surface enhanced Raman scattering, *J. Colloid Interface Sci.*, **499**, pp. 54–61.

120. Liu, Y., Kou, Q., Wang, D., Chen, L., Sun, Y., Lu, Z., Zhang, Y., Wang, Y., Yang, J. and Xing, S. G. (2017). Rational synthesis and tailored optical and magnetic characteristics of Fe_3O_4–Au composite nanoparticles, *J. Mater. Sci.*, **52**(17), pp. 10163–10174.

121. Miola, M., Ferraris, S., Pirani, F., Multari, C., Bertone, E., Rožman, K. Ž., Kostevšek, N. and Verné, E. (2017). Reductant-free synthesis of magnetoplasmonic iron oxide-gold nanoparticles, *Ceram. Int.*, **43**(17), pp. 15258–15265.

122. Najafishirtari, S., Guardia, P., Scarpellini, A., Prato, M., Marras, S., Manna, L. and Colombo, M. (2016). The effect of Au domain size on the CO oxidation catalytic activity of colloidal Au-FeOx dumbbell-like heterodimers, *J. Catal.*, **338**(Suppl C), pp. 115–123.

123. Sun, L., Joh, D. Y., Al-Zaki, A., Stangl, M., Murty, S., Davis, J. J., Baumann, B. C., Alonso-Basanta, M., Kao, G. D., Tsourkas, A. and Dorsey, J. F. (2016). Theranostic application of mixed gold and superparamagnetic iron oxide nanoparticle micelles in glioblastoma multiforme, *J. Biomed. Nanotechnol.*, **12**(2), pp. 347–356.

124. Song, J., Wu, B., Zhou, Z., Zhu, G., Liu, Y., Yang, Z., Lin, L., Yu, G., Zhang, F., Zhang, G., Duan, H., Stucky, G. D. and Chen, X. (2017). Double-layered plasmonic–magnetic vesicles by self-assembly of Janus amphiphilic gold–iron(II,III) oxide nanoparticles, *Angew. Chem. Int. Ed.*, **129**(28), pp. 8222–8226.

125. Xing, Y., Gao, Q., Zhang, Y., Ma, L., Loh, K. Y., Peng, M., Chen, C. and Cui, Y. (2017). The improved sensitive detection of C-reactive protein based on the chemiluminescence immunoassay by employing monodispersed PAA-Au/Fe_3O_4 nanoparticles and zwitterionic glycerophosphoryl choline, *J. Mater. Chem. B*, **5**(21), pp. 3919–3926.

126. Zhang, Q., Zhang, L., Li, S., Chen, X., Zhang, M., Wang, T., Li, L. and Wang, C. (2017). Designed synthesis of Au/Fe_3O_4@ C Janus nanoparticles for dual-modal imaging and actively targeted chemo-photothermal synergistic therapy of cancer cells, *Chemistry*, **23**(68), pp. 17242–17248.

127. Zhao, J., Tu, K., Liu, Y., Qin, Y., Wang, X., Qi, L. and Shi, D. (2017). Photo-controlled aptamers delivery by dual surface gold-magnetic nanoparticles for targeted cancer therapy, *Mater. Sci. Eng., C*, **80**, pp. 88–92.

128. Xu, C., Xie, J., Ho, D., Wang, C., Kohler, N., Walsh, E. G., Morgan, J. R., Chin, Y. E. and Sun, S. (2008). Au–Fe_3O_4 dumbbell nanoparticles as dual-functional probes, *Angew. Chem. Int. Ed.*, **47**(1), pp. 173–176.

129. Xu, C., Wang, B. and Sun, S. (2009). Dumbbell-like Au-Fe_3O_4 nanoparticles for target-specific platin delivery, *J. Am. Chem. Soc.*, **131**(12), pp. 4216–4217.

130. Abdulla-Al-Mamun, M., Kusumoto, Y., Zannat, T., Horie, Y. and Manaka, H. (2013). Au-ultrathin functionalized core-shell (Fe_3O_4@ Au)

monodispersed nanocubes for a combination of magnetic/plasmonic photothermal cancer cell killing, *RSC Adv.*, **3**(21), pp. 7816–7827.

131. Cheng, K., Yang, M., Zhang, R., Qin, C., Su, X. and Cheng, Z. (2014). Hybrid nanotrimers for dual T1 and T2-weighted magnetic resonance imaging, *ACS Nano*, **8**(10), pp. 9884–9896.
132. Felber, M. and Alberto, R. (2015). 99m Tc radiolabelling of Fe_3O_4-Au core–shell and Au-Fe_3O_4 dumbbell-like nanoparticles, *Nanoscale*, **7**(15), pp. 6653–6660.
133. Gan, N., Wang, L., Li, T., Sang, W., Hu, F. and Cao, Y. (2013). A novel signal-amplified immunoassay for myoglobin using magnetic core-shell Fe_3O_4@ Au-multi walled carbon nanotubes composites as labels based on one piezoelectric sensor, *Integr. Ferroelectr.*, **144**(1), pp. 29–40.
134. Hu, F., Lin, H., Zhang, Z., Liao, F., Shao, M., Lifshitz, Y. and Lee, S.-T. (2014). Smart liquid SERS substrates based on Fe_3O_4/Au nanoparticles with reversibly tunable enhancement factor for practical quantitative detection, *Sci. Rep.*, **4**, p. 7204.
135. Hu, Y., Yang, J., Wei, P., Li, J., Ding, L., Zhang, G., Shi, X. and Shen, M. (2015). Facile synthesis of hyaluronic acid-modified Fe_3O_4/Au composite nanoparticles for targeted dual mode MR/CT imaging of tumors, *J. Mater. Chem. B*, **3**(47), pp. 9098–9108.
136. Lin, F.-H., Peng, H.-H., Yang, Y.-H. and Doong, R.-A. (2013). Size and morphological effect of Au-Fe_3O_4 heterostructures on magnetic resonance imaging, *J. Nanopart. Res.*, **15**(12), p. 2139.
137. Felix, L. L., Coaquira, J. A. H., Martínez, M. A. R., Goya, G. F., Mantilla, J., Sousa, M. H., de los Santos Valladares, L., Barnes, C. H. W. and Morais, P. C. (2017). Structural and magnetic properties of core-shell Au/Fe_3O_4 nanoparticles, *Sci. Rep.*, **7**, p. 41732.
138. Zhao, H. Y., Liu, S., He, J., Pan, C. C., Li, H., Zhou, Z. Y., Ding, Y., Huo, D. and Hu, Y. (2015). Synthesis and application of strawberry-like Fe_3O_4-Au nanoparticles as CT-MR dual-modality contrast agents in accurate detection of the progressive liver disease, *Biomaterials*, **51**(Suppl C), pp. 194–207.
139. Liu, J., Zhang, W., Zhang, H., Yang, Z., Li, T., Wang, B., Huo, X., Wang, R. and Chen, H. (2013). A multifunctional nanoprobe based on Au-Fe_3O_4 nanoparticles for multimodal and ultrasensitive detection of cancer cells, *Chem. Commun.*, **49**(43), pp. 4938–4940.
140. Daniel, M.-C. and Astruc, D. (2004). Gold nanoparticles: assembly, supramolecular chemistry, quantum-size-related properties, and

applications toward biology, catalysis, and nanotechnology, *Chem. Rev.*, **104**(1), pp. 293–346.

141. Venditti, I. (2017). Gold nanoparticles in photonic crystals applications: a review, *Materials*, **10**(2), p. 97.

142. Kim, G.-C., Li, Y.-Y., Chu, Y.-F., Cheng, S.-X., Zhuo, R.-X. and Zhang, X.-Z. (2008). Nanosized temperature-responsive Fe_3O_4-UA-g-P(UA-co-NIPAAm) magnetomicelles for controlled drug release, *Eur. Polym. J.*, **44**(9), pp. 2761–2767.

143. Shi, W., Liu, X., Wei, C., Xu, Z. J., Sim, S. S. W., Liu, L. and Xu, C. (2015). Micro-optical coherence tomography tracking of magnetic gene transfection via Au-Fe_3O_4 dumbbell nanoparticles, *Nanoscale*, **7**(41), pp. 17249–17253.

144. Han, L., Xia, J.-M., Hai, X., Shu, Y., Chen, X.-W. and Wang, J.-H. (2017). Protein-stabilized gadolinium oxide-gold nanoclusters hybrid for multimodal imaging and drug delivery, *ACS Appl. Mater. Interfaces*, **9**(8), pp. 6941–6949.

145. Salunkhe, A. B., Khot, V. M. and Pawar, S. H. (2014). Magnetic hyperthermia with magnetic nanoparticles: a status review, *Curr. Top. Med. Chem.*, **14**(5), pp. 572–594.

146. Kumar, C. S. S. R. and Mohammad, F. (2011). Magnetic nanomaterials for hyperthermia-based therapy and controlled drug delivery, *Adv. Drug Delivery Rev.*, **63**(9), pp. 789–808.

147. Javed, Y., Akhtar, K., Anwar, H. and Jamil, Y. (2017). MRI based on iron oxide nanoparticles contrast agents: effect of oxidation state and architecture, *J. Nanopart. Res.*, **19**(11), p. 366.

148. Jia, D. and Liu, J. (2010). Current devices for high-performance whole-body hyperthermia therapy, *Expert Rev. Med. Devices*, **7**(3), pp. 407–423.

149. Fathi Karkan, S., Mohammadhosseini, M., Panahi, Y., Milani, M., Zarghami, N., Akbarzadeh, A., Abasi, E., Hosseini, A. and Davaran, S. (2017). Magnetic nanoparticles in cancer diagnosis and treatment: a review, *Artif. Cells Nanomed. Biotechnol.*, **45**(1), pp. 1–5.

150. Pitt, W. G., Husseini, G. A. and Staples, B. J. (2004). Ultrasonic drug delivery–a general review, *Expert Opin. Drug Delivery*, **1**(1), pp. 37–56.

151. Thiesen, B. and Jordan, A. (2008). Clinical applications of magnetic nanoparticles for hyperthermia, *Int. J. Hyperthermia*, **24**(6), pp. 467–474.

152. Hergt, R., Dutz, S. and Röder, M. (2008). Effects of size distribution on hysteresis losses of magnetic nanoparticles for hyperthermia, *J. Phys.: Condens. Matter*, **20**(38), p. 385214.

153. Burrows, F., Parker, C., Evans, R. F. L., Hancock, Y., Hovorka, O. and Chantrell, R. W. (2010). Energy losses in interacting fine-particle magnetic composites, *J. Phys. D: Appl. Phys.*, **43**(47), p. 474010.

154. Motoyama, J., Hakata, T., Kato, R., Yamashita, N., Morino, T., Kobayashi, T. and Honda, H. (2010). Size dependent heat generation of magnetic nanoparticles under AC magnetic field for cancer therapy, in *Animal Cell Technology: Basic & Applied Aspects* (Springer), pp. 415–421.

155. Lima, E., Torres, T. E., Rossi, L. M., Rechenberg, H. R., Berquo, T. S., Ibarra, A., Marquina, C., Ibarra, M. R. and Goya, G. F. (2014). Size dependence of the magnetic relaxation and specific power absorption in iron oxide nanoparticles, *J. Nanopart. Res.*, **15**(5), p. 1654.

156. Guardia, P., Di Corato, R., Lartigue, L., Wilhelm, C., Espinosa, A., Garcia-Hernandez, M., Gazeau, F., Manna, L. and Pellegrino, T. (2012). Water-soluble iron oxide nanocubes with high values of specific absorption rate for cancer cell hyperthermia treatment, *ACS Nano*, **6**(4), pp. 3080–3091.

157. Lartigue, L., Hugounenq, P., Alloyeau, D., Clarke, S. P., Lévy, M., Bacri, J. C., Bazzi, R., Brougham, D. F., Wilhelm, C. and Gazeau, F. (2012). Cooperative organization in iron oxide multi-core nanoparticles potentiates their efficiency as heating mediators and MRI contrast agents, *ACS Nano*, **6**(12), pp. 10935–10949.

158. Hilger, I. and Kaiser, W. A. (2012). Iron oxide-based nanostructures for MRI and magnetic hyperthermia, *Nanomedicine*, **7**(9), pp. 1443–1459.

159. Hiergeist, R., Andrä, W., Buske, N., Hergt, R., Hilger, I., Richter, U. and Kaiser, W. (1999). Application of magnetite ferrofluids for hyperthermia, *J. Magn. Magn. Mater.*, **201**(1), pp. 420–422.

160. Sharifi, I., Shokrollahi, H. and Amiri, S. (2012). Ferrite-based magnetic nanofluids used in hyperthermia applications, *J. Magn. Magn. Mater.*, **324**(6), pp. 903–915.

161. Mohammad, F., Balaji, G., Weber, A., Uppu, R. M. and Kumar, C. S. S. R. (2010). Influence of gold nanoshell on hyperthermia of superparamagnetic iron oxide nanoparticles, *J. Phys. Chem. C*, **114**(45), pp. 19194–19201.

162. Chung, R.-J., Wang, H.-Y. and Wu, K.-T. (2014). Preparation and characterization of Fe-Au alloy nanoparticles for hyperthermia application, *J. Med. Biol. Eng.*, **34**(3), p. 251.

163. Hoskins, C., Min, Y., Gueorguieva, M., McDougall, C., Volovick, A., Prentice, P., Wang, Z., Melzer, A., Cuschieri, A. and Wang, L. (2012). Hybrid gold-iron oxide nanoparticles as a multifunctional platform for biomedical application, *J. Nanobiotechnol.*, **10**(1), p. 27.

164. Chung, R.-J. and Shih, H.-T. (2014). Preparation of multifunctional Fe@Au core-shell nanoparticles with surface grafting as a potential treatment for magnetic hyperthermia, *Materials*, **7**(2), p. 653.

165. Khosroshahi, M. E. and Tajabadi, M. (2017). Multifunctional nanoplatform for targeted laser-induced hyperthermia and microscopy of breast cancer cells using SPION-based gold and folic acid conjugated nanodendrimers: an in vitro assay, *J. Nanomed. Nanotechnol.*, **8**(432), p. 2.

166. Bell, G., Bogart, L. K., Southern, P., Olivo, M., Pankhurst, Q. A. and Parkin, I. P. (2017). Enhancing the magnetic heating capacity of iron oxide nanoparticles through their postproduction incorporation into iron oxide–gold nanocomposites, *Eur. J. Inorg. Chem.*, **2017**(18), pp. 2386–2395.

167. Davis, M. E. and Shin, D. M. (2008). Nanoparticle therapeutics: an emerging treatment modality for cancer, *Nat. Rev. Drug Discovery*, **7**(9), pp. 771–782.

168. Pissuwan, D., Valenzueka, S. M. and Cortie, M. B. (2006). Therapeutic possibilities of plasmonically heated gold nanoparticles, *Trends Biotechnol.*, **24**, pp. 62–67.

169. Petros, R. A. and DeSimone, J. M. (2010). Strategies in the design of nanoparticles for therapeutic applications, *Nat. Rev. Drug Discovery*, **9**(8), pp. 615–627.

170. Huang, X., Jain, P. K., El-Sayed, I. H. and El-Sayed, M. A. (2008). Plasmonic photothermal therapy (PPTT) using gold nanoparticles, *Lasers Med. Sci.*, **23**(3), p. 217.

171. Qiao, R., Yang, C. and Gao, M. (2009). Superparamagnetic iron oxide nanoparticles: from preparations to in vivo MRI applications, *J. Mater. Chem.*, **19**(35), pp. 6274–6293.

172. Smolensky, E. D., Park, H. Y. E., Berquo, T. S. and Pierre, V. C. (2011). Surface functionalization of magnetic iron oxide nanoparticles for MRI applications–effect of anchoring group and ligand exchange protocol, *Contrast Media Mol. Imaging*, **6**(4), pp. 189–199.

173. Fortin, J.-P., Wilhelm, C., Servais, J., Ménager, C., Bacri, J.-C. and Gazeau, F. (2007). Size-sorted anionic iron oxide nanomagnets as colloidal

mediators for magnetic hyperthermia, *J. Am. Chem. Soc.*, **129**(9), pp. 2628–2635.

174. Lee, D.-E., Koo, H., Sun, I.-C., Ryu, J. H., Kim, K. and Kwon, I. C. (2012). Multifunctional nanoparticles for multimodal imaging and theragnosis, *Chem. Soc. Rev.*, **41**(7), pp. 2656–2672.

175. Dickerson, E. B., Dreaden, E. C., Huang, X., El-Sayed, I. H., Chu, H., Pushpanketh, S., McDonald, J. F. and El-Sayed, M. A. (2008). Gold nanorod assisted near-infrared plasmonic photothermal therapy (PPTT) of squamous cell carcinoma in mice, *Cancer Lett.*, **269**(1), pp. 57–66.

176. Kim, J.-W., Galanzha, E., Shashkov, E. V., Moon, H.-M. and Zharov, V. P. (2009). Golden carbon nanotubes as multimodal photoacoustic and photothermal high-contrast molecular agents, *Nat. Nanotechnol.*, **4**(10), pp. 688–694.

177. Ji, X., Shao, R., Elliott, A. M., Stafford, R. J., Esparza-Coss, E., Bankson, J. A., Liang, G., Luo, Z.-P., Park, K. and Markert, J. T. (2007). Bifunctional gold nanoshells with a superparamagnetic iron oxide–silica core suitable for both MR imaging and photothermal therapy, *J. Phys. Chem. C*, **111**(17), pp. 6245–6251.

178. Huang, X., El-Sayed, I. H., Qian, W. and El-Sayed, M. A. (2006). Cancer cell imaging and photothermal therapy in the near-infrared region by using gold nanorods, *J. Am. Chem. Soc.*, **128**(6), pp. 2115–2120.

179. Syed, S. H., Coughlin, A. J., Garcia, M. D., Wang, S., West, J. L., Larin, K. V. and Larina, I. V. (2015). Optical coherence tomography guided microinjections in live mouse embryos: high-resolution targeted manipulation for mouse embryonic research, *J. Biomed. Opt.*, **20**(5), pp. 051020–051020.

180. Reguera, J., Jimenez de Aberasturi, D., Winckelmans, N., Langer, J., Bals, S. and Liz-Marzan, L. M. (2015). Synthesis of Janus plasmonic-magnetic, star-sphere nanoparticles, and their application in SERS detection, *Faraday Discuss.*, **191**, pp. 47–59.

181. Ju, Y., Zhang, H., Yu, J., Tong, S., Tian, N., Wang, Z., Wang, X., Su, X., Chu, X. and Lin, J. (2017). Monodisperse Au-Fe2C Janus nanoparticles: an attractive multifunctional material for triple-modal imaging-guided tumor photothermal therapy, *ACS Nano*, **11**(9), pp. 9239–9248.

Chapter 8

Importance of Hexagonal Boron Nitride (hBN) Layers and Boron Nitride Nanotubes (BNNTs)

Nabanita Dutta
Nano Brik, Sunnyvale, CA 94086, USA
go2nabanita@gmail.com

8.1 Introduction

Nowadays, nanotechnology covers a broad spectrum of applications all over the globe. However, human welfare will be aided only after having been facilitated by their ease of technological benefits. Boron nitride offers a broad realm of the physics spectrum to investigate its novel behaviors. Development of high-quality 2D hexagonal boron nitride (hBN) layers and boron nitride nanotubes (BNNTs) in large-scale production brings opportunities to investigate the implications of their remarkable properties, which, in turn, facilitates tremendous technological benefits.

The physical property of hBN is analogous to graphite. hBN is basically an insulator and can be made semiconducting after doping

Hybrid Nanocomposites: Fundamentals, Synthesis, and Applications
Edited by Kaushik Pal

ISBN 978-981-4800-34-1 (Hardcover), 978-0-429-00096-6 (eBook)
www.panstanford.com

by Be, Mg, etc. It is a good dielectric. It possesses astounding chemical inertness and thermal stability up to a very high temperature. It gets a new dimension in materials science, while being developed in nanostructured forms like 2D atomic layers and nanotubes. hBN comprises alternating boron and nitrogen atoms in a honeycomb arrangement consisting of sp^2-bonded 2D layers held together by weak van der Waals forces. hBN layers can be peeled off from a bulk boron nitride crystal by micromechanical cleavage, but it is very hard to achieve. These layers structurally resemble graphene (Fig. 8.1).

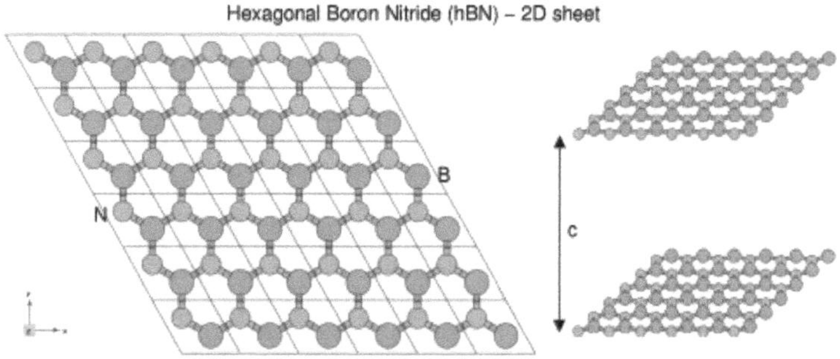

Figure 8.1 Stoichiometry of 2D hexagonal boron nitride. Image: http://www.yambo-code.org/wiki/index.php?title=First_steps:_a_walk_through_from_DFT_to_optical_properties.

BNNTs are structurally similar to carbon nanotubes (CNTs) (Fig. 8.2). Unlike CNTs, the bandgap of BNNTs does not depend upon helicity. Hydrogenated hBN/BNNTs are obtained by incorporation of hydrogen, where it is covalently bonded to boron, nitrogen, or both. Previous experiments and modeling suggest that the intercalation of molecules into a boron nitride interlayer space is much more difficult compared to that in graphite. Bandgap engineering is being pursued in hydrogenated hBN to obtain some desired physical properties. The uniqueness of hBN nanosheets/BNNTs makes them useful in various fields such as optoelectronic nanodevices, field emitters, hydrogen accumulators, electrically insulating substrates, etc., and their composites have a range of applications starting from wafers to spacecraft material. Moreover, BNNTs also have promising biomedical applications [1–4].

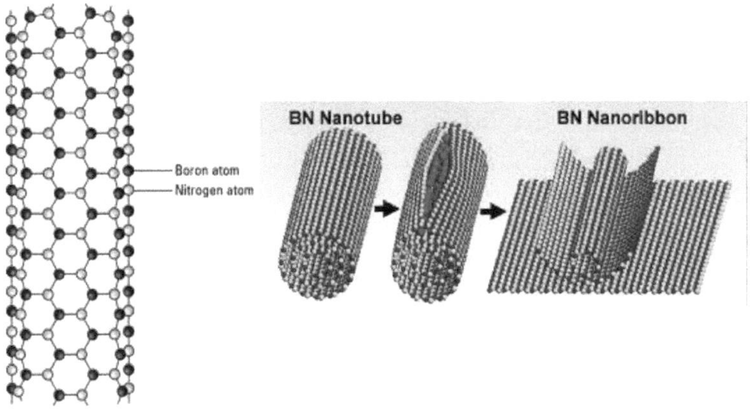

Figure 8.2 Stoichiometry of boron nitride nanotubes. (Left) *Nanotechnology for Dummies*, 2nd edition, courtesy of Earl Boysen et al. (Right) Courtesy of Alex Zettl, Department of Physics, University of California at Berkeley.

8.2 Development Methodology

Various fabrication processes have been adopted to develop 2D hBN and BNNTs such as ball milling, arc discharge, substitution, laser-based method, chemical vapor deposition (CVD), metal organic chemical vapor deposition. (MOCVD), and electron beam in situ deposition. Ajayan et al. at Rice University reported successful large-scale production of hBN films consisting of two to five atomic layers with a very high 2D elastic modulus, a wide optical energy bandgap, and high transparency over a broad wavelength range. A few layers of hBN can also be made by ultrasonication and also by high-energy electron beam irradiation. Furthermore, chemical decomposition of various precursors has been exploited to develop single-layer hBN domains over small areas. A group from NIMS, Japan, and NUAA, China, reported a kind of thermally and chemically stable nanoribbon production that shows insulator-semiconductor electrical transition behavior. A group from the University of California, Berkeley, developed a technique where defect-free boron nitride nanoribbons of uniform length and thickness are fabricated with similar advantages like graphene nanoribbons, along with an additional array of electronic, optical, and magnetic properties.

A group from Texas Tech University reported epitaxially grown semiconducting hBN layers as a deep ultraviolet photonic material. Monolayer hBN, so-called white graphene, could be useful as a complementary 2D dielectric substrate. Hydrogen incorporation into boron nitride layers is investigated through a neutron-scattering experiment. The wafer-scale semiconducting hBN epitaxial layers with high crystalline quality and electrical conductivity are highly desirable for optoelectronics to graphene electronics. P-type conductivity control is attained by in situ Mg doping. Dean et al. demonstrated such applications of hBN wafers with mobilities and carrier inhomogeneities that are almost an order of magnitude better than devices on SiO_2.

More consistent electronic properties than CNTs made BNNTs arouse significant interest. Tang et al. developed an effective CVD method for the large-scale synthesis of highly pure multiwalled BNNTs. Analysis of BNNTs indicates that they are growing by addition of atoms to the exposed ends but not at the substrate–nanostructure interface. Ma et al. found that the temperature gradient is the determining factor for the production of BNNTs. Huo et al. also proposed a stress-induced sequential growth model. Ming Xie et al. reported a mechanism for low-temperature growth (600°C–700°C) of BNNTs by plasma-enhanced pulsed laser deposition (PEPLD). Researchers from the Helsinki University of Technology and the Tampere University of Technology have attempted to calculate the energy cost to form bonds between individual boron and nitrogen atoms on a metal surface and summarized into a phase-selective growth model (vapor-liquid-solid [VLS]), which is guided by the theory of nucleation. Golberg et al. have shown that BNNTs have high oxidation resistance. Researchers at Northwestern University prepared single-walled BNNTs on tungsten substrates and are also trying to fabricate tailored BNNTs for technological applications. Radosavljevic et al. fabricated field-effect transistors exploiting the thermionic emission of single-walled BNNTs [5–11].

Two-dimensional hBN/BNNTs are characterized through various characterization tools like X-ray diffraction (XRD), scanning electron microscopy (SEM), high-resolution transmission electron microscopy (HRTEM) (Figs. 8.3 and 8.4), atomic force microscopy (AFM), X-ray photoelectron spectroscopy (XPS), and Raman spectroscopy. Basic characterization techniques like thermogravimetry and differential thermal analysis (TG/DTA), Fourier transform infrared (FTIR), and nuclear magnetic resonance (NMR) are generally performed for routine characterization. The optical phenomenon is evaluated through optical absorption and photoluminescence. Electronic properties are studied in a probe station inside an ultraclean room, which is essential to probe BNNTs individually by making contact with a focused ion beam. Mechanical properties of 2D hBN/BNNTs are mostly investigated through nanoindentation experiments. Moreover, radiation-shielding measurements are performed by the tools of neutron radiation exposure. However, these 2D hBN/BNNT nanostructures suffer from defects that furnish some additional properties to the material whereby substantial analysis of these defects is extremely desirable. These studies will give a platform of a new research domain in materials science.

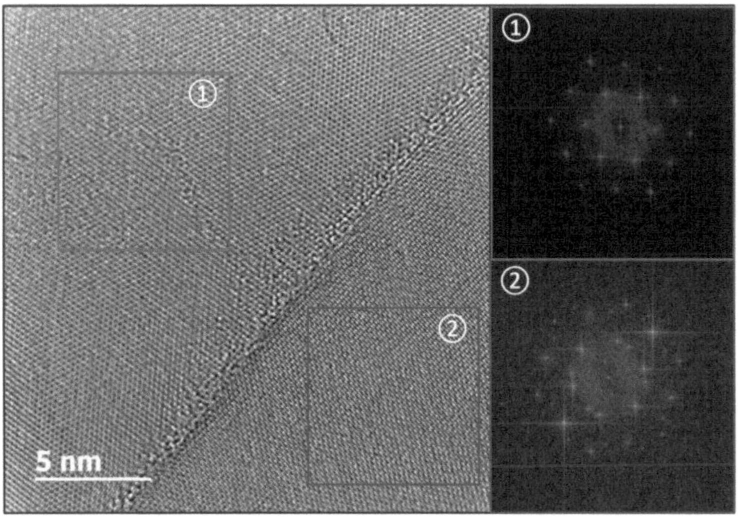

Figure 8.3 HRTEM of 2D hexagonal boron nitride. Courtesy: Andrieux et al., la Hunière, Palaiseau, France.

Being an indirect-bandgap material, phonon contribution is also expected to determine its performance. Boron compounds are known as Lewis acids since they accept a lone pair of electrons from a donor (Lewis base), which is basically a nitrogen compound such as ammonia (NH_3); thereby, using this strategy, it is possible to design some nanovectors for targeted drug delivery or gene therapy by decorating the surface of BNNTs with a desired drug or antibodies specific to the surface-expressed antigens [12].

Figure 8.4 SEM of boron nitride nanotubes [13]. Courtesy: Dr. Zhi Chunyi City University of Hong Kong.

It is worth mentioning that among all development protocols, ambient pressure chemical vapor deposition (APCVD) is considered one of the best fabrication techniques for the growth of hBN/BNNTs. APCVD seems to be ideal for boron nitride, it being a binary compound, whereby few synthesis parameters are needed to optimize the growth condition as well as stoichiometry. APCVD offers some possibilities to reduce the growth temperature. It is essential for hBN growth to overcome the substrate contamination effect that may occur at high temperature. APCVD offers significant advantages obtaining large-area hBN films/BNNTs with more purity compared to other wet-chemistry methods. The key feature of these processes is the reaction of nitrogen with boron to produce boron nitride with a small amount of hydrogen incorporation. The growth of these layers also depends on the orientation of the substrate and stress factors. In this process, generally the synthesis of 2D hBN is conducted in a split tube furnace with a fused quartz processing tube in an inert gas atmosphere, which is better compared to vacuum for hBN growth. Since the deposition rate is affected by the molar ratio of boron and nitrogen, in this method, the use of a single precursor like ammonia borane (NH_3-BH_3) shows many advantages over other precursors due to the 1:1 B/N ratio to obtain stoichiometric hBN layers. Moreover, it is relatively less toxic. Furthermore, in the case of BNNTs also, the same precursor borazine is preferable. However, a metal catalyst and a pristine high-vacuum environment are needed in the case of BNNTs. The role of a metal catalyst is to provide nucleation centers for boron/nitrogen in order to grow BNNTs at a large temperature gradient by creating some condensed spots. The growth temperature and catalyst concentration affect the morphology of BNNTs. Besides, the most crucial challenge lies in the purification of hBN films/BNNTs.

8.3 Utilization and Applications

Low-*Z* elements possess a significantly large neutron absorption cross section, whereby a low-*Z*, structural, and lightweight material is needed to build a radiation-shielding coating to protect spacecraft from high-energy secondary neutrons generated from the interactions of galactic cosmic rays and solar energy particles with the nuclei in the earth's atmosphere. The absorption cross section is governed by the following factors: the number of electrons per unit volume, electronic mean excitation energy, and tight binding corrections for the inner shell electrons. As a consequence of interplay of these factors, hydrogen has the highest neutron absorption cross section, boron has a large neutron absorption cross section, and nitrogen has also a fairly high neutron absorption cross section compared to carbon. Accordingly the shielding property of hydrogenated 2D hBN/BNNTs is expected to be very high due to the coordinated participation of hydrogen, boron, and nitrogen. The addition of hydrogenated hBN into a polymer matrix leads to a composite with minimal weight penalty and improved structural integrity. Basically, these are transparent, high-temperature aromatic polyamides. This type of composite is ideal for a radiation-shielding coating owing to its huge heat absorption capacity and inflammability. Due to a large hydrogen content, high-density polyethylene, a polyamide composite, is extensively used in such coatings. Another advantage of these composites is that they can forbid the generation of high-energy X-rays. These X-rays are generated while high-energy electrons hit the metallic body of a spacecraft [14, 15]. The radiation-shielding measurements are performed with the tools of neutron radiation exposure followed by Radiation Shielding Materials Evaluation

Software (RSMES). Furthermore, these composite films are needed to be tested in order to optimize the ratio of BNNTs/polyamide to obtain maximum protection and mechanical strength.

The inhibition of cancer cells in boron neutron capture cancer therapy based on molecular recognition of boron–cancer cell interaction hypothesizes the idea of clubbing BNNTs with the irreversible electroporation (IRE) therapeutic process. IRE is an electrical system device methodology whereby a standard value of electrical field is applied to induce a characteristic nanopore over the target cell (biological system), leading to the death of the target cell. In an IRE-BNNT clubbed system, the resulting pore over the cell surface aids BNNTs to be inserted into the desired target cell to achieve the characteristic change owing to boron–target cell interaction upon application of a desired high voltage. Exploitation of BNNTs as nanovectors of boron atoms to address targeted cancer cells in the IRE process can enhance the effective inhibition of cancer cells in IRE therapy up to a large extent. It is likely to obtain a concentration-dependent activity of BNNTs toward cancer cells. To address particularly Rb, p53 gene of a chosen cancer cell line, BNNT nanovectors could be a promising candidate. This class of genes remains activated in normal healthy cells, whereas in cancerous cells it gets altered and thus cancer cells undergo unrestricted growth and proliferation [16]. This application is proposed to herald a medically significant novel therapeutic system for cancer intercalating BNNTs with IRE. This study is focused on the application of BNNTs over one of the easily cultured and maintained cancer cell lines (HeLa; commercially available) in vitro. It is hypothesized that along with higher values of the applied electric field of the IRE device, a selected dosage of BNNTs will assist in achieving marked therapeutic efficacy toward the targeted cancer cells. It is desirable to obtain a specific interaction of BNNTs with the surface-expressed antigens lying over the altered tumor suppressor genes (TSGs), which are present in the HeLa cell line. It is assumed that after the antigen–BNNT recognition is built, BNNTs may thereby initiate their potential work by turning the switched-off condition of TSGs to the switched-on condition and activating it. In a row, activated TSGs could, in turn, assist the immortal cancer cells of HeLa to be mortalized. Research on cancer therapy using IRE experiments is currently going on at MD Anderson Cancer Institute at Houston, Texas.

8.4 Conclusions and Outlook

Theoretical studies on hydrogenation of hBN suggest its stable conformers, which opens up tremendous opportunities of bandgap engineering to derive desired physical properties However, there is not much experimental verification of these results. Under these circumstances, methodical research on hydrogenated hBN to explore its physical properties would be highly appreciated.

A strong radiation-shielding coating that can overcome the adverse effects of high-energy neutron beam encounters is always desired for space missions. So far it is known that the highly energetic charged particles themselves have radiation hazards to humans (or any living tissue). Moreover another issue is that as high-energy electrons hit the metal surface of a spacecraft, they produce very strong X-rays and any living tissue cannot withstand this. NASA is looking for designing some materials to resolve this issue. The studies on physical properties of hBN/polyamide provide genuine information to fabricate such materials in the future. This composite is expected to have appreciable technological benefits.

Moreover, IRE-mediated delivery of antibody-coated BNNTs could emerge as a potential candidate for effective cancer treatment. The primary studies on BNNT/IRE-clubbed experiments will guide to design such cancer therapeutic systems that can address late-stage cancer. Attempts will be highly appreciated to analyze the underlying mechanism of this event to decode the interaction of BNNT/cancer genes.

References

1. Alem, N., Erni, R., Kisielowski, C., Rossell, M. D., Gannett, W. and Zettl, A. (2009). *Phys. Rev. B*, **80**, p. 155425.
2. Rubio, A., Corkill, J. L. and Cohen, M. L. (1994). *Phys. Rev. B*, **49**, pp. 5081–5084.
3. Hod, O., Barone, V. and Scuseria, G. E. (2008). *Phys. Rev. B*, **77**, p. 035411.
4. Ciofani, G., Raffaa, V., Menciassia, A. and Cuschieria, A. (2009). *Nano Today*, **4**, pp. 8–10.
5. Kobayaashi, Y., Kumakura, K., Akasaka, T. and Makimoto, T. (2012). *Nature*, **484**, p. 223.

6. Zeng, H., Zhi, C., Zhang, Z., Wang, X., Guo, W., Bando, Y. and Golberg, D. (2010). *Nano Lett.*, **10**, pp. 5049–5055.
7. Dean, C. R., Young, A. F., Meric, I., Lee, C., Wang, L., Sorgenfrei, S., Watanabe, K., Taniguchi, T., Kim, P. and Shepard, K. L. (2010). *J. Hone Nat. Nanotechnol.*, **5**, p. 722.
8. Song, L., Ci, L., Lu, H., Sorokin, P. B., Jin, C., Ni, J., Kvashnin, A. G., Kvashnin, D. G., Lou, J., Yakobson, B. I. and Ajayan, P. M. (2010). *Nano Lett.*, **10**(8), pp. 3209–3215.
9. Dahal, R., Li, J., Majety, S., Pantha, B. N., Cao, X. K., Lin, J. Y. and Jiang, H. X. (2011). *Appl. Phys. Lett.*, **98**, p. 211110.
10. Lee, Y.-H., Liu, K.-K., Lu, A.-Y., et al. (2012). *RSC Adv.*, **2**, p. 111.
11. Huo, K. F., Hu, Z., Fu, J. J., Xu, H., Wang, X. Z. and Lu, Y. N. (2003). *J. Phys. Chem. B*, **107**, p. 11316.
12. Ferreira, T. H., da Silva, P. R. O., dos Santos, R. G. and Barros de Sousa, E. M. (2011). *J. Biomater. Nanobiotechnol.*, **2**, pp. 426–434.
13. Zhi, C. Y. (2013). *JSM Nanotechnol. Nanomed.*, **1**(1), p. 1005.
14. Smith, M. W., Jordan, K. C., Park, C., Kim, J. W., Lillehei, P. T., Crooks, R. and Harrison, J. S. (2009). *Nanotechnology*, **20**, p. 505604.
15. Harrison, C., Weaver, S., Bertelsen, C., Burgett, E., Hertel, N. and Grulke, E. (2008). *J. Appl. Polym. Sci.*, **109**, pp. 2529–2538.
16. Raffa, V., Riggio, C., Smith, M. W., Jordan, K. C., Cao, W. and Cuschieri, A. (2012). *Technol. Cancer Res. Treat.*, **5**, pp. 459–465.

Chapter 9

Natural Polymer-based Bionanocomposites as Smart Adsorbents for Removal of Metal Contaminants from Water

Anamika Kalita[a] and Pranjal Barman[b]
[a]*Physical Science Division, Institute of Advanced Study in Science and Technology, Pachim Boragaon, Guwahati 781035, Assam, India*
[b]*Department of Electronics and Communication Engineering, Tezpur University, Napam, Tezpur 784028, Assam, India*
anamik.kalita01@gmail.com; pranjalele@gmail.com

9.1 Introduction

Pollution by heavy metals is one of the major environmental concerns of modern society because of their persistent and bioaccumulative nature unlike organic contaminants. They are nonbiodegradable and highly water solubile that facilitates their interactions and accumulation by living organisms, threatening human health as well as ecosystems. Furthermore, many of these heavy metals are known to have carcinogenic nature [1]. Thus, there is a high demand for

Hybrid Nanocomposites: Fundamentals, Synthesis, and Applications
Edited by Kaushik Pal

ISBN 978-981-4800-34-1 (Hardcover), 978-0-429-00096-6 (eBook)
www.panstanford.com

effective removal and treatment of heavy metals from wastewater to protect the human population and the environment. Several heavy metals like Zn, Cu, Ni, Hg, Cd, Pb, As, and Cr are particularly important in the treatment of industrial wastewaters [2–4]. Various methods have already been utilized for the removal of harmful metal ions. These conventional methods like chemical precipitation, ion-exchange filtration, flotation, electrochemical treatment, reverse osmosis, membrane technologies, and evaporation [5–7] suffer from a major disadvantage, production of toxic chemical sludge, whose treatment becomes a costly affair and is not eco-friendly. Therefore, considerable attention has been devoted to study the removal of heavy metal ions from water by adsorption using eco-friendly materials in an environmentally safe level with cost-effectiveness [8].

The major challenges of modern advancement in polymer technology is to find solutions to the problems of plastic waste generation using the diminishing natural resources that are nonrenewable [9]. Naturally occurring polymers (biopolymers) offering interesting properties of biocompatibility and biodegradability represent a new approach to obtain structural and functional materials that are provided with exceptional properties in the development of nanocomposites. Among various biopolymers, starch, cellulose, chitin, and chitosan (CS) (Fig. 9.1) are an abundant and widespread group of biomacromolecules acquired from renewable sources that can be used for preparation of a large variety of composite systems. These nanocomposite–polymer matrix materials are of great interest for removal of metals due to the functional groups of the polymeric matrixes that provide specific bindings to target pollutants.

Bionanocomposites (BNCs) are considered as a 21st-century emerging group of nanostructured hybrid materials designed from natural biodegradable polymers and organic/inorganic fillers having dimensions in the nanorange [10, 11]. Apart from conventional nanocomposites, based on synthetic polymers, biohybrid materials exhibit improved structural and functional properties of great interest with a wide number of applications, including drug delivery, biosensors, tissue engineering, environmental remediation, etc. [12–15]. Due to the accessibility of a large variety of biopolymers, fillers, and ease of their processing, these hybrid composites display versatility in nature. Biopolymers have inherent properties like biocompatibility and biodegradability that open up a new prospect for these hybrid materials with special incidence in environmentally friendly materials. Research on BNCs can be regarded as a new interdisciplinary field representing a promising research topic that takes advantage of the synergistic assembling of biopolymers with inorganic solids, introducing multifunctionality in a single system. This chapter explores the new avenues of BNC hybrid materials where they may be exploited to address certain unanswered issues that are pertaining to environmental remediation, that is, water treatment via removing pollutants from water bodies.

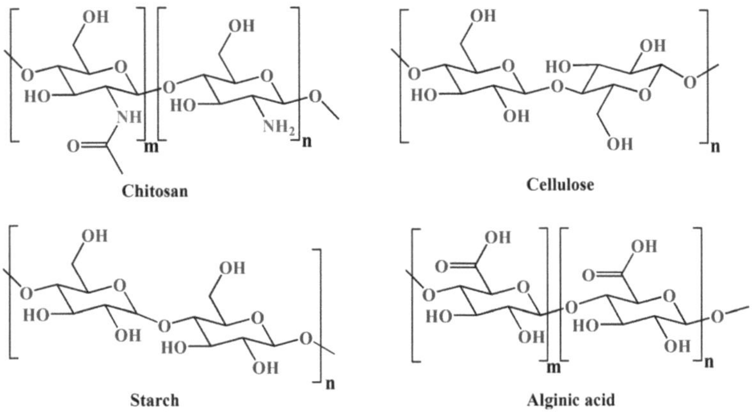

Figure 9.1 Structures of some selected biopolymers.

9.2 Removal of Metal Contaminants from Water Using Bionanocomposites

9.2.1 Chitosan-Based Bionanocomposites

Two-dimensional materials based on graphene have been explored for treatment of wastewater containing heavy metals. Lee et al. [16] reported ethylenediaminetetraacetic acid (EDTA)-functionalized magnetic chitosan (MCS) graphene oxide (GO) nanocomposites (EDTA-MCS/GO) using a reduction precipitation method, and this hybrid system is applied to the efficient removal of divalent (Pb(II) and Cu(II) and trivalent (As(III)) metal ions from aqueous solutions.

GO, a carbonaceous nanomaterial, has attracted the attention of researchers because of its unique properties like the presence of consecutive oxygen functional groups and a large surface area, that make GO an excellent adsorbent for the removal of heavy metal contaminants [17]. Moreover, GO with a large surface area provides abundant binding sites for other compounds, in this case CS and EDTA, which ultimately leads to an increased number of surface functional groups, and that might boost its metal adsorption activity. The primary amino groups of CS are easily functionalized with different organic ligands, like GO and EDTA, to improve its adsorption capacity [18, 19]. Moreover, EDTA is considered as a good candidate for the adsorption of heavy metals via chelation and can be functionalized with other materials such as GO and CS. Magnetic properties were also incorporated into the nanocomposites to allow the used absorbent to be easily recovered from an aqueous solution after the adsorption of heavy metals. Owing to the large specific surface area, hydrophilic behavior, and functional moieties, the magnetic hybrid nanocomposite demonstrated excellent removal ability with a maximum adsorption capacity of 206.52, 207.26, and 42.75 mg/g for Pb(II), Cu(II), and As(III), respectively. The nanocomposite was reused in four successive adsorption–desorption cycles, revealing a good regeneration capacity. Efficient removal of metal ions in real wastewater was also achieved by using this adsorbent system.

Development of magnetic nanoparticles, particularly iron oxide, provides a convenient approach toward exploring magnetic separation techniques. They have the capability to treat large amounts of wastewater within a short time. Moreover, they can be functionally tuned by using polymers, novel molecules, or inorganic materials to further impart surface reactivity [20, 21]. Liu et al. [22] reported the fabrication of MCS nanoparticles, as shown in Fig. 9.2, and their adsorption capacity for heavy metal ions such as Pb(II), Cu(II), and Cd(II) ions.

Magnetic nanoparticles were added as a glutaraldehyde solution and the suspension was irradiated by ultrasonic waves to obtain carbonyl-magnetic nanoparticles. MCS nanoparticles were achieved by adding the carbonyl-magnetic nanoparticles to a CS solution with intensive stirring. Finally the prepared MCS nanoparticles were investigated for efficient removal of Pb(II) using an external magnetic field, as shown in Fig. 9.3.

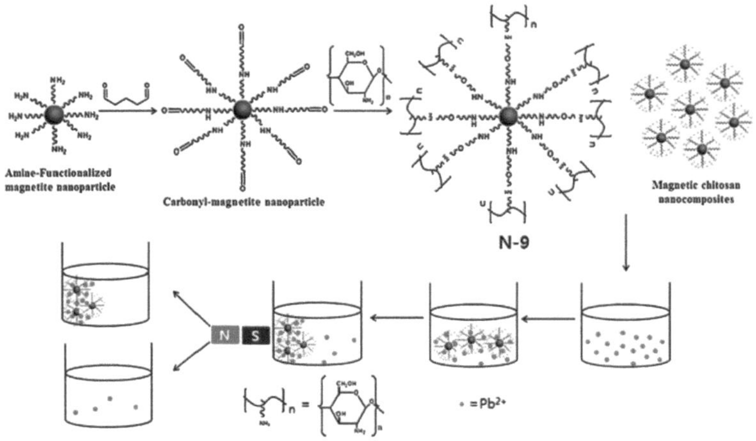

Figure 9.2 Synthetic route of magnetic chitosan nanocomposites (N-9) and their use as a facile tool for Pb(II) removal with the help of an external magnetic field. Reprinted with permission from Ref. [22]. Copyright (2009) American Chemical Society.

Figure 9.3 Photographs of a magnetic chitosan nanocomposite colloidal solution containing 10 mL/L of Pb(II) (a) before and (b) after magnetic separation by an external magnetic field. Reprinted with permission from Ref. [22]. Copyright (2009) American Chemical Society.

To stimulate rapid interaction between the MCS nanoparticles and Pb(II) ions, ultrasound radiation was employed to disperse the MCS nanoparticles into the Pb(II) ion solution. The concentration of free Pb(II) remaining in solution after 10 min. of ultrasound radiation declined from 10 to 0.54 mg/L. The efficiency of lead ion removal was ~94.6%. The concentration of residual Pb(II) could be decreased further by increasing the duration of sonication. In addition, the adsorption capacities of CS-immobilized nanoparticles for Cu(II) and Cd(II) were 1.09 and 0.79 mg/L, respectively. The CS-immobilized nanoparticles were also treated with deionized water to neutralize the solution and then were tested for Pb(II) removal in subsequent cycles.

Again Tran and coworkers [23] reported a new platform, CS/magnetite nanocomposite beads that were proven to be an effective adsorbent for removal of toxic metal ions such as Pb(II), Ni(II), etc. The adsorption parameters demonstrated good compatibility with the Langmuir model and adsorption capacities of CS/magnetite composite beads reached a maxima at pH 6.0 for both Pb(II) and Ni(II). The metals ion adsorption on the surface of the CS/magnetite composite was observed via experimental techniques like scanning electron microscopy (SEM) and energy-dispersive X-ray spectroscopy (EDS). The high saturation magnetization value (M_s) after coating with CS made these beads advantageous for heavy metal ion removal from water with the help of an external magnet.

CS is a cost-effective, biocompatible, biodegradable, and nontoxic biopolymer and carries amino and hydroxyl groups along its backbone that make it a good sorbent for heavy metals. Therefore, it can be used as a matrix for the preparation of BNCs [24, 25], while MnO_2, the most attractive inorganic material, acts as a filler because of its considerable ion-exchange and molecular adsorption properties [26, 27]. Among various methods, the hypothermal method has been facilely used to synthesize various shapes of MnO_2 nanostructures (e.g., nanoparticles, nanocubes, nanorods, and nanotubes) and different crystallographic forms (e.g., α, β, γ, and δ forms) [28, 29]. To uniformly disperse α-MnO_2 nanorods within the CS matrix, α-MnO_2 nanorods were modified with L-valine by a solvothermal approach. Among various morphologies and crystallographic forms, the α-MnO_2 nanorod structure is very promising according to the literature for the intercalation phenomenon, that is, a potential candidate for the adsorption of heavy metal ions [30, 31]. By simultaneously taking advantage of the CS and MnO_2 nanorod as well as its tunnel structure,

Mallakpour et al. [32] developed an ideal BNC, that is, CS/MnO_2, as shown in Fig. 9.4, with great potential for use in environmental remediation. Using this adsorbent, it is possible to take advantage of both the natural adsorbent and the nanostructure for heavy metal removal, while also controlling nanomaterial emission to the environment. The CS/α-MnO_2-valine BNC was used as a potential adsorbent for removal of Pb(II) ions from aqueous solutions.

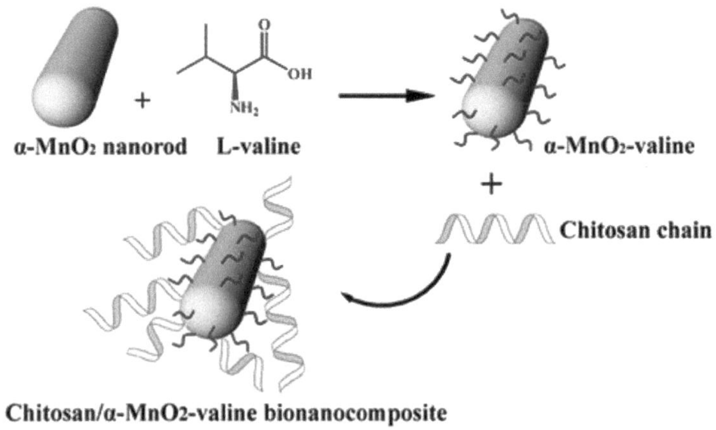

Figure 9.4 Schematics of prepared chitosan/α-MnO_2-valine bionanocomposite. Reprinted with permission from Ref. [32]. Copyright (2016) American Chemical Society.

Tao et al. [33] demonstrated removal of Pb(II) from an aqueous solution utilizing a CS/TiO_2 hybrid film (CTF). The CTF was synthesized by the sol–gel method where the CTF was grafted into the Ti–O group on the CS backbone. The CTF exhibited high adsorption ability toward Pb(II). It illuminates that this model is reliable to optimize the adsorption process and the CTF is suitable for adsorbing Pb(II) from an aqueous solution. Similarly Zimmerman and coworkers reported a novel biobased sorbent, TiO_2-impregnated chitosan beads (TICBs), as an arsenic adsorbent, shown in Fig. 9.5. TICBs are well reported for removal of arsenite and arsenate and oxidize arsenite to arsenate in the presence of UV light and oxygen [34].

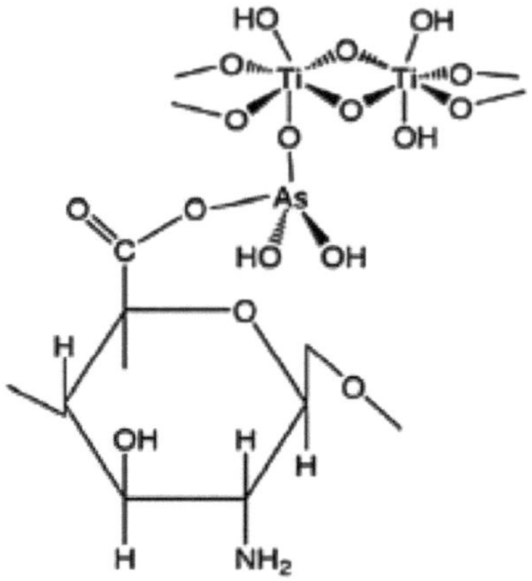

Figure 9.5 Schemitic representation showing complexation of arsenate by TiO_2 and oxidation product of chitosan. Reprinted from Ref. [34], Copyright (2011), with permission from Elsevier.

Though natural polymers have an eco-friendly nature and they show poor performance compare to synthetic polymers. Therefore, the development of an organic–inorganic hybrid BNC is required that can possibly have the best properties like a hydrophilic matrix of CS and the adsorption capacity of the silicate layers of the organoclay filler. Besides, CS possesses an extraordinary ability to uptake heavy metal ions, whereas layered silicates have a high specific area for adsorption with a limitation of low metal-binding constant [35, 36]. Mishra et al. [37] developed a BNC, as shown in Fig. 9.6, based on CS and clay (Cloisite 10A) with combined properties of hydrophilicity of an organic polycation and adsorption capacity of inorganic polyanion. The chitosan/clay nanocomposite (CCN) was prepared by the solvent-casting method. Addition of small amounts of montmorillonite–Na+ (MMT–Na+) to the CS matrix makes the composite highly efficient for the removal of Cr(VI) from an aqueous solution under optimized conditions.

Charlet et al. [38] reported a novel material, chitosan goethite (α-FeOOH), that is, chitosan-iron (oxyhydr)oxide (CGB), BNC beads for effective removal of both inorganic As(III) and As(V) from water. A small amount of CGB can purify high arsenic–contaminated water to a potable level by forming inner-sphere complexes between arsenic species and goethite nanoparticles. Both As(III) and As(V) can penetrate the entire CGB material, which indicates that CGB has a higher arsenic removal capacity by comparison with conventional adsorbents that can only be partially removed. CGB also showed other outstanding properties like facile one-pot synthesis without using toxic chemicals and low difficulty and cost in filtration; the large-scale bead (1 mm) is more easily removed than fine nanoparticle powders that can be handled easily without high-pressure membrane filtration or other energy-consuming separation techniques and finally desirable mechanical properties. When considering a cost-effective facile (green) synthetic route, CGB is a promising material for arsenic remediation, particularly in developing countries, which suffer with regard to water purification and sanitation.

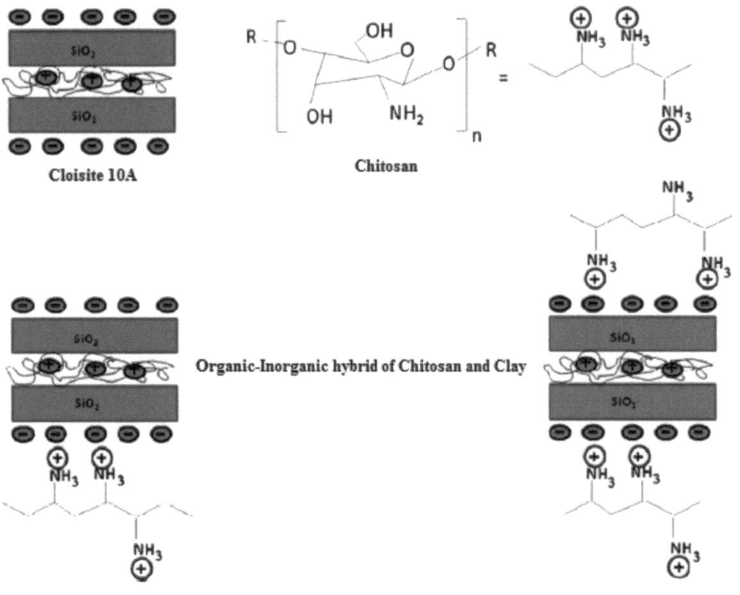

Figure 9.6 Organic–inorganic hybrid of chitosan and clay. Reprinted from Ref. [37], Copyright (2011), with permission from Elsevier.

Yang et. al. [39] reported a novel polyacrylamide-grafted chitosan magnetic composite microsphere (CS-PAM-MCM) prepared by a simple method and studied its application as an efficient adsorbent for the removal of Cu(II), Pb(II), and Hg(II) ions from aqueous solutions. In addition, a chitosan magnetic composite microsphere (CS-MCM) without modification was prepared for comparison. During the synthesis of CS-PAM-MCM, the magnetic Fe_3O_4 nanoparticles were further coated by a layer of silica to improve its acid duration to generate silica-coated nanoparticles ($Fe_3O_4@SiO_2$). Compared to CS-MCM without modification, CS-PAM-MCM showed improved adsorption capacity for each metal ion and highly selective adsorption for Hg (Fig. 9.7) from Pb and Cu. This improvement is attributed to the formation of stronger interactions between Hg and the amide groups of PAM branches for chelating effects. Furthermore, these MCS-based adsorbents could be easily regenerated in an EDTA aqueous solution and reused/recycled, which displayed its importance in real applications.

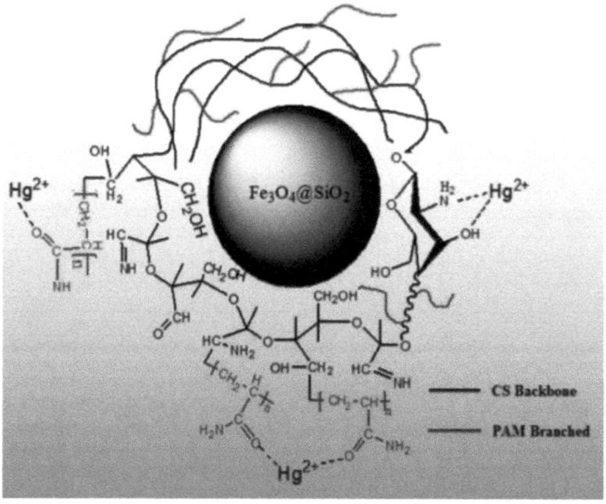

Figure 9.7 Schematic representation showing adsorption of Hg (II) onto CS-PAM-MCM. Reprinted from Ref. [39], Copyright (2015), with permission from Elsevier.

9.2.2 Cellulose-Based Bionanocomposites

Tong et al. [40] established a facile, green pathway to prepare composite materials containing magnetic nanoparticles and cellulose by the one-step co-precipitation method using a NaOH–thiourea–urea aqueous solution for cellulose dissolution. The Fe_2O_3 nanoparticles were uniformly dispersed in the cellulose matrix due to the strong interaction between cellulose and Fe_2O_3 nanoparticles (Fig. 9.8). The resultant cellulose@Fe_2O_3 composites exhibited outstanding adsorption efficiency of arsenic compared to other magnetic materials reported. The nanocomposite exhibited a sensitive magnetic response and superparamagnetic behavior with an external magnetic field and could be easily separated from an aqueous solution using an external magnetic field. The Langmuir adsorption capacities of the composites for the removal of As(III) and As(V) were 23.16 and 32.11 mg/g, respectively. Moreover, the adsorption capacities of arsenic were less affected by coexisting ions.

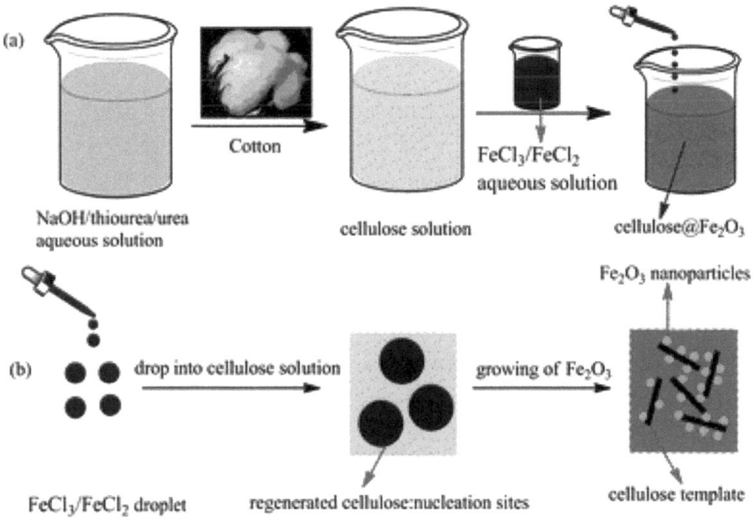

Figure 9.8 Schematic representation of formation procedure for (a) cellulose@ Fe_2O_3 composites and (b) growing of Fe_2O_3 nanoparticles. Reproduced from Ref. [40] with permission of The Royal Society of Chemistry.

Similarly Yu et al. [41] reported environmentally friendly, magnetic, millimeter-scale cellulose-based beads with micro- and nanopore structures fabricated via an optimal extrusion dropping technology from a NaOH/urea aqueous solution. Functional fillers of carboxyl-decorated Fe_3O_4 nanoparticles and nitric acid modification of activated carbon imparted to the beads a convenient, operating-based, sensitive magnetic response and highly effective adsorption performance for Cu(II), Pb(II), and Zn(II). Adsorption experiments showed that these adsorption processes were spontaneous endothermic reactions controlled by combining physical and chemical adsorptive mechanisms. The adsorption of Cu(II), Pb(II), and Zn(II) by MCBs is predominantly by electrostatic attraction between the adsorbent's surface and heavy metals. A simple preparation procedure, cheap cellulose feedstock, their availability on an industrial scale, fast adsorption speed, great adsorption capacity, and good reusability make the beads an economically viable and environmentally friendly cellulose-based adsorbent for highly efficient removal of the tested heavy metal ions from the aqueous environment. For the removal of heavy metal ions from water, new resources should be exploited to design more efficient, environmentally friendly adsorbents.

Sarkar and coworkers [42] successfully prepared a novel biobased adsorbent cerium-loaded cellulose nanocomposite bead (CCNB) and efficiently employed it for adsorptive removal and recovery of As(V) from synthetic and field samples of arsenic-affected areas. The adsorbent CCNB was synthesized via the sol–gel method. The validation of the experimental design for adsorptive recovery of aqueous As(V), applicability in real samples, and recycling of CCNBs were demonstrated along with the mechanistic pathway of As(V)–CCNB interaction, as shown in Fig. 9.9.

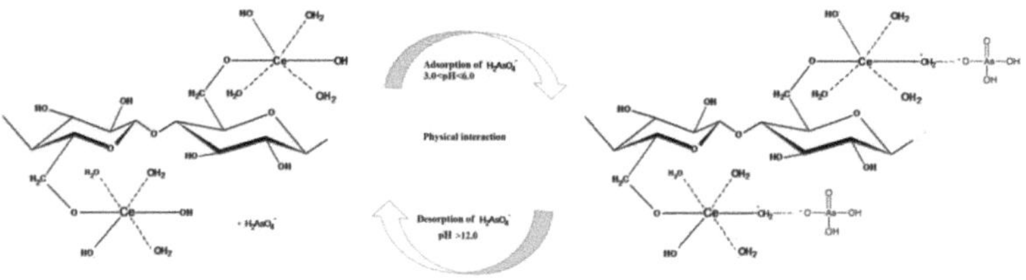

Figure 9.9 Schematic representation showing the adsorption desorption pathway. Reprinted from Ref. [42], Copyright (2016), with permission from Elsevier.

9.2.3 Starch-Based Bionanocomposites

Ruiz-Hitzky et al. [43] developed functional BNCs shown in Fig. 9.10, based on cationic starch (CST)/clay by assembling the modified biopolymer to two-layered silicates, commercial Cloisite®Na and a purified bentonite. Neutral polymers like starch are commonly used to prepare green nanocomposites due to their cost-effectiveness and abundant nature. Structural modifications can be done by covalently grafting functional groups, which provides improvement in adsorption capacity [44, 45]. Moreover the presence of silicate layers within the BNC material is essential to increase the stability of the CST in water and to allow its easy recovery from an aqueous solution after the adsorption process. Introduction of quaternary ammonium groups offers anion-exchange properties and it is also helpful to produce more stable adsorbents, favoring the intercalation of the polysaccharide chains in the layered silicates compared to the neutral starch [46, 47].

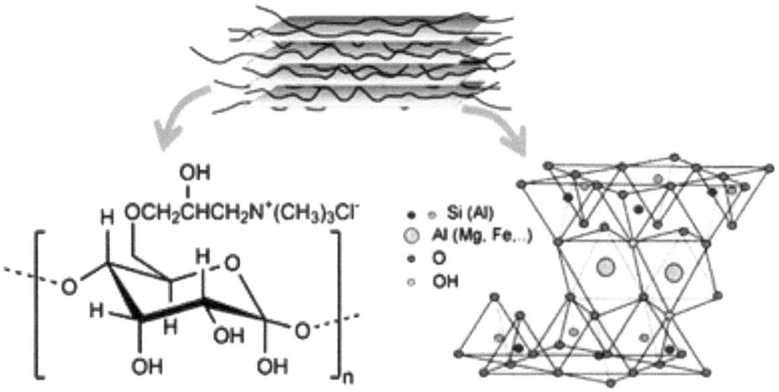

Figure 9.10 Bionanocomposites based on the intercalation of cationic starch functionalized with quaternary ammonium groups in the layered silicate montmorillonite. Reproduced from Ref. [43] with permission of The Royal Society of Chemistry.

The developed platform was utilized to evaluate the adsorption capacity of hexavalent chromium anions, Cr(VI), from an aqueous solution. The adsorption process was also well described by the Langmuir isotherm model, and the kinetic data were fitted by the pseudo-second-order model. The efficiency of the adsorption process was also successfully proved in the presence of competing anions, such as such as NO_3^-, ClO_4^-, SO_4^{2-}, and Cl^-, as shown in Fig. 9.11. The degree of interference follows the sequence $SO_4^{2-} > H_2PO_4^- > ClO_4^- > Cl^- > NO_3^-$, which is in agreement with the hydrated radii and charges of the different anions except for the phosphate species. The regenerating of the adsorbents for reusing them in several adsorption cycles was also possible with this system. These results show the potential interest of this type of low-cost biosorbents based on layered silicates of different origin in the removal of pollutants with potential interest in countries with the tannery industry.

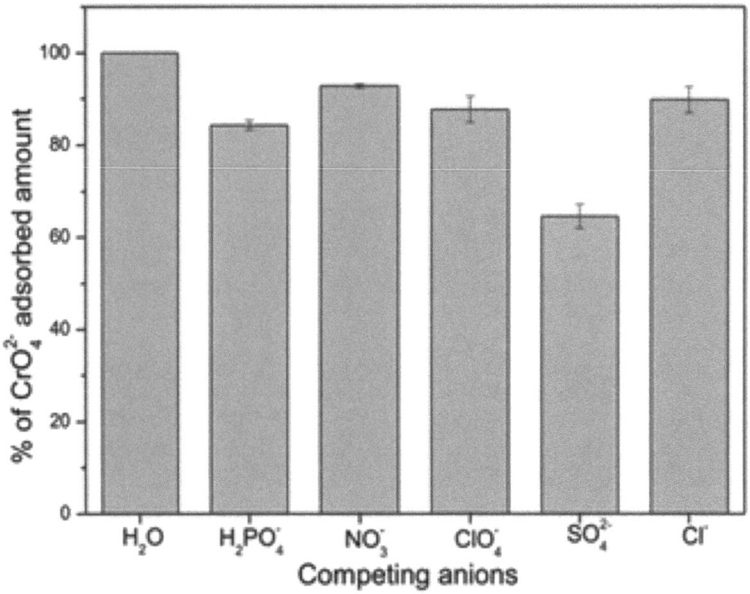

Figure 9.11 Effect of competing anions on chromate adsorption. Reproduced from Ref. [43] with permission of The Royal Society of Chemistry.

Naushad and coworkers [48] reported a starch-based nanocomposite (starch/SnO_2), as shown in Fig. 9.12, synthesized via a simple sol–gel method. The prepared starch/SnO_2 nanocomposite was used for the removal of Hg(II) from an aqueous medium. The experimental results showed that the starch/SnO_2 nanocomposite had a high ability to remove Hg(II) ions from an aqueous medium. The adsorption of Hg(II) was maximum at pH 6 and the maximum adsorption capacity was found to be 333 mg/g at 25°C. The adsorbed Hg(II) metal ions could be successfully desorbed using 0.1 M HCl solution.

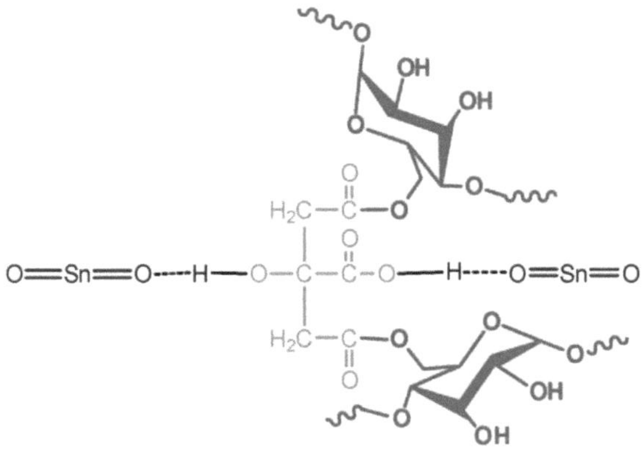

Starch-SnO_2 composite (STO)

Figure 9.12 Schematic representation of starch/SnO_2nanocomposite material. Reprinted from Ref. [48], Copyright (2016), with permission from Elsevier.

9.2.4 Alginate-Based Bionanocomposites

Alginate (G) is a nontoxic, biocompatible, and biodegradable biopolymer that is extracted from brown seaweeds. It consists of blocks of 1–4 linked α-l-guluronic and β-d-mannuronic acids [49]. In the presence of divalent cations, especially Ca^{2+} ions, G can easily form crosslinked gel matrices. Therefore, these Ca-crosslinked G matrices can be used to prepare adsorbents in gel phase, which are easier to handle than the powder materials [50]. The use of G-based formulations as efficient adsorbents is related particularly to the presence of carboxylic groups in the G structure, which enable it to form complexes with metal ions in aqueous solutions [51, 52]. El-Sherbiny et al. [53] synthesized cobalt ferrite (CF) nanoparticles and titanate (T) nanotubes and developed a new series of G-based nanocomposite microparticles (CF/G and T/G). The developed nanomaterials and their nanocomposite microparticles were investigated as potential adsorbents for efficient removal of Cu(II), Fe(III), and As(III) ions from water. It is worth mentioning that both CF and T were synthesized and selected as nanofillers upon preparing G-based nanocomposites, not only because of a lack of previous studies reporting their use in removal of metal ions, particularly Fe(III), but also because CF and T were found to have tunable surface charges at different pH values, which enables the improvement of adsorption efficiency toward metal ions. The removal efficiencies for Cu^{2+} using G, CF, T, CF/G, and T/G were found to be 91%, 100%, 99.9%, 95%, and 98%, respectively, while those of Fe^{3+} removal were 60%, 100%, 100%, 60%, and 82%, respectively. Efficient removal of As^{3+} ions was also attained (98% upon using T nanoadsorbents).

Naturally occuring nontoxic biopolymer adsorbents prepared by G and its derivatives has been broadly investigated because of their renewability, sustainability, and biodegradability. Biopolymer composites incorporating inorganic nanoparticle such as Au nanoparticles have been receiving considerable attention as they improve the performance properties owing to their large specific surface area, excellent biocompatibility, easy preparation, and huge applications. Au nanoparticles are synthesized via green synthesis using glutathione and oxalic acid due to an environmentally benign nature and greater stability of nanoparticles. Glutathione acts as a capping and stabilizing agent, whereas oxalic acid acts as a reducing agent in the formation of gold nanoparticles. An environmentally friendly mineral nanofiller, mica, which is a layered aluminum silicate with reactive groups on its surface, has been incorporated to enhance the mechanical strength of BNCs. Ahmad and coworkers [54] presented the successful synthesis of eco-friendly novel G-Au-mica BNCs that were found to exhibit good adsorption capacity for the adsorption of Pb(II) and Cu(II) metal ions in a single-component system and Pb(II) in a Pb(II)+Cu(II) binary-component system from an aqueous solution. The adsorption capacity of Pb(II) is comparatively better than Cu(II). BNCs have proved to be an excellent adsorbent for the removal of heavy metal ions from industrial wastewater (electroplating and battery manufacturing wastewater) as well. The overall results suggest that the present novel BNCs proved to be a potential adsorbent for the removal of Pb(II)and Cu(II) from an aquatic environment. This study could also provide innovative insight into the removal of toxic heavy metals from wastewater.

Zirconium-based oxides are stable, nontoxic, and water insoluble; they are therefore attractive sorption materials in the field of water purification. G is known to have a strong affinity for metal ions. The encapsulation of particles using G beads by crosslinking with calcium ions is an eco-friendly method. This composite adsorbent can be used to remove cationic and anionic contaminants simultaneously from an aqueous solution. Kim et al. [55] developed a new adsorbent for the efficient removal of arsenic species (AsO_3^{3-}/AsO_4^{3-}) and Cu(II) by immobilizing zirconium oxide on alginate beads (ZOAB). The mechanism of Cu(II) and As(III,V) adsorption by ZOAB is shown in Fig. 9.13. Kinetics, isotherm sorption experiments, and effects of contact time, initial adsorbate concentration, and pH on the adsorption performance of ZOAB were examined using the developed sorbent. The sorption behavior of selected anions and cations in the binary system is also reported. The developed sorbent can potentially be utilized to simultaneously treat wastewater contaminated with cationic and anionic contaminants.

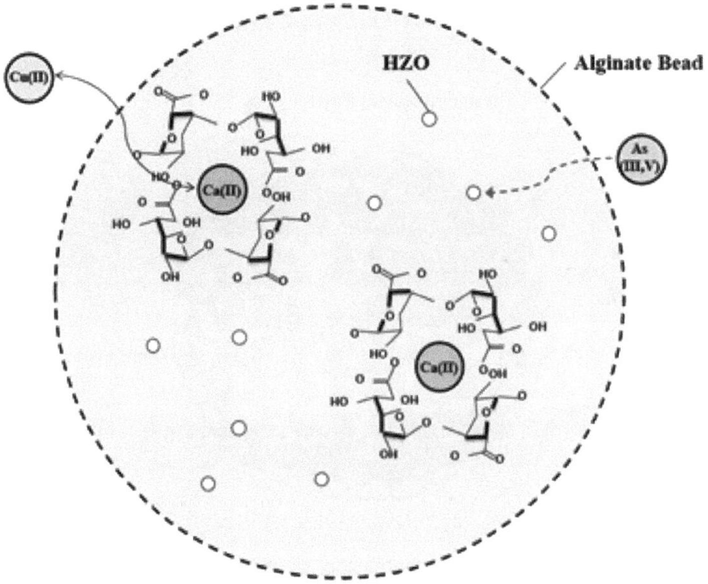

Figure 9.13 Schematic representation showing the adsorption of Cu(II) and As(III,V) by ZOAB. Reprinted from Ref. [55], Copyright (2016), with permission from Elsevier.

The chemical precipitation method was utilized for preparing amorphous hydrous iron oxides that were homogeneously dispersed into a G gel matrix to form composite beads. Park and coworkers [56] developed HIO-G beads for effective treatment of arsenic-contaminated water. Studies shows that As(III) removal by adsorption onto HIO-G beads was maximized at pH 6–9, while adsorption of As(V) was higher in an acidic solution than an alkaline solution. Studies were also performed to explore the effects of interfering ions on arsenic adsorption. The competitive effects of other anions, such as sulfate, bicarbonate, chloride, and nitrate, were insignificant, whereas phosphate showed a pronounced interfering effect, especially at high concentration. The regeneration studies showed that beads could be regenerated and reused for multiple cycles. Overall, HIO-G beads have potential for use as an effective adsorbent for arsenic removal from water.

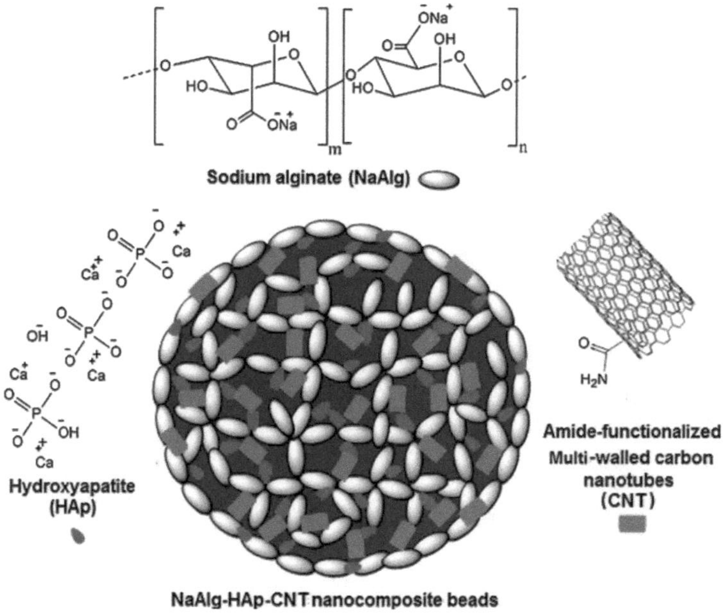

Figure 9.14 Schematic representation of NaAlg-Hap-CNT nanocomposite beads. Reprinted from Ref. [57], Copyright (2016), with permission from Elsevier.

The novel nanocomposite bead containing an amide group–functionalized multiwalled carbon nanotube (MWCNT-$CONH_2$) imprinted in the network of sodium alginate containing hydroxyapatite, (NaAlg-HAp-CNT) was prepared by Samandari et al. [57], as shown in Fig. 9.14, for effective removal of Co(II) ions from aqueous solutions. The combined advantages of both polymer and ceramic materials by incorporating CNTs on the network of NaAlg-HAp were investigated. The results show that the adsorption capacity of the nanocomposite beads was increased due to introducing CNTs with a large surface area into the network of NaAlg-HAp. The maximum adsorption capacity for Co(II) ions by prepared nanocomposite beads with the largest surface area of 163.4 m^2/g was reported to be 347.8 mg/g in the optimized condition. Therefore, the overall results proposed that the prepared nanocomposite adsorbent beads are highly efficient and cost effective and indicate their potentiality of practical application for metal removal in water treatment industries.

9.3 Conclusion

The concept of BNCs has been recognized as a trigger for the construction of novel and innovative materials with improved properties. This field of BNCs benefits from the functionality provided by the biopolymer/inorganic host solid/carbonaceous materials, with the possibility to develop synergistic interactions between all the components. Owing to their abundance, high strength and stiffness, low weight, and biodegradability, biopolymers like cellulose, CS, starch, etc., serve as promising candidates for the preparation of BNCs. In this chapter, our focus was on bionanohybrid materials with functionalities suited to playing an active part in the removal of metal contaminants from water sources. However, the present level of improvements is not enough to overcome issues like environmental remediation. Therefore, further development of BNCs is desirable to obtain ideal properties as well as to reduce cost in the production and processing of BNCs. In addition, significant research is still required to estimate the toxicity of the nanomaterials used, as well as their detrimental effects on the environment.

References

1. Turhanen, P. A., Vepsäläinen, J. J. and Peräniemi, S. (2015). Advanced material and approach for metal ions removal from aqueous solutions, *Sci. Rep.*, **5**, p. 8992.
2. Fu, F. and Wang, Q. (2011). Removal of heavy metal ions from wastewaters: a review, *J. Environ. Manage.*, **92**, pp. 407–418.

3. Sharma, S. K., Sanghi, R. and Mudhoo, A. (2012). *Advances in Water and Pollution Prevention*, Sharma, S. K. and Sanghi, R., eds. (Springer), pp. 1–36.
4. Huang, M.-R., Li, S. and Li, X.-G. (2010). Longan shell as novel biomacromolecular sorbent for highly selective removal of lead and mercury Ions, *J. Phys. Chem. B*, **114**, pp. 3534–3542.
5. Kurniawan, T. A., Chan, G. Y. S., Lo, W.-H. and Babel, S. (2006). Physico-chemical treatment techniques for wastewater laden with heavy metals, *Chem. Eng. J.*, **118**, pp. 83–98.
6. Puget, F. P., Melo, M. V. and Massarani, G. (2008). Comparative study of flotation techniques for the treatment of liquid effluents, *Environ. Technol.*, **25**, pp. 79–87.
7. Vaaramaa, K. and Lehto, J. (2003). Removal of metals and anions from drinking water by ion exchange, *Desalination*, **155**, pp. 157–170.
8. Al-Qahtani, K. M. (2016). Water purification using different waste fruit cortexes for the removal of heavy metals, *J. Taibah Univ. Sci.*, **10**, pp. 700–708.
9. Idumah, C. I. and Hassan, A. (2016). Emerging trends in flame retardancy of biofibers, biopolymers, biocomposites, and bionanocomposites, *Rev. Chem. Eng.*, **32**, pp. 115–148.
10. Ruiz-Hitzky, E., Aranda, P. and Darder, M. (2009). Polymer and biopolymer-layered solid nanocomposite: organic-inorganic assembling in two-dimensional hybrid systems, in *Bottom-Up Nanofabrication: Supramolecules, Self-Assemblies, and Organized Films*, Ariga, K., ed. (American Scientific, Valencia, California), pp. 39–76.
11. Ruiz-Hitzky, E., Darder, M. and Aranda, P. (2005). Functional biopolymer nanocomposites based on layered solids, *J. Mater. Chem.*, **15**, pp. 3650–3662.
12. Gibson, L. J. and Ashby M. F., eds. (1997). *Cellular Solids: Structure and Properties*, 2nd ed. (Cambridge University Press, Cambridge, UK).
13. Nussinovitch, A. (2005). Production, properties, and applications of hydrocolloid cellular solids, *Mol. Nutr. Food Res.*, **49**, pp. 195–213.
14. Bhowmick, A., Pramanik, N., Mitra, T., Gnanamani, A., Das, M. and Kundu, P. P. (2017). Mechanical and biological investigations of chitosan–polyvinyl alcohol based ZrO_2 doped porous hybrid composites for bone tissue engineering applications, *New J. Chem.*, **41**, pp. 7524–7530.
15. Levengood, S. K. L. and Zhang, M. (2014). Chitosan-based scaffolds for bone tissue engineering, *J. Mater. Chem. B*, **2**, pp. 3161–3184.

16. Shahzad, A., Miran, W., Rasool, K., Nawaz, M., Jang, J., Lim, S. R. and Lee, D. S. (2017).Heavy metals removal by EDTA-functionalized chitosan graphene oxide nanocomposites, *RSC Adv.*, **7**, pp. 9764–9771.
17. Galashev, A. E. and Polukhin, V. A. (2014). *Phys. Met. Metallogr.*, **115**, pp. 697–704.
18. Cui, L., Wang, Y., Gao, L., Hu, L., Yan, L., Wei, Q. and Du, B. (2015). Removal of copper from graphene by bombardment with argon clusters: Computer experiment, *Chem. Eng. J.*, **281**, pp. 1–10.
19. Repo, E., Koivula, R., Harjula, R. and Sillanpää, M. (2013).Effect of EDTA and some other interfering species on the adsorption of Co(II) by EDTA-modified chitosan, *Desalination*, **321**, pp. 93–102.
20. Rocher, V., Siaugue, J. M., Cabuil, V. and Bee, A. (2008). Removal of organic dyes by magnetic alginate beads, *Water Res.*, **42**, pp. 1290–1298.
21. Banerjee, S. S. and Chen, D. H. (2007). Fast removal of copper ions by gum arabic modified magnetic nano-adsorbent, *J. Hazard. Mater.*, **147**, pp. 792–799.
22. Liu, X., Hu, Q., Fang, Z., Zhang, X. and Zhang, B. (2009). Magnetic chitosan nanocomposites: a useful recyclable tool for heavy metal ion removal, *Langmuir*, **25**, pp. 3–8.
23. Tran, H. V., Tran, L. D. and Nguyen, T. N. (2010). Preparation of chitosan/ magnetite composite beads and their application for removal of Pb(II) and Ni(II) from aqueous solution, *Mater. Sci. Eng., C*, **30**, pp. 304–310.
24. O'Toole, M. G., Soucy, P. A., Chauhan, R., Raju, M. V. R., Patel, D. N., Nunn, B. M., Keynton, M. A., Ehringer, W. D., Nantz, M. H., Keynton, R. S. and Gobin, A. S. (2016). Release-modulated antioxidant activity of a composite curcumin-chitosan polymer, *Biomacromolecules*, **17**, pp. 1253–1260.
25. Ayoub, A., Venditti, R. A., Pawlak, J. J., Salam, A. and Hubbe, M. A. (2013). Novel hemicellulose–chitosan biosorbent for water desalination and heavy metal removal, *ACS Sustainable Chem. Eng.*, **1**, pp. 1102–1109.
26. He, Y., Chen, W., Li, X., Zhang, Z., Fu, J., Zhao, C. and Xie, E. (2013). Freestanding three-dimensional graphene/MnO_2 composite networks as ultralight and flexible supercapacitor electrodes, *ACS Nano*, **7**, pp. 174– 182.
27. Yin, B., Zhang, S., Jiang, H., Qu, F. and Wu, X. (2015). Phase-controlled synthesis of polymorphic MnO_2 structures for electrochemical energy storage, *J. Mater. Chem. A*, **3**, pp. 5722– 5729.

28. Su, D., Ahn, H.-J. and Wang, G. (2013). Hydrothermal synthesis of α-MnO_2 and β-MnO_2 nanorods as high capacity cathode materials for sodium ion batteries, *J. Mater. Chem. A*, **1**, pp. 4845–4850.

29. Su, X., Yang, X., Yu, L., Cheng, G., Zhang, H., Lin, T. and Zhao, F.-H. (2015). A facile one-pot hydrothermal synthesis of branched α-MnO_2 nanorods for supercapacitor application, *CrystEngComm*, **17**, pp. 5970–5977.

30. Liang, S., Teng, F., Bulgan, G., Zong, R. and Zhu, Y. (2008). Effect of phase structure of MnO_2 nanorod catalyst on the activity for CO Oxidation, *J. Phys. Chem. C*, **112**, pp. 5307– 5315.

31. Zhang, L., Tian, Y., Guo, Y., Gao, H., Li, H. and Yan, S. (2015). Introduction of α-MnO_2 nanosheets to NH_2 graphene to remove Cr^{6+} from aqueous solutions, *RSC Adv.*, **5**, pp. 44096–44106.

32. Mallakpour, S. and Madani, M. (2016). Use of valine amino acid functionalized α-MnO_2/chitosan bionanocomposites as potential sorbents for the removal of lead(II) ions from aqueous solution, *Ind. Eng. Chem. Res.*, **55**, pp. 8349–8356.

33. Tao, Y., Ye, L., Pan, J., Wang, Y. and Tang, B. (2009). Removal of Pb(II) from aqueous solution on chitosan/TiO_2 hybrid film, *J. Hazard. Mater.*, **161**, pp. 718–722.

34. Miller, S. M., Spaulding, M. L. and Zimmerman, J. B. (2011). Optimization of capacity and kinetics for a novel bio-based arsenic sorbent, TiO_2-impregnated chitosan bead, *Water Res.*, **45**, pp. 5745–5754.

35. Li, F., Du, P., Chen, W. and Zhang, S. (2007). Preparation of silica-supported porous sorbent for heavy metal ions removal in wastewater treatment by organic-inorganic hybridization combined with sucrose and polyethylene glycol imprinting, *Anal. Chim. Acta*, **585**, pp. 211–218.

36. Badshah, S. and Airlodi, C. (2011). Layered organoclay with talc-like structure as agent for thermodynamics of cations sorption at the solid/liquid interface, *Chem. Eng. J.*, **166**, pp. 420–427.

37. Pandey, S. and Mishra, S. B. (2011). Organic-inorganic hybrid of chitosan/organoclay bionanocomposites for hexavalent chromium uptake, *J. Colloid Interface Sci.*, **361**, pp. 509–520.

38. He, J., Bardelli, F., Gehin, A., Silvester, E. and Charlet, L. (2016). Novel chitosan goethite bionanocomposite beads for arsenic remediation, *Water Res.*, **101**, pp. 1–9.

39. Li, K., Wang, Y., Huang, M., Yan, H., Yang, H., Xiao, S. and Li, A. (2015). Preparation of chitosan-*graft*-polyacrylamide magnetic composite

microspheres for enhanced selective removal of mercury ions from water, *J. Colloid Interface Sci.*, **455**, pp. 261–270.

40. Yu, X., Tong, S., Ge, M., Zuo, J., Cao, C. and Song, W. (2012). One-step synthesis of magnetic composites of cellulose@iron oxide nanoparticles for arsenic removal, *J. Mater. Chem. A*, **1**, pp. 959–965.
41. Luo, X., Lei, X., Cai, N., Xie, X., Xue, Y. and Yu, F. (2016). Removal of heavy metal ions from water by magnetic cellulose-based beads with embedded chemically modified magnetite nanoparticles and activated carbon, *ACS Sustainable Chem. Eng.*, **4**, pp. 3960–3969.
42. Santra, D. and Sarkar, M. (2016). Optimization of process variables and mechanism of arsenic (V) adsorption onto cellulose nanocomposite, *J. Mol. Liq.*, **224**, pp. 290–302.
43. Koriche, Y., Darder, M., Aranda, P., Semsari, S. and Ruiz-Hitzky, E. (2014). Bionanocomposites based on layered silicates and cationic starch as eco-friendly adsorbents for hexavalent chromium removal, *Dalton Trans.*, **43**, pp. 10512–10520.
44. Xing, G.-X., Zhang, S.-F., Ju, B.-Z. and Yang, J.-Z. (2006). Study on adsorption behavior of crosslinked cationic starch maleate for chromium(VI), *Carbohydr. Polym.*, **66**, pp. 246–251.
45. Xie, G., Shang, X., Liu, R., Hu, J. and Liao, S. (2011). Synthesis and characterization of a novel amino modified starch and its adsorption properties for Cd(II) ions from aqueous solution, *Carbohydr. Polym.*, **84**, pp. 430–438.
46. Koriche, Y., Darder, M., Aranda, P., Semsari, S. and Ruiz-Hitzky, E. (2013). Efficient and ecological removal of anionic pollutants by cationic starch-clay bionanocomposites, *Sci. Adv. Mater.*, **5**, pp. 994– 1005.
47. Chivrac, F., Pollet, E., Schmutz, M. and Avérous, L. (2008). New approach to elaborate exfoliated starch-based nanobiocomposites, *Biomacromolecules*, **9**, pp. 896–900.
48. Naushad, M., Ahamad, T., Sharma, G., Al-Muhtaseb, A. H., Albadarin, A. B., Alam, M. M., ALOthman, Z. A., Alshehri, S. M. and Ghfar, A. A. (2016). Synthesis and characterization of a new starch/SnO_2 nanocomposite for efficient adsorption of toxic Hg^{2+} metal ion, *Chem. Eng. J.*, **300**, pp. 306–316.
49. Davis, T. A., Llanes, F., Volesky, B., Diaz-Pulido, G., Mccook, L. and Mucci, A. (2003). 1H-NMR study of Na alginates extracted from Sargassum spp. in relation to metal biosorption, *Appl. Biochem. Biotechnol.*, **110**, pp. 75–90.

50. Cataldo, S., Gianguzza, A., Milea, D., Muratore, N. and Pettignano, A. (2016). Pb(II) adsorption by a novel activated carbon - alginate composite material. A kinetic and equilibrium study, *Int. J. Biol. Macromol.*, **92**, pp. 769–778.

51. Cataldo, S., Muratore, N., Orecchio, S. and Pettignano, A. (2015). Enhancement of adsorption ability of calcium alginate gel beads towards Pd(II) ion. A kinetic and equilibrium study on hybrid Laponite and Montmorillonite–alginate gel beads, *Appl. Clay Sci.*, **118**, pp. 162–170.

52. Crist, R. H., Oberholser, K., Shank, N. and Nguyen, M. (1981). Nature of bonding between metallic ions and algal cell walls, *Environ. Sci. Technol.*, **15**, pp. 1212–1217.

53. Esmat, M., Farghali, A. A., Khedr, M. H. and El-Sherbiny, I. M. (2017). Alginate-based nanocomposites for efficient removal of heavy metal ions, *Int. J. Biol. Macromol.*, **102**, pp. 272–283.

54. Ahmad, R. and Mirza, A. (2017). Adsorption of Pb(II) and Cu(II) by alginate-Au-mica bionanocomposite: kinetic, isotherm and thermodynamic studies, *Process Saf. Environ. Prot.*, **109**, pp. 1–10.

55. Kwon, O. H., Kim, J.-O., Cho, D. W., Kumar, R. l., Baek, S. H., Kurade, M. B. and Jeon, B.-H. (2016). Adsorption of As(III), As(V) and Cu(II) on zirconium oxide immobilized alginate beads in aqueous phase, *Chemosphere*, **160**, pp. 126–133.

56. Sigdel, A., Park, J., Kwak, H. and Park, P.-K. (2016). Arsenic removal from aqueous solutions by adsorption onto hydrous iron oxide-impregnated alginate beads, *J. Ind. Eng. Chem.*, **35**, pp. 277–286.

57. Karkeh-Abadi, F., Samandari, S. S. and Samandari, S. S. (2016). The impact of functionalized CNT in the network of sodium alginate-based nanocomposite beads on the removal of Co(II) ions from aqueous solutions, *J. Hazard. Mater.*, **312**, pp. 224–233.

Chapter 10

Processing of Nanocomposite Solar Cells in Optical Applications

Khuram Ali and Yasir Javed
Nano-optoelectronics Research Laboratory, Department of Physics, University of Agriculture Faisalabad, Faisalabad, Pakistan
khuram_uaf@yahoo.com

10.1 Introduction

Nanocomposites are of great interest for the past five decades. Nanocomposite materials have been considered the most admirable alternatives in order to overcome the limitations of microcomposites [1]. Design uniqueness and property combinations are the reasons owing to which they are considered to be the materials of the 21st century. Conventional composites do not possess such properties. However, a complete understanding of these properties demands more research yet. Nanocomposites made their first appearance in the early 1950s [2].

Hybrid Nanocomposites: Fundamentals, Synthesis, and Applications
Edited by Kaushik Pal

ISBN 978-981-4800-34-1 (Hardcover), 978-0-429-00096-6 (eBook)
www.panstanford.com

The most commonly used clay mineral in nanocomposites is montmorillonite. It is commonly referred to as nanoclay and sometimes regarded as bentonite. Bentonite is an absorbent aluminum phyllosilicate clay and it is natural clay. It is formed by two ways: first is the hydrothermal alteration of volcanic rocks and the second is by the in situ alteration of volcanic ash. Another benefit is that this clay is vastly available and relatively cheap. This clay is widely used in nanocomposite applications. The work on nanocomposites started in the 1950s but the true start of polymer nanocomposites never started before the 1990s [2, 3]. It was Toyota that first used clay/nylon-6 nanocomposites for Toyota car-timing belt covers. In 1991, the idea of structural ceramic nanocomposites was first introduced. In current perspectives, these materials are going to be used in future space missions and other interesting applications. Composites are generally considered to be environment friendly. Nanocomposites have opened new avenues for technology and business in multiple sectors of the biotechnology, automotive, electronics, and aerospace industries [4, 5].

The combination of two or more simple materials results in a solid composite material that develops a dispersed phase and a continuous phase. In other words, a nanocomposite is a multiphased material, one phase of which can have one, two, or three dimensions (normally less than 100 nm). One can also define it as structures that have a repetition of nanoscale distances between different phases that make up the material [6].

Nanocomposites are nanomaterials that consolidate one or more distinct components for obtaining the best properties of each composite. The working idea behind the development of nanocomposites is developing new materials by using the materials whose dimensions are in the range of nanometers [7]. These new materials have unprecedented flexibility and improved physical properties. It is because of the fact that physical properties of particles witness an alteration after achieving a size less than a specific level (known critical size) of the particle. It is an undeniable fact that dimensionality plays a vital role in defining the properties of matter [8, 9]. Dimensionality controls the structure at the nanolevel. Additionally, extensive improvement in the interactions at phase boundries has been witnessed when dimensions extend to the nanometer level. It can be undoubtedly said that it is a particularly important factor to increase the material's properties. The surface-to-volume ratio of reinforcement materials that is used during their preparation is helpful to demonstrate the structure–property relationships of nanocomposites [10].

10.1.1 Nanocomposite Materials in Solar Cells

The sun, being the major source of energy, transmits about 3×10^{24} J of energy every year. This huge amount of energy is 1000 times more than the actual need of requirement of the earth. The demand of energy is estimated to increase at a rate of 2% each year for the next 2.5 decades. If 0.1% of the earth's surface would be covered with solar cells that have 10% efficiency, then the energy needs of the present world would be fulfilled [11, 12]. On the other hand, reality seems to be quite opposite as less than 0.1% of the world's total energy demand for electricity is obtained via solar cells. For this purpose, photovoltaics (PVs) are being developed that are based on single or multicrystalline p-n junctions. These PVs have a high manufacturing cost with relatively less efficiency. This drawback is overcome by using nanocomposites (which include molecular assemblies, nanosemiconductors, organic–inorganic hybrid assemblies), whose aim is to provide high efficiency at a relatively low cost [13].

Semiconductor nanocomposites are a huge group of nanomaterials. These have physical, chemical, and optoelectrical properties. These properties are tuned by altering the composition of the semiconductor in a mixture to obtain specific requirements of semiconductor devices. To enhance the power conversion efficiency (PCE), cyclic stability, and electrocatalytic activity in solar cells, nanocomposites are combined with transition metal oxides and carbonaceous material (carbon nanofibers, graphene, and carbon nanotubes) [14, 15]. Nanocomposites manufactured by using metal oxides have important applications for electrochemical energy storage. These materials have low conductivity and poor stability. So it becomes a necessity to add conductive phases so that electron transport and electrical contact of the active material in the electrodes of a Li-ion battery would be enhanced [16].

Optical properties are also affected by particle size. Variations of semiconductor nanoparticle size results in bandgap alteration, which allows tenability of their optical properties. On the other hand, the surface-to-volume ratio and particle size are inversely related. A decrease in particle size sharply affects the material's electronic structure, owing to the quantum confinement effect. Because of this effect the electronic states become discrete, settle at higher energies, and ultimately alter the optical properties of the nanoparticle. Photoluminescence, high or low refractive index, plasmon resonance, and high transparency are some optical properties of nanocomposite materials [17, 18].

Dye-sensitized solar cells (DSSCs) are considered to be next-generation devices because of their reasonable PCE and manufacturing ease. To enhance efficiency and minimize the device cost, DSSCs are modified. For this purpose different materials and their nanocomposites for dye, photoanode, electrolyte, and cathode are used [19]. Organic and inorganic nanocomposites are considered to be a promising choice for applications in devices like photodiodes, gas sensors, light-emitting diodes (LEDs), and PV cells [20]. The exciton dissociation efficiency is determined by the morphology and charge transport properties of composites in solar cells. They are also responsible for the performance of bulk heterojunction solar cells. The fabrication conditions of solar cells have an impact on device performance and carrier mobility. Additionally, properties of bulk heterojunction devices are extremely dependent on film morphology. However, film morphology does not affect the performance of nanocomposite-based devices to a large extent [21].

Nanocomposite materials give a good option to optimize the electrical junction characteristics, carrier transport, and absorption of the solar spectrum. All this would be helpful to upgrade PV energy conversion efficiency. Such approaches generally include the use of semiconductor (groups IV, II–VI, and IV–VI) quantum dots (QDs) to provide quantum-size-tuned and compositional electronic structures [22]. The spatial distribution of the semiconductor and the inserted material are defined by the phase assembly of semiconductor-based nanocomposites. These characteristics are important in governing the transport over long length scales.

The most extensively used material among functional metal oxide semiconductors is TiO_2 [23]. A wide optical bandgap (~3.2 eV), high conversion efficiency, strong UV absorptivity, and good photocatalysis are distinctive properties of TiO_2, which make it a better choice [24]. There exist some limitations as well, of which major drawbacks are weak absorption in the visible region and high recombination of photogenerated electron–hole pairs. These drawbacks can be overcome by using various techniques. A major technique in this regard is to utilize the advantage of the comparatively large surface area of TiO_2 nanostructures like nanorods, nanotubes, thin-film structures, nanosheets, and nanoparticle to upgrade its performance. The use of TiO_2 with incorporation of suitable semiconductors is considered to be another alternate technique to overcome this drawback [25, 26].

It is reported that doping of TiO_2 with aluminum (Al) and tungsten (W) results in increasing the short-circuit current and open-circuit voltage of DSSCs [27]. It is also reported that in photoreactivity, the recombination of a electron–hole pair decreases by using nanotube composites of SnO_2/TiO_2. Scientists profitably upgraded visible light absorption of TiO_2 using a ZnS/TiO_2 nanocomposite. It was also reported that the use of hybrid nanocomposites like poly(phenylene vinylene) (PPV)/TiO_2 in solar cells can increase the photoelectric conversion efficiency. To reduce the charge carrier recombination rate and to increase light absorption in the visible region, metal nanoparticles (especially Au) are introduced. Fabrication of composites (TiO_2-based) with other oxides of metal provided a beneficial impact on the reduction of charge carrier recombination [28].

Perovskite solar cells (PSCs) are considered to be next-generation devices in solar energy conversion devices. The most important advantage of PSCs is that they have a simple production method and high energy conversion efficiency. The maximum efficiency of PSCs is greater than that of traditional DSSCs. Up to now, the best calculated power conversion efficiencies for PSCs, depending on $CH_3NH_3PbI_3$ and $CH_3NH_3PbI_{3-x}Cl_x$ sensitizers, is 15% and 19.3% respectively [29].

10.2 Dye-Sensitized Solar Cells

10.2.1 Dye-Sensitized vs. Conventional Solar Cells

DSSCs are preferred over conventional solar cells nowadays because they are more flexible and transparent. Printable DSSCs have a share in the development of optoelectronics of high efficiency in upcoming years. DSSCs are considered to be better than amorphous silicon solar cells in terms of efficiency and cost [30]. DSSCs are considered to be long lasting and can work at wide solar angles. Additionally, the indoor light efficiency of DSSCs is much better because the dye of DSSCs is a good absorber of fluorescent light and diffuse sunlight.

DSSCs are different from conventional semiconductor devices because the function of light absorption is different from charge carrier transportation. The dye sensitizer takes in the striking sunlight, which in turn, induces the vectorial electron transfer reaction. DSSCs have several advantages over Si-based PV. First of all, they have a low cost as compared to Si-based solar cells. Second, there exists a great possibility of the transfer of direct energy (photons) to chemical energy. It can be done by the use of nanoporous structures. These structures have an enormous surface area, which can adsorb dye molecules. Another name for DSSCs is Gratzel cells. Third, it is easy to form the semiconductor–electrolyte interface (SEI) in DSSCs and its production is cost-effective. Fourth, DSSCs are preferred due to less sensitivity to semiconductor defects [31].

10.2.2 Basic Principle

The principle of operation of DSSCs is analogous to photosynthesis in plants. DSSCs consist of four parts: a dye sensitizer, a transparent conducting electrode (photoanode), a counterelectrode (cathode), and an electrolyte. A semiconductor material (TiO_2) with a wide bandgap is used for the anode of the cell. The cathode is made by using materials having carbon and platinum. As light hits, a photon moves into the solar cell via the anode, while positive and negative carriers are generated in the cell when the layer of dye sensitizer absorbs the light [32]. Photons that are absorbed by the dye molecules have wavelengths according to the energy difference between the lowest unoccupied molecular orbit (LUMO) and the highest occupied molecular orbit (HOMO) of the dye sensitizer. This absorption causes the electrons to be shifted to the excited state of the dye from the ground electronic state. This is known as photoexcitation of the dye. The electrons of the excited state are introduced into the conduction band of TiO_2. It is a diffusion process that transports the electrons from the conduction band of TiO_2 through the semiconductor. Afterward, these electrons reach the conducting layer of fluorine tin oxide (FTO) glass. Electrons then flow to the cathode through an external circuit and work is performed. Electrons then drive a reduction oxidation process in the electrolyte solution after re-entering through the cathode. At anode, the electrons can reach the oxidized dye from the tri-iodide electrolyte [33, 34]. A dye molecule is regenerated in order to continue the process. The current continuously flows through the circuit until the light hits the solar cell once again. The performance of DSSCs can be calculated by measuring two important factors, energy conversion efficiency (η) and fill factor (FF).

10.2.3 Fabrication of DSSCs

Figure 10.1 shows three main parts of a DSSC. First, the layer of transparent conducting oxide (indium tin oxide [ITO]- or FTO-coated glass) is followed by an electrode (a mesoscopic porous structure) with a very high surface area (normally TiO_2). The TiO_2 is usually annealed at a certain temperature. Then it is left to cool at room temperature [36]. The electrode is then soaked with a high absorbent called the sensitizer or dye, which is usually a photosensitive ruthenium-polypyridine dye [37]. This electrode is sandwiched with a counterelectrode usually made of platinum or graphite. Afterward, dips of electrolyte are used between the sandwiched solar cell to jump-start the electron flow.

There are many oxide semiconductors that are used as photocatalysts in DSSCs but titania is proven to be a good option for solar energy conversion. Its biological and chemical inertness, strong oxidizing power, cost-effectiveness, and long-term stability against photocorrosion and chemical corrosion make it the most suitable one [38]. It is an established fact that morphology, porous structure, and crystallinity are the properties due to which titania plays a vital role in the photoelectric conversion efficiency of DSSCs. Many mesoporous TiO_2 powders have also been used to make the porous electrodes for DSSCs [39]. By now, the total efficiency of DSSCs has increased up to 11.1% due to an increase in the number of absorbed dye molecules on the surface of TiO_2. The photogenerated electrons in a TiO_2 film are transported and transferred very rapidly, which results in increased photoelectric conversion efficiency and decreased recombinations of electron–hole pairs. To enhance the performance of DSSCs, the most effective way is to fabricate the films by using 1D nanostructures [40]. These nanostructures make easy electron transport as well as enhance light harvesting due to light scattering. In most conventional DSSCs, the use of nanometer-size TiO_2 is commonly observed [41, 42]. The sizes of nanoparticles are much smaller as compared to the wavelength of visible light, so little light is scattered. On the other hand, more light is scattered when the optical length of the film is increased by incorporating large nanoparticles (100–400 nm).

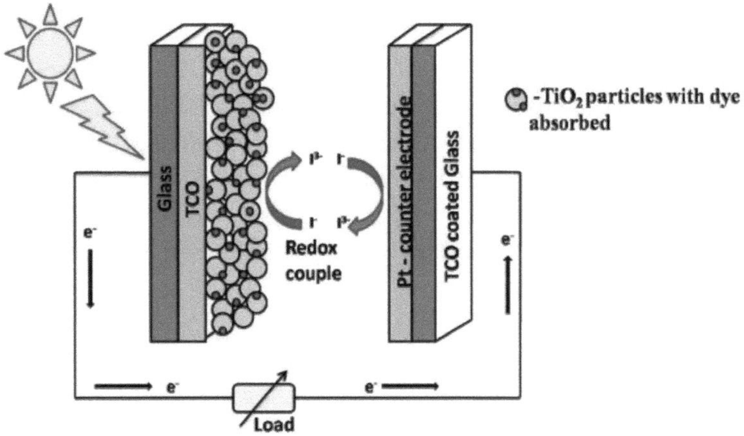

Figure 10.1 Schematic of a dye-sensitized solar cell. Reprinted from Ref. [35], Copyright (2015), with permission from Elsevier.

It is normally assumed that the overall device efficiency of DSSCs is higher than predicted sum of the properties of constituents of the cells [43]. On the other hand, N_3 dye degrades under light within a few hours after dissolving in the solution. It is worth mentioning that when both of them are combined in a device, their properties alter. In a device, the solar cell conducts current (20 mA/cm^2) and the N_3 dye shows stability for more than 1.5 decades in outdoor solar radiation. Hence it is considered that the PV function is the most important and promising property of the device. These results provide a promising future for the development of DSSCs. For this purpose use of organic dyes can extend light absorption as well as synthesis and modification of various types of TiO_2 [44]. The optical absorption of light (visible region) can be extended by modifying the physical properties of TiO_2 nanostructures [38].

10.2.4 Photocatalysis and Photoelectric Conversion in DSSCs

In metal oxide semiconductors, photocatalytic and PV performance is enhanced by separating photogenerated charge carriers (holes and electrons). TiO_2 is used as a scattering layer in order to improve the optical path in DSSCs. TiO_2 is widely used in different applications because it is chemical stable, nontoxic, and odorless. TiO_2-based photocatalysts also have the ability to capture the UV part of sunlight. TiO_2 is also used to increase the separation of photogenerated charge carriers in photocatalytic applications (Fig. 10.2). In DSSCs the sensitizer (dye) acts as a light-harvesting component [45].

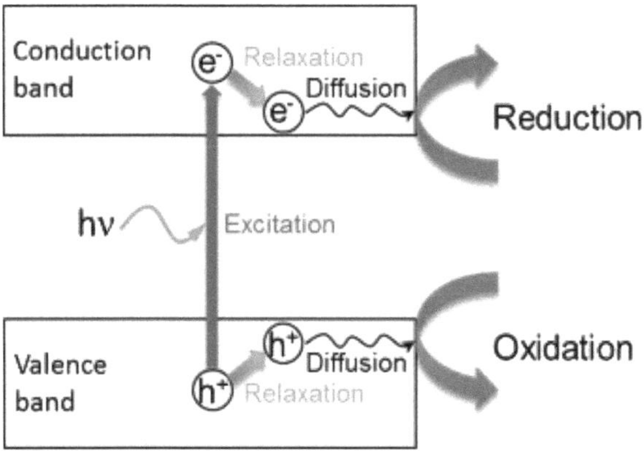

Figure 10.2 Schematic of photocatalysis of DSSCs. Reprinted from Ref. [46], Copyright (2012), with permission from Elsevier.

If absorption of sunlight is a matter of concern, TiO_2 is sensitized by visible-light-active photocatalysts. The build in the electric field is generated by different energy bandgaps between the sensitizer and TiO_2, which ultimately generates the photoexcited charge carriers. These carriers are then injected from one semiconductor to the other. This causes the generated charge carriers to recombine. Meantime the photogenerated electrons and holes can be well separated and improve the photocatalytic activity. So combined semiconductors have a higher photocatalytic activity as compared to single-component semiconductors. The final photocatalytic activity is also influenced by the procedures by which these combined semiconductors are prepared [47].

10.2.5 Enhanced Optical Properties in DSSCs

The underlying electronic structure of any material is responsible for the optical response. On the other hand, the electronic structure of a nanomaterial depends on its arrangement, chemical composition, and physical dimensions. The conversion efficiency is highly dependent on the light absorption by the photoanodes of the sensitizers. Parameters that determine the efficiency of DSSCs include dye molecules adsorbed on the photoanode, thickness of the photoanode, and response of dye molecules [48].

Raising the light absorption is considered as one approach to enhance the efficiency of DSSCs. The light absorption is enhanced by increasing the thickness of the TiO_2 layer in DSSCs. Much attention has been directed recently toward the surface plasmon resonance (SPR) of nanoparticles of noble metals such as copper, silver, and gold because of their unique magnetic, optical, and electronic properties [49].

There is another strategy by which the efficiency of DSSCs is further improved. If the thickness of the photoanode is increased, the diffusion path length of electrons also increases. By increasing the diffusion path length, electrons are recombined instead of being collected at the electrode. Recently, the plasmon resonance concept has been introduced to the DSSCs that use noble metals like gold or silver. Light harvesting efficiency is enhanced by the localized surface plasmon resonance (LSPR) phenomenon of metal nanoparticles [50].

Recently, metal nanoparticles have been incorporated into DSSCs with strong SPR. This strategy overcomes the main disadvantages produced by the addition of metal nanoparticles into the bulk of DSSCs. Photocurrent generation and light absorption are enhanced in DSSCs due to this new plasmonic PV system. Various processes like stability of interfaces and interfacial charge transfer can greatly influence the competency of a semiconductor–metal composite to prolong the charge separation [51].

10.3 Quantum Dot–Based Nanocomposite Solar Cells

Quantum dot solar cells (QDSCs) are a field of great interest in the area of solar energy technology. QDs are very useful regarding the production of energy-efficient solar cells, owing to their small size [52]. QDs have bandgaps that are easily tunable. The unique property that draws attention toward QDs is multiple excitation generation (MEG). In this process, the absorption of a single high-energy photon causes the generation of multiple bound charge–carrier pairs. The effects of charge–carrier multiplication are particularly beneficial for solar cells where they increase the photocurrent significantly. Theoretically, due to MEG, QDSCs can be used to obtain an efficiency of about 60%. But currently it seems harder to achieve it experimentally as the highest efficiency achieved is 8.6% yet [53].

10.3.1 Quantum Confinement

QDs are nanocrystals that are made up of materials that are in groups II–VI, III–V, and IV–VI. QDs have the property of confining electrons in the zero dimension. Electrons can be considered as free in the material whose dimensions are very large as compared to the electrons' wavelength [54]. On the other hand, quantum confinement comes into effect when the size of the particle and the wavelength of the electron are comparable. It indicates the confinement or restriction of random motion of electron indiscrete energy levels. As the particle attains a nanoscale size, the confinement dimension decreases, so the energy levels become discrete. This ultimately increases the bandgap and hence the bandgap energy also increases.

Due to the quantum confinement effect there is a change in the electronic structure of the material [55]. It ultimately changes the electronic and optical properties of the material. The following equation interrelates the dimensions of a material and its energy gap.

$$E_{n_x,n_y,n_z} = \frac{\hbar^2 \Pi^2}{2m}\left[\left(\frac{n_x}{L_x}\right)^2 + \left(\frac{n_y}{L_y}\right)^2 + \left(\frac{n_z}{L_z}\right)^2\right] \tag{10.1}$$

In the transition from the bulk semiconductor to the QD, a blue shift is observed with an increase in the energy gap. It also explains the effect of the size of the particle on the bandgap and ultimately on the bandgap energy [56].

10.3.2 Absorbance in the Quantum Dot Layer

As light is absorbed in the QD layer of a solar cell, excited bound charge–carrier pairs (i.e., excitons) are generated. Before these can be extracted through an external circuit they often encounter loss processes that reduce the efficiency of the solar cell. Normally, incomplete light trapping or angle restrictions account for up to 20% of all energy losses [58].

DSSCs and quantum dot–sensitized solar cells (QDSSCs) share almost the same working principle and structure. QDs are a source of current injection and represent the only difference between both types of the solar cells. Figure 10.3 shows the structure of a QDSC. Important elements of the cell include transparent conductive glass, a TiO_2 layer (nanostructured), an electrolyte, a QD layer, and a counterelectrode [59].

In QDSCs, metal oxides like TiO_2 and ZnO, having wide bandgaps, are mostly used as photoelectrodes. To achieve maximum power conversion efficiencies from solar cells, morphologies of such metal oxides have been actively explored. The widely used electrolyte for QDSSCs is aqueous polysulfied solution [60].

From Fig. 10.3 it can be observed that in QDSSCs, excitons are generated when optical absorption occur. Electrons (which are photoexcited) are then injected into the TiO_2 layer. The regeneration of oxidized QDs by the electrolyte occurs when these electrons are transported to the transparent conducting electrode [62].

Further, oxidized species of redox couples are regenerated at the counterelectrode.

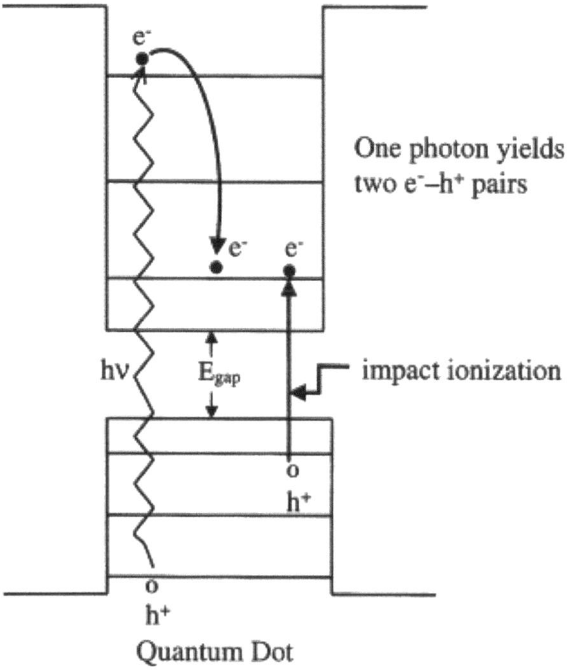

Figure 10.3 Impact ionization in quantum dot solar cells.Reprinted from Ref. [61], Copyright (2002), with permission from Elsevier.

Three possible charge transfer processes can be observed in Fig. 10.4: injection, recombination, and trapping of holes and excited electrons. All these processes occur at the interfaces between the TiO_2 layer, the QD layer, and the electrolyte [64].

There are four paths for injection of holes and photoexcited electrons: injection of electrons from the LUMO to TiO_2, from the electron-trapping level to TiO_2, hole injection from the HOMO to the electrolyte, and injection of the hole-trapping level to the electrolyte. On the other hand, there are five possible recombinations of holes and electrons, such as recombination of holes and electrons in QDs and through trapping levels. Electrons that were injected into TiO_2 transfer back to QDs and then recombine in the electrolyte. In the fifth type of recombination, excited electrons in QDs recombine with the oxidized species in the electrolyte [65].

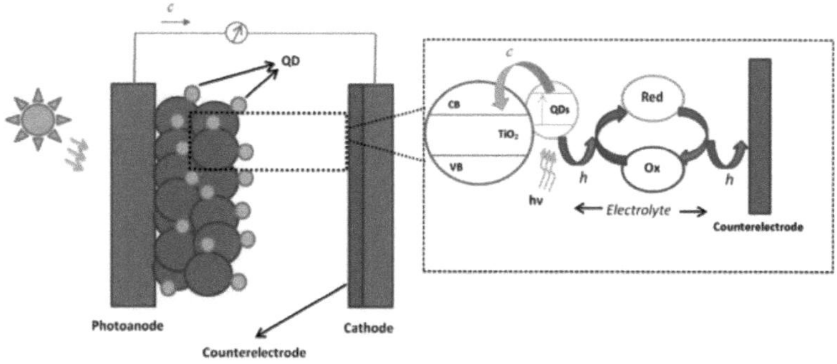

Figure 10.4 Schematic of operating process of a quantum dot–sensitized solar cell.Reprinted from Ref. [63], Copyright (2014), with permission from Elsevier.

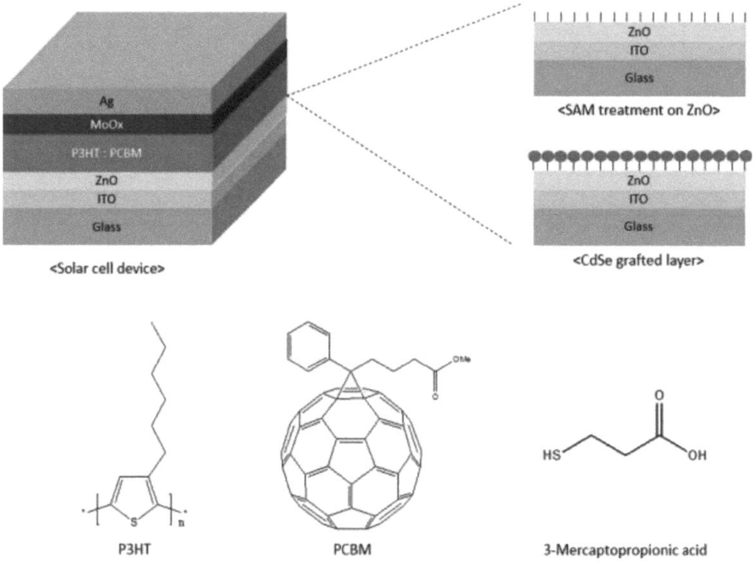

Figure 10.5 Schematic of fabrication process for a CdSe quantum dot solar cell. Reprinted from Ref. [66], Copyright (2015), with permission from Elsevier.

These charge transfer processes or recombinations lower the efficiency of QDSCs. Recombination of charges at each interface results in a reduction of charge collection efficiency and charge separation efficiency. Consequently, lower values of short-circuit current, FF. and open-circuit voltage are obtained with overall poor performance of a solar cell.

10.4 Nanocomposite Materials in Organic Solar Cells

Organic photovoltaic (OPV) cells have good flexibility, low cost, and light weight. Owing to these qualities OPVs have attracted great attention. OPV cells consist of two electrodes having organic photoactive materials packed between them [67]. ITO glass is mainly used in OPVs due to high transparency, good electrical conductivity, and ease of patterning. Efficient hole-extracting layers of poly-(3,4-ethylenedioxythiophene):polystyrene sulfonic acid (PEDOT:PSS) are mainly uses and show good optical transparency [68]. PEDOT:PSS is doped with an MWCNT film. After doping PEDOT:PSS acts as a hole extraction layer for OPV cells [69]. The MWCNT doping enhances the PCE, FF, and short-circuit current density of OPV cells. PEDOT:PSS is a conducting dispersion of PEDOT nanoparticles (conducting species) dispersed in a solution of PSS (film-forming species). In organic solar cells (OSCs) it is used extensively. Its main function is to act as a hole-transporting "buffer layer" between the ITO anode and the active organic layer [70].

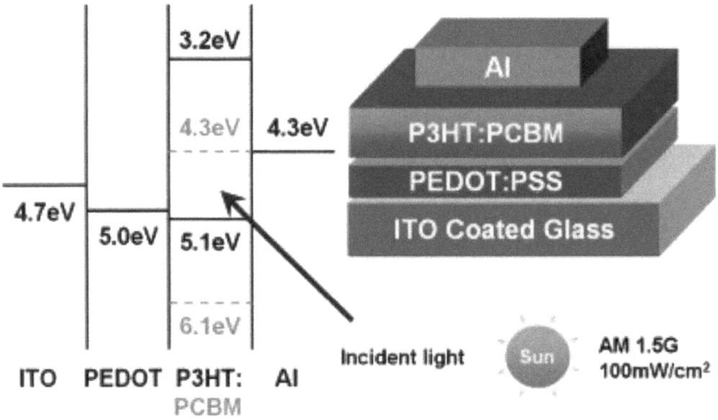

Figure 10.6 Schematic of an OPV cell and its energy band diagram. Reprinted from Ref. [71], Copyright (2009), with permission from Elsevier.

Generally, OSCs are fabricated on a glass substrate with ITO electrodes, as shown in Fig. 10.6. It contains two electrodes that have different work functions. An active layer is packed between these two electrodes. One of the electrodes must absorb the light in the active layer of the cell, so this electrode must be transparent.

This electrode is often a conductive oxide that can be a solution processed from a precursor material [50]. The second electrode is a metal. It can easily be evaporated on the active layer. This metal contact reflects off all the light that was not absorbed and thus helps to maximize the exciton generation in the active layer. The choice of electrodes, charge transport layers, and the morphology of the photoactive layer plays a vital role in determining the overall performance of such type of solar cells.

10.4.1 NiO_x-Based Heterojunction Perovskite Solar Cell

Organic–inorganic hybrid PSCs have excellent light harvesting, a long carrier lifetime, and high charge carrier mobility. These are the characteristics owing to which organic–inorganic hybrid PSC have seen great progress in recent decades. As a result the PCE of organic–inorganic hybrid PSCs has exceeded up to 20% [72].

The absorption layer is packed between the electron contact layer and the hole contact layer in a planar heterojunction PSC. In PSCs, NiO_x is used due to its distinctive features. NiO_x is considered a good hole-selective contact [73]. Second, NiO_x has high conduction band-edge position, which helps to block electrons and transmit holes efficiently. NiO_x is a chemically and thermally stable material, as well as having good optical transparency. It has a valence band that is well aligned with inverted hybrid PSCs [74].

Also the low-temperature deposition process of NiO_x will make the film very relevant and attractive for flexible devices because it can be deposited efficiently at a temperature as low as 130°C without any posttreatments. NiO_x-based PSCs have the potential to demonstrate a preliminary PCE of >17% and >14% on ITO–glass and ITO–polyethylene naphthalate (PEN) substrates, respectively [75].

On the basis of their structures the reported PSCs can be mainly divided into two types. The first type emerges from solid-state DSSCs and consists of mesoporous films such as TiO_2 [76, 77] and Al_2O_3 films [78, 79]. The other type has a similar structure to p-i-n silicon solar cells and is known as a planar heterojunction solar cell.

10.4.2 Fabrication of NiO_x-Based Solar Cells

A perovskite absorption layer is sandwiched between a hole contact layer and an electron contact layer in a typical planar heterojunction PSC. In n-i-p-type devices compact TiO_2 and ZnO films usually work as electron contacts [80, 81]. For hole contact, 2,2′,7,7′-tetrakis[*N,N*-di(4-methoxyphenyl)amino]-9,9′-spirobifluorene (spiro-OMeTAD) is most frequently used in these devices but it has some drawbacks in terms of a very high price and poor stability against moisture and temperature [82]. On the other hand, phenyl-C61-butyric acid methylester (PCBM) and PEDOT:PSS are frequently used as electron and hole contacts, respectively, in p-i-n-type devices.

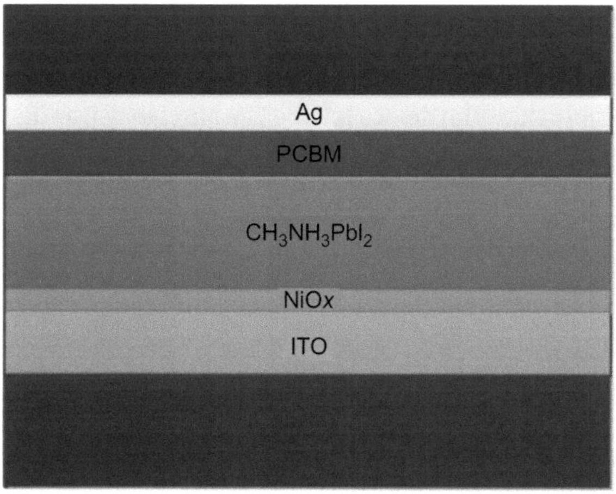

Figure 10.7 Schematic of NiO_x-based inverted planar heterojunction perovskite solar cell.

A schematic of a NiO_x-based inverted planar heterojunction perovskite solar cell is shown in Fig. 10.7. Such type of configuration can give a PCE as high as 18.1% [82, 83] and is named as "inverted planar heterojunction solar cell." But due to its high acidity and hygroscopicity, PEDOT:PSS is not good for long-term stability of the device [84]. Therefore, many inorganic semiconductors, which include MoO_x, NiO_x, V_2O_3, and WO_3, have been engaged to take over PEDOT:PSS. Of these, NiO_x is a low-price material with better chemical and thermal stability. In addition, due to a suitable work function and high conduction band-edge position that can block electrons and transport holes effectively, NiO_x has also been proven to be a good hole-selective contact for PSCs [85].

Research on NiO_x-based PSCs has gained a great achievement [86], but an expensive pulse laser deposition method has been used to deposit NiO_x films. This type of deposition is not suitable for large-scale production of NiO_x-based PSCs. In addition, an essential annealing process at 300°C–500°C to enhance the quality of the NiO_x films makes them unsuitable for flexible substrates [87, 88].

Recently, Jen et al. achieved a PCE as high as 17.74% by using a combustion method to prepare a Cu-doped NiO_x hole contact for PSCs [89]. High-quality perovskite films were made by Burschka et al. (2013), who identified the value of perovskite films for highly efficient PSCs. They developed spin coating of methylammonium iodide with a two-step deposition technique and subsequent submerging into a lead halide solution.

A flexible thin-film device having 7% efficiency was developed by Roldán-Carmona et al. [90]. They used a thermally evaporated lead iodide perovskite (CH_3NH_3PbI) layer sandwiched between two thin holes (PCBM) and electron poly(4-butylphenyl-diphenyl-amine) (poly-TPD) blocking layers. Flexible organometallic PSCs have an advantage due to their flexibility, but the hunt for a flexible structure is accompanied with deterioration in performance, as demonstrated by Cui et al. [91].

Irrespective of the compatibility of low-temperature sputtered NiO_x films with flexible devices, their low PCE (below 10%) makes them unappealing to the research society [91]. Therefore, for flexible PSCs, it is very purposeful to explore low-temperature processed NiO_x films with effective hole extraction capabilities. It is recorded that on an ITO glass substrate a solution-derived NiO_x hole contact layer–based inverted planar heterojunction PSC can gain a PCE of as high as 16.47% [92].

NiO_x is also very relevant and attractive for flexible devices because it can be deposited efficiently at a temperature as low as 130°C without any posttreatment. A prior PCE of 13.43% was reported with a NiO_x-based flexible PSC employing an ITO-PEN substrate. It was reported that a p-i-n structure of PSCs with an inverted planar heterojunction have less serious hysteresis compared to n-i-p structured planar heterojunction solar cells. For example, a PCE of 18.1% has been recorded with hysteresis-less inverted planar heterojunction PSCs with PEDOT:PSS hole contact films [84].

10.4.3 Absorption Gap and Optimization in Organic Solar Cells

Today's OSCs are efficient and their efficiency depends on the ability of the active layer to absorb the maximum amount of sunlight. OSCs have a photoactive layer [93]. This layer contains n-doped (electron acceptor) and p-doped (electron donor) semiconducting materials. The organic semiconductors are substances that help to absorb light in polymers of the solar cell [94]. The benefit of organic semiconductor polymers is that their physical properties such as absorption spectrum and bandgap can be altered by tuning their chemical structure. The high absorption coefficients of thin layers of semiconducting polymers make it possible to absorb an adequate amount of light. Such substances contain a bandgap having a particular energy gap (E_g) [95]. E_g depicts the energetic division among the nearest free electronic state and the valence electrons. While considering organic semiconductors, the energetic separation can be termed as the difference between the HOMO and the LUMO [96].

The performance of OSCs is closely related to the choice of material. Carrier mobility, molecular energy levels, and bandgaps are the properties of a material that greatly affect the performance of the device [97, 98]. The photons that have energy greater than the bandgap of materials are only absorbed. This property is important in the context that greater absorption leads to a larger number of photogenerated carriers. Hence a large number of carriers is collected at the electrode. In turn, a high external quantum efficiency value is achieved. Poly(3-hexylthiophene-2,5-diyl) (P3HT) is the first ever polymer that showed high current density and hole mobility values of about 8.7 mA/cm^2 and 0.1 cm^2/Vs, respectively [99]. This high hole mobility allows the inclusion of a thicker active layer for effective charge transport to the electrode.

Third-generation solar cells are mainly multijunction devices. The purpose of a multijunction structure is to get maximum absorption of the solar spectrum of materials having different bandgaps [100]. A junction having a high bandgap absorbs high-energy photons and the lower-bandgap junction absorbs low-energy photons. It is observed that the sunlight conversion efficiency for a three-junction solar cell is about 44%. The absorption bandgap is a strong limitation in the case of OSCs, so extension in the absorption range results in improved efficiency of the device [101].

10.5 Novel Nanocomposites for Efficient Optical Solar Cell Applications

In recent years, nanocomposites of inorganic compounds and conjugated polymers are an area of great development. It is because of their applications in devices (photodiodes, LEDs, PV cells, and sensors). Nanocomposites can show novel synergic effects along with their electronic and optical properties [102]. Solar cells containing PPV/TiO_2 composites show enhanced PV activity. In TiO_2 and PPV nanocomposites, TiO_2 is used as an electron-accepting material [103]. The PPV/TiO_2 composites are investigated for solar cell applications. Nanoscale CdS particles and poly-3-octyithiophene (PTO) are combined to form a novel inorganic/organic composite material. POT/CdS nanocomposites are encouraging materials with extremely good performance characteristics in PV applications [104–106].

Graphene/TiO_2 nanocomposite photoanodes are fabricated for DSSC applications. This structure significantly increases the PV performance as compared to conventional TiO_2-based solar cells [107]. On the other hand, light absorption is increased because of a higher surface area. Incorporation of CNTs into the TiO_2 matrix increases the electron transport, which results in the reduction of charge recombination. High-performance CNT–TiO_2 nanocomposites are widely used in DSSC applications [75]. A composite photoanode of porous TiO_2 with CNTs increases the surface area of TiO_2 nanoparticles to achieve high efficiency of DSSCs.

A novel nanocomposite comprising TiO_2, cuprous oxide (Cu_2O) nanoparticles, and reduced graphene oxide (RGO) sheets has been reported recently [108, 109]. By using the combination of these three materials, an effective methylene blue photodegradation process was observed. The nanocomposite helps to harvest a broad portion of the solar spectrum containing visible and UV light. It also plays a key role in the separation of photogenerated electron–hole pairs throughout the photocatalytic process. Mostly, n-type ZnO is paired with p-type materials like CuO and Cu_2O to make a p-n junction for optoelectronic devices [110, 111]. Low-cost solar cells are made by a combination of ZnO nanoparticles with CuO. It is observed that FF, short-circuit current density, open-circuit voltage, and PCE of the solar cell is improved.

The efficiency of photoelectrochemical (PEC) solar cells can be enhanced by metal–semiconductor composite films. Improvements in charge transfer processes are taking place by a combination of semiconductor substrates (i.e., ZnO and TiO_2) and metal clusters like Au, Ag, Pt, and Pd. These nanocomposites help in enhancing photocatalytic activity by trapping photoinduced charge carriers [112, 113].

Noble metals such as platinum and gold have high electron affinity behavior. So these metals are used with ZnO to form metal–ZnO composites. The electrons are transferred from Au to ZnO up to dynamic equilibrium [114]. Where Au works like a sink for photoinduced charge carriers and upgrades the interfacial charge transfer processes. In PSCs, the electron collection layer (ECL) plays an important role in decreasing recombination. The use of nanocomposite ECLs in PSCs is of considerable interest for further development of PV cells. ZnO–SnO_2 nanocomposite thin films are prepared as ECLs to fabricate PSCs, which results in increased FF and short-circuit current density [115].

10.6 Conclusions

Solar energy is considered the most suitable energy source for the 21st century as nonrenewable resources are becoming scarce with each passing day. Solar cells are used to trap available sunlight. It is an area of great interest for the past five decades. Research has gone far in recent years. Conventional solar cells have been largely replaced by nanocomposites due to a number of advantages. DSSCs are preferred over conventional solar cells because of their efficiency, flexibility, and transparency. Low cost and durability are the properties that attract much attention toward DSSCs. QD-based nanocomposite solar cells are also common these days. The small size of QDs makes them energy efficient. Quantum confinement is the property of QDs that makes them most suitable for energy conversion. OSCs have proven their effectiveness owing to many reasons. Organic cells have an active layer to absorb the maximum amount of incident sunlight so that maximum output may be achieved. All these developments give a ray of hope for future energy needs. Further developments in this field would help in space missions and other technological developments.

References

1. Thakur, V. K. and Kessler, M. R. (2015). Self-healing polymer nanocomposite materials: a review, *Polymer*, **69**, pp. 369–383.
2. Camargo, P. H. C., Satyanarayana, K. G. and Wypych, F. (2009). Nanocomposites: synthesis, structure, properties and new application opportunities, *Mater. Res.*, **12**(1), pp. 1–39.
3. Hanemann, T. and Szabó, D. V. (2010). Polymer-nanoparticle composites: from synthesis to modern applications, *Materials*, **3**, pp. 3468–3517.
4. Sanchez, C., Julián, B., Belleville, P. and Popall, M. (2005). Applications of hybrid organic–inorganic nanocomposites, *J. Mater. Chem.*, **15**, pp. 3559–3592.
5. Voevodin, A. A. and Zabinski, J. (2005). Nanocomposite and nanostructured tribological materials for space applications, *Compos. Sci. Technol.*, **65**, pp. 741–748.
6. Jancar, J., Douglas, J., Starr, F. W., Kumar, S., Cassagnau, P., Lesser, A., Sternstein, S. S. and Buehler, M. (2010). Current issues in research on structure–property relationships in polymer nanocomposites, *Polymer*, **51**, pp. 3321–3343.
7. Wang, Z., Gu, P. and Zhang, Z. (2010). Indentation and scratch behavior of nano-SiO 2/polycarbonate composite coating at the micro/nano-scale, *Wear*, **269**, pp. 21–25.
8. Wang, Z. Z., Gu, P., Zhang, Z., Gu, L. and Xu, Y. Z. (2011). Mechanical and tribological behavior of epoxy/silica nanocomposites at the micro/nano scale, *Tribol. Lett.*, **42**, pp. 185–191.
9. Wang, Y. and Herron, N. (1991). Nanometer-sized semiconductor clusters: materials synthesis, quantum size effects, and photophysical properties, *J. Phys. Chem.*, **95**, pp. 525–532.

10. Roychowdhury, A., Pati, S. P., Kumar, S. and Das, D. (2015). Tunable properties of magneto-optical Fe_3O_4/CdS nanocomposites on size variation of the magnetic component, *Mater. Chem. Phys.*, **151**, pp. 105–111.

11. Gençer, E., Miskin, C., Sun, X., Khan, M. R., Bermel, P., Alam, M. A. and Agrawal, R. (2017). Directing solar photons to sustainably meet food, energy, and water needs, *Sci. Rep.*, **7**, p. 3133.

12. Ali, K. and Javed, Y. (2017). Radiation-resistant solar cells: recent updates and future prospective, in *Handbook of Ecomaterials*, Martínez, L. M. T., Kharissova, O. V. and Kharisov, B. I., eds. (Springer International, Cham), pp. 1–26.

13. Small, C. E., Chen, S., Subbiah, J., Amb, C. M., Tsang, S.-W., Lai, T.-H., Reynolds, J. R. and So, F. (2012). High-efficiency inverted dithienogermole-thienopyrrolodione-based polymer solar cells, *Nat. Photonics*, **6**, pp. 115–120.

14. Siwach, B., Sharma, S. and Mohan, D. (2017). Structural, optical and morphological properties of ZnO/MWCNTs nanocomposite photoanodes for dye sensitized solar cells (DSSCs) application, *J. Integr. Sci. Technol.*, **5**, pp. 1–4.

15. Jeon, N. J., Noh, J. H., Kim, Y. C., Yang, W. S., Ryu, S. and Seok, S. I. (2014). Solvent engineering for high-performance inorganic–organic hybrid perovskite solar cells, *Nat. Mater.*, **13**, pp. 897–903.

16. Demirocak, D. E., Srinivasan, S. S. and Stefanakos, E. K. (2017). A review on nanocomposite materials for rechargeable Li-ion batteries, *Appl. Sci.*, **7**, p. 731.

17. Pourabdollah, M., Zeynali, H. and Akbari, H. (2017). Controlled synthesis, characterization, and optical properties of ZnIn2S4 and CdIn2S4 nanostructures with enhanced performance for solar cell applications, *Mater. Lett.*, **196**, pp. 312–315.

18. Zang, Z., Zeng, X., Wang, M., Hu, W., Liu, C. and Tang, X. (2017). Tunable photoluminescence of water-soluble AgInZnS–graphene oxide (GO) nanocomposites and their application in-vivo bioimaging, *Sens. Actuators, B*, **252**, pp. 1179–1186.

19. Grätzel, M. (2014). The light and shade of perovskite solar cells, *Nat. Mater.*, **13**, pp. 838–842.

20. Shih, C.-C., Chen, P.-C., Lin, G.-L., Wang, C.-W. and Chang, H.-T. (2014). Optical and electrochemical applications of silicon–carbon dots/ Silicon dioxide nanocomposites, *ACS Nano*, **9**, pp. 312–319.

21. Khuspe, G., Chougule, M., Navale, S., Pawar, S. and Patil, V. (2014). Camphor sulfonic acid doped polyaniline-tin oxide hybrid nanocomposites: synthesis, structural, morphological, optical and electrical transport properties, *Ceram. Int.*, **40**, pp. 4267–4276.

22. Labiadh, H., Chaabane, T. B., Balan, L., Becheik, N., Corbel, S., Medjahdi, G. and Schneider, R. (2014). Preparation of Cu-doped ZnS QDs/TiO_2 nanocomposites with high photocatalytic activity, *Appl. Catal., B*, **144**, pp. 29–35.

23. Sun, M., Ma, X., Chen, X., Sun, Y., Cui, X. and Lin, Y. (2014). A nanocomposite of carbon quantum dots and TiO_2 nanotube arrays: enhancing photoelectrochemical and photocatalytic properties, *RSC Adv.*, **4**, pp. 1120–1127.

24. Bian, J., Huang, C., Wang, L., Hung, T., Daoud, W. A. and Zhang, R. (2014). Carbon dot loading and TiO_2 nanorod length dependence of photoelectrochemical properties in carbon dot/TiO_2 nanorod array nanocomposites, *ACS Appl. Mater. Interfaces*, **6**, pp. 4883–4890.

25. Li, Z., Cui, X., Hao, H., Lu, M. and Lin, Y. (2015). Enhanced photoelectrochemical water splitting from Si quantum dots/TiO_2 nanotube arrays composite electrodes, *Mater. Res. Bull.*, **66**, pp. 9–15.

26. Ren, P., Fu, X. and Zhang, Y. (2017). Carbon quantum dots-TiO_2 nanocomposites with enhanced catalytic activities for selective liquid phase oxidation of alcohols, *Catal. Lett.*, **147**, pp. 1679–1685.

27. Abdullah, H., Razali, M. Z., Shaari, S. and Taha, M. R. (2014). Enhancement of dye-sensitized solar cell efficiency using carbon nanotube/TiO_2 nanocomposite thin films fabricated at various annealing temperatures, *Electron. Mater. Lett.*, **10**, pp. 611–619.

28. Pandikumar, A., Suresh, S., Murugesan, S. and Ramaraj, R. (2015). Dual functional TiO_2–Au nanocomposite material for solid-state dye-sensitized solar cells, *J. Nanosci. Nanotechnol.*, **15**, pp. 6965–6972.

29. Boro, B., Rajbongshi, B. and Samdarshi, S. (2016). Synthesis and fabrication of TiO_2–ZnO nanocomposite based solid state dye sensitized solar cell, *J. Mater. Sci.: Mater. Electron.*, **27**, pp. 9929–9940.

30. Kim, H., Lee, E.-J. and Sun, Y.-K. (2014). Recent advances in the Si-based nanocomposite materials as high capacity anode materials for lithium ion batteries, *Mater. Today*, **17**, pp. 285–297.

31. Inoue, I., Yamauchi, H., Okamoto, N., Toyoda, K., Horita, M., Ishikawa, Y., Yasueda, H., Uraoka, Y. and Yamashita, I. (2015). Thermo-stable carbon nanotube-TiO_2 nanocompsite as electron highways in dye-sensitized solar cell produced by bio-nano-process, *Nanotechnology*, **26**, p. 285601.

32. Eshaghi, A. and Aghaei, A. A. (2015). Effect of TiO_2–graphene nanocomposite photoanode on dye-sensitized solar cell performance, *Bull. Mater. Sci.*, **38**, pp. 1177–1182.

33. Han, G. S., Song, Y. H., Jin, Y. U., Lee, J.-W., Park, N.-G., Kang, B. K., Lee, J.-K., Cho, I. S., Yoon, D. H. and Jung, H. S. (2015). Reduced graphene oxide/mesoporous TiO_2 nanocomposite based perovskite solar cells, *ACS Appl. Mater. Interfaces*, **7**, pp. 23521–23526.

34. Calogero, G., Calandra, P., Irrera, A., Sinopoli, A., Citro, I. and Di Marco, G. (2011). A new type of transparent and low cost counter-electrode based on platinum nanoparticles for dye-sensitized solar cells, *Energy Environ. Sci.*, **4**, pp. 1838–1844.

35. Shalini, S., Balasundara Prabhu, R., Prasanna, S., Mallick, T. K. and Senthilarasu, S. (2015). Review on natural dye sensitized solar cells: operation, materials and methods, *Renewable Sustainable Energy Rev.*, **51**, pp. 1306–1325.

36. Wang, Z., Xu, H., Zhang, Z., Zhou, X., Pang, S. and Cui, G. (2014). High-performance cobalt selenide and nickel selenide nanocomposite counter electrode for both iodide/triiodide and cobalt (II/III) redox couples in dye-sensitized solar cells, *Chin. J. Chem.*, **32**, pp. 491–497.

37. Wu, W.-Q., Xu, Y.-F., Su, C.-Y. and Kuang, D.-B. (2014). Ultra-long anatase TiO_2 nanowire arrays with multi-layered configuration on FTO glass for high-efficiency dye-sensitized solar cells, *Energy Environ. Sci.*, **7**, pp. 644–649.

38. Jung, H.-Y., Yeo, I.-S., Kim, D.-G., Oh, B.-Y., Gu, H.-B. and Ki, H.-C. (2017). Solar conversion efficiency improvement of dye-sensitized solar cells via plasma treatment of transparent conducting oxide substrate, *J. Nanosci. Nanotechnol.*, **17**, pp. 3323–3327.

39. Xu, P., Tang, Q., Chen, H. and He, B. (2014). Insights of close contact between polyaniline and FTO substrate for enhanced photovoltaic performances of dye-sensitized solar cells, *Electrochim. Acta*, **125**, pp. 163–169.

40. Park, K.-H. and Dhayal, M. (2014). Simultaneous growth of rutile TiO_2 as 1D/3D nanorod/nanoflower on FTO in one-step process enhances electrochemical response of photoanode in DSSC, *Electrochem. Commun.*, **49**, pp. 47–50.

41. Liao, M., Fang, L., Xu, C., Wu, F., Huang, Q. and Saleem, M. (2014). Effect of seed layer on the growth of rutile TiO_2 nanorod arrays and their performance in dye-sensitized solar cells, *Mater. Sci. Semicond. Process*, **24**, pp. 1–8.

42. Gong, J., Sumathy, K., Qiao, Q. and Zhou, Z. (2017). Review on dye-sensitized solar cells (DSSCs): Advanced techniques and research trends, *Renewable Sustainable Energy Rev.*, **68**, pp. 234–246.
43. Zhao, X., Li, M., Song, D., Cui, P., Zhang, Z., Zhao, Y., Shen, C. and Zhang, Z. (2014). A novel hierarchical Pt-and FTO-free counter electrode for dye-sensitized solar cell, *Nanoscale Res. Lett.*, **9**, p. 202.
44. Gong, H. H., Park, S. H., Lee, S.-S. and Hong, S. C. (2014). Facile and scalable fabrication of transparent and high performance Pt/reduced graphene oxide hybrid counter electrode for dye-sensitized solar cells, *Int. J. Precs. Eng. Manuf.*, **15**, pp. 1193–1199.
45. Petronella, F., Truppi, A., Ingrosso, C., Placido, T., Striccoli, M., Curri, M., Agostiano, A. and Comparelli, R. (2017). Nanocomposite materials for photocatalytic degradation of pollutants, *Catal. Today*, **281**, pp. 85–100.
46. Nakata, K. and Fujishima, A. (2012). TiO_2 photocatalysis: design and applications, *J. Photochem. Photobiol., C*, **13**, pp. 169–189.
47. Vafaei, S., Manseki, K., Horita, S., Matsui, M. and Sugiura, T. (2017). Controlled assembly of nanorod TiO_2 crystals via a sintering process: photoanode properties in dye-sensitized solar cells, *Int. J. Photoenergy*, **2017**, p. 7686053.
48. Gregg, B. A. (2004). Interfacial processes in the dye-sensitized solar cell, *Coord. Chem. Rev.*, **248**, pp. 1215–1224.
49. Yu, H., Zhang, S., Zhao, H., Will, G. and Liu, P. (2009). An efficient and low-cost TiO_2 compact layer for performance improvement of dye-sensitized solar cells, *Electrochim. Acta*, **54**, pp. 1319–1324.
50. Gong, J., Sumathy, K., Qiao, Q. and Zhou, Z. (2017). Review on dye-sensitized solar cells (DSSCs): advanced techniques and research trends, *Renewable Sustainable Energy Rev.*, **68**, pp. 234–246.
51. Guan, J., Zhang, J., Yu, T., Xue, G., Yu, X., Tang, Z., Wei, Y., Yang, J., Li, Z. and Zou, Z. (2012). Interfacial modification of photoelectrode in ZnO-based dye-sensitized solar cells and its efficiency improvement mechanism, *RSC Adv.*, **2**, pp. 7708–7713.
52. Tan, H., Jain, A., Voznyy, O., Lan, X., de Arquer, F. P. G., Fan, J. Z., Quintero-Bermudez, R., Yuan, M., Zhang, B. and Zhao, Y. (2017). Efficient and stable solution-processed planar perovskite solar cells via contact passivation, *Science*, **355**, pp. 722–726.
53. Vitoreti, A. B. F., Corrêa, L. B., Raphael, E., Patrocinio, A. O. T., Nogueira, A. F. and Schiavon, M. A. (2017). Quantum dot-sensitized solar cells, *Quím. Nova*, **40**, pp. 436–446.

54. Sidhik, S., Esparza, D., Martínez-Benítez, A., Lopez-Luke, T., Carriles, R., Mora-Sero, I. and de la Rosa, E. (2017). Enhanced photovoltaic performance of mesoscopic perovskite solar cells by controlling the interaction between CH3NH3PbI3 films and CsPbX3 perovskite nanoparticles, *J. Phys. Chem. C*, **121**, pp. 4239–4245.

55. Krishna, M. V. R. and Friesner, R. A. (1991). Quantum confinement effects in semiconductor clusters, *J. Chem. Phys.*, **95**, pp. 8309–8322.

56. Jeong, W.-S., Lee, J.-W., Jung, S., Yun, J. H. and Park, N.-G. (2011). Evaluation of external quantum efficiency of a 12.35% tandem solar cell comprising dye-sensitized and CIGS solar cells, *Sol. Energy Mater. Sol. Cells*, **95**, pp. 3419–3423.

57. Koulentianos, D. (2014). Quantum confinement effect in materials for solar cell applications, Uppsala University.

58. Anta, J. A. (2012). Electron transport in nanostructured metal-oxide semiconductors, *Curr. Opin. Colloid Interface Sci.*, **17**, pp. 124–131.

59. Mora-Seró, I., Giménez, S., Moehl, T., Fabregat-Santiago, F., Lana-Villareal, T., Gómez, R. and Bisquert, J. (2008). Factors determining the photovoltaic performance of a CdSe quantum dot sensitized solar cell: the role of the linker molecule and of the counter electrode, *Nanotechnology*, **19**, p. 424007.

60. Chen, J., Zhao, D., Song, J., Sun, X., Deng, W., Liu, X. and Lei, W. (2009). Directly assembled CdSe quantum dots on TiO_2 in aqueous solution by adjusting pH value for quantum dot sensitized solar cells, *Electrochem. Commun.*, **11**, pp. 2265–2267.

61. Nozik, A. J. (2002). Quantum dot solar cells, *Physica E*, **14**, pp. 115–120.

62. Zhang, Q., Chen, G., Yang, Y., Shen, X., Zhang, Y., Li, C., Yu, R., Luo, Y., Li, D. and Meng, Q. (2012). Toward highly efficient CdS/CdSe quantum dots-sensitized solar cells incorporating ordered photoanodes on transparent conductive substrates, *Phys. Chem. Chem. Phys.*, **14**, pp. 6479–6486.

63. Kouhnavard, M., Ikeda, S., Ludin, N. A., Ahmad Khairudin, N. B., Ghaffari, B. V., Mat-Teridi, M. A., Ibrahim, M. A., Sepeai, S. and Sopian, K. (2014). A review of semiconductor materials as sensitizers for quantum dot-sensitized solar cells, *Renewable Sustainable Energy Rev.*, **37**, pp. 397–407.

64. Jovanovski, V., González-Pedro, V., Giménez, S., Azaceta, E., Cabañero, G. N., Grande, H., Tena-Zaera, R., Mora-Seró, I. N. and Bisquert, J. (2011). A sulfide/polysulfide-based ionic liquid electrolyte for quantum dot-sensitized solar cells, *J. Am. Chem. Soc.*, **133**, pp. 20156–20159.

65. Lee, Y.-L. and Chang, C.-H. (2008). Efficient polysulfide electrolyte for CdS quantum dot-sensitized solar cells, *J. Power Sources*, **185**, pp. 584–588.

66. Kim, K. M., Jeon, J. H., Kim, Y. Y., Lee, H. K., Park, O. O. and Wang, D. H. (2015). Effects of ligand exchanged CdSe quantum dot interlayer for inverted organic solar cells, *Org. Electron.*, **25**, pp. 44–49.

67. Xu, M., Feng, J., Fan, Z.-J., Ou, X.-L., Zhang, Z.-Y., Wang, H.-Y. and Sun, H.-B. (2017). Flexible perovskite solar cells with ultrathin Au anode and vapour-deposited perovskite film, *Sol. Energy Mater. Sol. Cells*, **169**, pp. 8–12.

68. Hong, W., Xu, Y., Lu, G., Li, C. and Shi, G. (2008). Transparent graphene/PEDOT–PSS composite films as counter electrodes of dye-sensitized solar cells, *Electrochem. Commun.*, **10**, pp. 1555–1558.

69. Vosgueritchian, M., Lipomi, D. J. and Bao, Z. (2012). Highly conductive and transparent PEDOT: PSS films with a fluorosurfactant for stretchable and flexible transparent electrodes, *Adv. Funct. Mater.*, **22**, pp. 421–428.

70. Kim, Y. H., Sachse, C., Machala, M. L., May, C., Müller-Meskamp, L. and Leo, K. (2011). Highly conductive PEDOT: PSS electrode with optimized solvent and thermal post-treatment for ITO-free organic solar cells, *Adv. Funct. Mater.*, **21**, pp. 1076–1081.

71. Baek, W.-H., Yang, H., Yoon, T.-S., Kang, C. J., Lee, H. H. and Kim, Y.-S. (2009). Effect of P3HT:PCBM concentration in solvent on performances of organic solar cells, *Sol. Energy Mater. Sol. Cells*, **93**, pp. 1263–1267.

72. Groeneveld, B. G., Najafi, M., Steensma, B., Adjokatse, S., Fang, H.-H., Jahani, F., Qiu, L., ten Brink, G. H., Hummelen, J. C. and Loi, M. A. (2017). Improved efficiency of NiOx-based pin perovskite solar cells by using PTEG-1 as electron transport layer, *APL Mater.*, **5**, p. 076103.

73. Assadi, M. K., Bakhoda, S., Saidur, R. and Hanaei, H. (2017). Recent progress in perovskite solar cells, *Renewable Sustainable Energy Rev.*, **81**, pp. 2812–2822.

74. Islam, M. B., Yanagida, M., Shirai, Y., Nabetani, Y. and Miyano, K. (2017). NiOx hole transport layer for perovskite solar cells with improved stability and reproducibility, *ACS Omega*, **2**, pp. 2291–2299.

75. Haruyama, J., Sodeyama, K., Hamada, I., Han, L. and Tateyama, Y. (2017). First-principles study of electron injection and defects at the TiO_2/CH3NH3PbI3 interface of perovskite solar cells, *J. Phys. Chem. Lett.*, **8**, pp. 5840–5847.

76. Jung, H. S. and Park, N. G. (2015). Perovskite solar cells: from materials to devices, *Small*, **11**, pp. 10–25.

77. Stranks, S. D., Eperon, G. E., Grancini, G., Menelaou, C., Alcocer, M. J., Leijtens, T., Herz, L. M., Petrozza, A. and Snaith, H. J. (2013). Electron-hole diffusion lengths exceeding 1 micrometer in an organometal trihalide perovskite absorber, *Science*, **342**, pp. 341–344.

78. Burschka, J., Pellet, N., Moon, S.-J., Humphry-Baker, R., Gao, P., Nazeeruddin, M. K. and Grätzel, M. (2013). Sequential deposition as a route to high-performance perovskite-sensitized solar cells, *Nature*, **499**, pp. 316–319.

79. Lee, M. M., Teuscher, J., Miyasaka, T., Murakami, T. N. and Snaith, H. J. (2012). Efficient hybrid solar cells based on meso-superstructured organometal halide perovskites, *Science*, **338**, pp. 643–647.

80. Liu, M., Johnston, M. B. and Snaith, H. J. (2013). Efficient planar heterojunction perovskite solar cells by vapour deposition, *Nature*, **501**, pp. 395–398.

81. Liu, D. and Kelly, T. L. (2014). Perovskite solar cells with a planar heterojunction structure prepared using room-temperature solution processing techniques, *Nat. Photonics*, **8**, pp. 133–138.

82. Liu, J., Wu, Y., Qin, C., Yang, X., Yasuda, T., Islam, A., Zhang, K., Peng, W., Chen, W. and Han, L. (2014). A dopant-free hole-transporting material for efficient and stable perovskite solar cells, *Energy Environ. Sci.*, **7**, pp. 2963–2967.

83. Chen, Y., Chen, T. and Dai, L. (2015). Layer-by-layer growth of CH3NH3PbI3− xClx for highly efficient planar heterojunction perovskite solar cells, *Adv. Mater.*, **27**, pp. 1053–1059.

84. Heo, J. H., Han, H. J., Kim, D., Ahn, T. K. and Im, S. H. (2015). Hysteresis-less inverted CH3 NH3 PbI3 planar perovskite hybrid solar cells with 18.1% power conversion efficiency, *Energy Environ. Sci.*, **8**, pp. 1602–1608.

85. Yin, X., Que, M., Xing, Y. and Que, W. (2015). High efficiency hysteresis-less inverted planar heterojunction perovskite solar cells with a solution-derived NiOx hole contact layer, *J. Mater. Chem. A*, **3**, pp. 24495–24503.

86. Park, J. H., Seo, J., Park, S., Shin, S. S., Kim, Y. C., Jeon, N. J., Shin, H. W., Ahn, T. K., Noh, J. H. and Yoon, S. C. (2015). Efficient CH3NH3PbI3 perovskite solar cells employing nanostructured p-type NiO electrode formed by a pulsed laser deposition, *Adv. Mater.*, **27**, pp. 4013–4019.

87. Zhu, Z., Bai, Y., Zhang, T., Liu, Z., Long, X., Wei, Z., Wang, Z., Zhang, L., Wang, J. and Yan, F. (2014). High-performance hole-extraction layer of sol–gel-processed NiO nanocrystals for inverted planar perovskite solar cells, *Angew. Chem. Int. Ed.*, **126**, pp. 12779–12783.

88. Kim, J. H., Liang, P. W., Williams, S. T., Cho, N., Chueh, C. C., Glaz, M. S., Ginger, D. S. and Jen, A. K. Y. (2015). High-performance and environmentally stable planar heterojunction perovskite solar cells based on a solution-processed copper-doped nickel oxide hole-transporting layer, *Adv. Mater.*, **27**, pp. 695–701.

89. Jung, J. W., Chueh, C. C. and Jen, A. K. Y. (2015). A low-temperature, solution-processable, Cu-doped nickel oxide hole-transporting layer via the combustion method for high-performance thin-film perovskite solar cells, *Adv. Mater.*, **27**, pp. 7874–7880.

90. Roldán-Carmona, C., Malinkiewicz, O., Soriano, A., Espallargas, G. M., Garcia, A., Reinecke, P., Kroyer, T., Dar, M. I., Nazeeruddin, M. K. and Bolink, H. J. (2014). Flexible high efficiency perovskite solar cells, *Energy Environ. Sci.*, **7**, pp. 994–997.

91. Cui, J., Meng, F., Zhang, H., Cao, K., Yuan, H., Cheng, Y., Huang, F. and Wang, M. (2014). CH3NH3PbI3-based planar solar cells with magnetron-sputtered nickel oxide, *ACS Appl. Mater. Interfaces*, **6**, pp. 22862–22870.

92. Yin, X., Chen, P., Que, M., Xing, Y., Que, W., Niu, C. and Shao, J. (2016). Highly efficient flexible perovskite solar cells using solution-derived NiOx hole contacts, *ACS Nano*, **10**, pp. 3630–3636.

93. Wang, Y., Wu, D., Fu, L. M., Ai, X. C., Xu, D. and Zhang, J. P. (2015). Correlation between energy and spatial distribution of intragap trap states in the TiO_2 photoanode of dye-sensitized solar cells, *ChemPhysChem*, **16**, pp. 2253–2259.

94. Huang, A., Lei, L., Zhu, J., Yu, Y., Liu, Y., Yang, S., Bao, S., Cao, X. and Jin, P. (2017). Achieving high current density of perovskite solar cells by modulating the dominated facets of room-temperature DC magnetron sputtered TiO_2 electron extraction layer, *ACS Appl. Mater. Interfaces*, **9**, pp. 2016–2022.

95. Su, T., Yang, Y., Na, Y., Fan, R., Li, L., Wei, L., Yang, B. and Cao, W. (2015). An insight into the role of oxygen vacancy in hydrogenated TiO_2 nanocrystals in the performance of dye-sensitized solar cells, *ACS Appl. Mater. Interfaces*, **7**, pp. 3754–3763.

96. Nelson, J. (1999). Continuous-time random-walk model of electron transport in nanocrystalline TiO_2 electrodes, *Phys. Rev. B*, **59**, p. 15374.

97. Li, C., Guo, Q., Wang, Z., Bai, Y., Liu, L., Wang, F., Zhou, E., Hayat, T., Alsaedi, A. and Tan, Z. A. (2017). Efficient planar structured perovskite solar cells with enhanced open-circuit voltage and suppressed charge recombination based on a slow grown perovskite layer from lead acetate precursor, *ACS Appl. Mater. Interfaces*, **9**, pp. 41937–41944.

98. He, Q., Yao, K., Wang, X., Xia, X., Leng, S. and Li, F. (2017). Room-temperature and solution-processable Cu-doped nickel oxide nanoparticles for efficient hole-transport layers of flexible large-area perovskite solar cells, *ACS Appl. Mater. Interfaces*, **9**, pp. 41887–41897.

99. Wang, W., Wu, H., Yang, C., Luo, C., Zhang, Y., Chen, J. and Cao, Y. (2007). High-efficiency polymer photovoltaic devices from regioregular-poly (3-hexylthiophene-2, 5-diyl) and [6, 6]-phenyl-C 61-butyric acid methyl ester processed with oleic acid surfactant, *Appl. Phys. Lett.*, **90**, p. 183512.

100. Pae, S. R., Byun, S., Kim, J., Kim, M., Gereige, I. and Shin, B. (2017). Improving uniformity and reproducibility of hybrid perovskite solar cells via a low-temperature, vacuum deposition process for NiOx hole transport layers, *ACS Appl. Mater. Interfaces*, **10**(1), pp. 534–540.

101. Liska, P., Thampi, K., Grätzel, M., Bremaud, D., Rudmann, D., Upadhyaya, H. and Tiwari, A. (2006). Nanocrystalline dye-sensitized solar cell/copper indium gallium selenide thin-film tandem showing greater than 15% conversion efficiency, *Appl. Phys. Lett.*, **88**, p. 203103.

102. Han, D., Wu, C., Zhao, Y., Chen, Y., Xiao, L. and Zhao, Z. (2017). Ion implantation modified fluorine-doped tin oxide by zirconium with continuously tunable work function and its application in perovskite solar cells, *ACS Appl. Mater. Interfaces*, **9**(48), pp. 42029–42034.

103. Yang, C., Yu, M., Chen, D., Zhou, Y., Wang, W., Li, Y., Lee, T.-C. and Yun, D. (2017). Annealing-free aqueous-processed anatase TiO_2 compact layer for efficient planar heterojunction perovskite solar cells, *Chem. Commun.*, **53**, pp. 10882–10885.

104. Song, Z., Tong, G., Li, H., Li, G., Ma, S., Yu, S., Liu, Q. and Jiang, Y. (2017). Three-dimensional architecture hybrid perovskite solar cells using CdS nanorod arrays as an electron transport layer, *Nanotechnology*, **29**, p. 025401.

105. Wang, J., Liu, L., Liu, S., Yang, L., Zhang, B., Feng, S., Yang, J., Meng, X., Fu, W. and Yang, H. (2017). Influence of a compact CdS layer on the photovoltaic performance of perovskite-based solar cells, *Sustainable Energy Fuels*, **1**, pp. 504–509.

106. Chen, J., Cai, X., Yang, D., Song, D., Wang, J., Jiang, J., Ma, A., Lv, S., Hu, M. Z. and Ni, C. (2017). Recent progress in stabilizing hybrid perovskites for solar cell applications, *J. Power Sources*, **355**, pp. 98–133.

107. Murakami, T. N., Miyadera, T., Funaki, T., Cojocaru, L., Kazaoui, S., Chikamatsu, M. and Segawa, H. (2017). Adjustment of conduction band edge of compact TiO_2 layer in perovskite solar cells through TiCl4 treatment, *ACS Appl. Mater. Interfaces*, **9**, pp. 36708–36714.

108. Nazari, P., Ansari, F., Abdollahi Nejand, B., Ahmadi, V., Payandeh, M. and Salavati-Niasari, M. (2017). Physicochemical interface engineering of CuI/Cu as advanced potential hole-transporting materials/metal contact couples in hysteresis-free ultralow-cost and large-area perovskite solar cells, *J. Phys. Chem. C*, **121**, pp. 21935–21944.

109. Nordseth, Ø., Kumar, R., Bergum, K., Fara, L., Foss, S. E., Haug, H., Drăgan, F., Crăciunescu, D., Sterian, P. and Chilibon, I. (2017). Optical analysis of a ZnO/Cu2O subcell in a silicon-based tandem heterojunction solar cell, *Green Sustainable Chem.*, **7**, p. 57.

110. Panigrahi, S., Nunes, D., Calmeiro, T., Kardarian, K., Martins, R. and Fortunato, E. (2017). Oxide-based solar cell: impact of layer thicknesses on the device performance, *ACS Comb. Sci.*, **19**, pp. 113–120.

111. Mitroi, M. R., Ninulescu, V. and Fara, L. (2017). Performance optimization of solar cells based on heterojunctions with Cu_2O: numerical analysis, *J. Energy Eng.*, **143**, p. 04017005.

112. Tai, Q. and Yan, F. (2017). Emerging semitransparent solar cells: materials and device design, *Adv. Mater.*, **29**(34), p. 1700192.

113. Sun, M., Hu, J., Zhai, C., Zhu, M. and Pan, J. (2017). A p-n heterojunction of CuI/TiO_2 with enhanced photoelectrocatalytic activity for methanol electro-oxidation, *Electrochim. Acta*, **245**, pp. 863–871.

114. Zhang, R., Fei, C., Li, B., Fu, H., Tian, J. and Cao, G. (2017). Continuous size tuning of monodispersed ZnO nanoparticles and its size effect on the performance of perovskite solar cells, *ACS Appl. Mater. Interfaces*, **9**(11), pp. 9785–9794.

115. Guo, Y., Kang, L., Zhu, M., Zhang, Y., Li, X. and Xu, P. (2017). A strategy toward air-stable and high-performance ZnO-based perovskite solar cells fabricated under ambient conditions, *Chem. Eng. J.*, **336**, pp. 732–740.

Index